Hair Care Products

Ashish Dwivedi • A. B. Pant •
Shiv Poojan • Malini Kotak •
Anurag Tripathi
Editors

Hair Care Products

Efficacy, Safety and Global Regulation

Springer

Editors
Ashish Dwivedi
Photobiology Division
Indian Institute of Toxicology Research
Lucknow, Uttar Pradesh, India

A. B. Pant
Toxicology and Health Risk Assessment
Indian Institute of Toxicology Research
Lucknow, Uttar Pradesh, India

Shiv Poojan
Dermatology and Cutaneous Biology
Thomas Jefferson University
Philadelphia, PA, USA

Malini Kotak
ITC Life Sciences and Technology Center
Bangalore, Karnataka, India

Anurag Tripathi
Toxicology and Health Risk Assessment
Indian Institute of Toxicology Research
Lucknow, Uttar Pradesh, India

ISBN 978-981-97-6739-7 ISBN 978-981-97-6740-3 (eBook)
https://doi.org/10.1007/978-981-97-6740-3

This Springer imprint is published by the registered company Springer Nature Singapore Pte Ltd.
The registered company address is: 152 Beach Road, #21-01/04 Gateway East, Singapore 189721, Singapore

Preface

The global hair care industry has flourished, offering various products like shampoo, conditioners, dyes, and styling products to consumers worldwide. These products promise to enhance the look and health of our hair. However, there is a paucity of proper regulation guidelines, and with abundance comes a critical need for robust regulation and safety assessments. This book, *Hair Care Products: Efficacy, Safety and Global Regulation*, tackles the complexities and opportunities within the ever-evolving regulatory framework for hair care products. Our aim is to deliver a thorough examination of the standards and guidelines that ensure the efficacy and safety of these products on a global scale. This book seeks to bridge that gap, offering a comprehensive and multidisciplinary exploration of hair care from a scientific and regulatory perspective.

We begin with the history of the regulations by exploring the fascinating world of the hair itself in Chaps. 1, 2, and 3. Understanding the anatomy, physiology, and genetic underpinnings of hair color equips us to appreciate the complexities of hair health and treatment. Chapter 4 unveils the raw materials that form the foundation of hair care products. We explore how regulations ensure their safety and ethical sourcing. Chapter 5 includes the different types of hair care products like oils, dyes, and serums, examining their functionalities and potential interactions. The subsequent chapters navigate the intricate web of global regulations governing hair care products. Chapters 6 and 7 shed light on the opportunities and challenges associated with efficacy and safety testing, highlighting the evolving landscape of traditional and novel methodologies. Chapters 8 and 9 explore the exciting frontiers of hair care science. This further explores the role of the microbiome in hair health and the potential of preclinical and clinical models to assess product safety. Chapter 10 delves into the power of artificial intelligence in predicting hair product toxicity, while Chap. 11 unveils the cutting-edge omics approaches for comprehensively evaluating product safety and efficacy. Looking ahead, Chap. 12 explores the next generation of hair products, emphasizing innovation for safe and healthy consumer choices. Finally, Chaps. 13 and 14 provide an in-depth analysis of global regulatory frameworks for the hair care industry, equipping readers with a clear understanding of key agencies and guidelines. This book is a valuable resource for scientists, formulators, regulators, and consumers alike. It empowers informed decision-making, fostering a future of safe and effective hair care for all.

This book caters to a diverse audience, including researchers, regulatory bodies, industry professionals, and consumers who prioritize the safety and efficacy of hair care products. Our objective is to provide a clear understanding of the current regulatory landscape for hair care products, while highlighting the challenges and opportunities in guaranteeing these products' safety and efficacy globally. In essence, *Hair Care Products: Efficacy, Safety and Global Regulation* serves as a timely and comprehensive resource, offering insights into the intricate world of global hair care product regulations. We trust this book will be a valuable contribution to the field, ultimately promoting the safe and effective use of hair care products worldwide.

Lucknow, Uttar Pradesh, India — Ashish Dwivedi
Lucknow, Uttar Pradesh, India — A. B. Pant
Philadelphia, PA, USA — Shiv Poojan
Bangalore, Karnataka, India — Malini Kotak
Lucknow, Uttar Pradesh, India — Anurag Tripathi

Contents

Editors and Contributors

About the Editors

Ashish Dwivedi graduated from CSIR-Indian Institute of Toxicology Research, Lucknow, India, in Phototoxicology. Thereafter, he enriched himself in the same domain of research as Post-Doctoral Fellow at Israel Institute of Technology, Israel, Colorado University, USA, and Banaras Hindu University, Varanasi, India. Based on his outstanding research experience, he got an opportunity to return to his parenteral research organization-CSIR-IITR, where he is currently a Scientist in the Drug and Chemical Toxicology Group. In a decade-long research career, Dr. Dwivedi has established himself as one of the pioneers in the area of phototoxicity/photo safety with special reference to the photo safety of cosmetics ingredients (hair dyes, colorants, preservatives, and fragrances), therapeutic agents, nanotized phytochemicals, environmental pollutants, etc. Besides the laboratory work, he has also been involved in fostering contemporary science to post-graduate, PhD students, researchers, and university teachers. Dr. Dwivedi's research efforts have been rewarding in terms of the quantifiable number of research publications in journals of high repute. He has also edited three books for Springer Nature in the area of Photo Protection, Skin Aging, and Neuroprotection. He is a member of various scientific societies, including the Society of Toxicology, USA; the American Society for Photobiology, USA; the Indian Photobiology Society, India; the Society of Toxicology, India, etc.

A. B. Pant is a seasoned toxicologist with over 30 years of active research career. Presently serving as "Senior Principal Scientist" at CSIR-Indian Institute of Toxicology Research, Lucknow, and also as a "Professor" of the Academy of Scientific and Innovative Research, an Institution of National Importance set up by the Act of Parliament, Government of India. In the scientific fraternity, Professor Pant is renowned for his elegant research on the application of human stem cells and iPSCs in developmental neurotoxicity and for establishing the number of in vitro model systems as an alternative to laboratory animals for biomedical research. He also founded a new discipline of "In-Vitro Toxicology" in the country. He also dedicated himself to fostering contemporary science to post-graduate, PhD students, researchers, and university teachers. He is a lead GLP inspector for the Government of India and also serves as an expert for regulatory bodies, such as BIS, CDSCO, NABL, FSSAI, NGCMA, ICMR, DST, and Pharmacovigilance. He has won several awards and honors, including Vigyan Ratna Award-2010 (CST UP); Shakuntala Amir Chand Prize-2007 (ICMR); National Bioscience Award-2012 (DBT); Prof. KT Shetty Memorial Oration Award-2017 (Indian Academy of Neurosciences); Toxicology Promotion Award-2018 (National Academy of Sciences, Allahabad); STOX/ASAW Gold Medal (STOX, India), etc. He has been elected Fellow of the Academy of Toxicological Sciences, USA; the Society of Toxicology, India; the Indian Academy of Neurosciences, the Academy of Sciences for Animal Welfare, India; the Academy of Environmental Biology India, etc. In view of outstanding credibility in toxicology, Professor Pant has been included as a global toxicologist in the European and United Kingdom Register of Toxicologists and has also been conferred the DSc (h.c.) degree by HNB Medical Education University, Dehradun, Uttarakhand, India. He currently serves as Chairman and Member of several National and International Committees/Boards of Government Institutes and Bodies. He also serves as a member of the editorial board of several reputable journals.

Shiv Poojan, Founder and CEO of bioTox Solutions, is a visionary entrepreneur and academician with over 15 years of experience in scientific research. Prior to bioTox, Dr. Shiv held prestigious positions at research institutions worldwide as the Head of R&D at MS Clinical Research Pvt. Ltd., Bangalore (2022–2023, as a Scientist at Eliksa Therapeutics, Philadelphia, USA (2021–2022), Research Scientist at the Sidney Kimmel Cancer Center, Thomas Jefferson University in the USA (2018–2021), and Post-Doctoral Fellow at the National Cancer Center in South Korea (2014–2018). He has received his PhD (Biochemistry) from CSIR-Indian Institute of Toxicology Research, India. He is a leading expert in media formulation and tissue culture, extending the lifespan of 3D cell cultures and ex-plant growth beyond 2 weeks—a groundbreaking achievement. Dr. Shiv's expertise extends to drug discovery and development. He has experience in preclinical assessments, developing knock-in/knock-out transgenic mice for cancer research, and creating cell lines for drug testing. His research has been recognized with awards like the Rambhau Kulkarni Best Young Scientist Award from ICAR and the prestigious Carl Storm International Diversity Fellowship. Notably, he was the first in India to establish natural human skin in a lab setting. Dr. Shiv's unwavering commitment to innovation is evident in his developing unique media compositions for ex-plant cultures, significantly impacting tissue engineering research. His diverse skillset in molecular biology, biochemistry, and omics research makes him a true leader in scientific advancement.

Malini Kotak is an Associate Scientist in Health and Hygiene segment at ITC Ltd., India's leading fast moving consumer goods company. During her doctoral training at the University of Texas at Arlington and post-doctoral training at ITC Ltd., she has extensively worked on gut, hand, skin, and vaginal microbiomes and their association with human health. She has more than 10 years of work experience in industry and academia and has done cutting-edge research in fields like biotherapeutics, human microbiome, precision medicine, genomics, and cancer biology. She also has extensive experience in research and development in the personal care industry. She has published several research papers in reputed journals, book chapters, and is a recipient of several research fellowships in India and abroad. Her current research involves the development of novel products which are of relevance in the Health and Hygiene Industry.

Anurag Tripathi is working as Principal Scientist at the CSIR-Indian Institute of Toxicology Research. He obtained his Ph.D. from the School of Biotechnology, Banaras Hindu University, on the immunomodulatory effects of prolactin and growth hormone on macrophages. His work contributes to an in-depth understanding of the immunomodulatory and immunosuppressive effects of common food additives, drugs, and cosmetic ingredients. He has also worked on the tumorigenic potential of mycotoxins and the molecular mechanisms underlying skin cancer. Dr. Tripathi has over 16 years of research experience and has published 50 articles in various peer-reviewed international journals. His publications are highly cited, with an h-index of 25. He is a life member of the Indian Nanoscience Society and serves as an editorial board member for several prestigious international journals.

Contributors

Samreen Abbas Department of Biotechnology, IET Lucknow, Lucknow, India

M. Akshay Anand Regulatory Affairs, ITC Limited—ITC Life Sciences and Technology Centre, Bengaluru, Karnataka, India

Abhay Bajaj Environmental Toxicology Group, FEST Division, CSIR-Indian Institute of Toxicology Research (CSIR-IITR), Lucknow, Uttar Pradesh, India

Tanmayee Basu Department of Biotechnology, Thapar Institute of Engineering and Technology, Patiala, Punjab, India

Anugya Bhatt Division of Thrombosis Research, Department of Applied Biology, Biomedical Technology Wing, Sree Chitra Tirunal Institute for Medical Sciences and Technology, Trivandrum, Kerala, India

Shishir Bobate Environmental Toxicology Group, FEST Division, CSIR-Indian Institute of Toxicology Research (CSIR-IITR), Lucknow, Uttar Pradesh, India

Amrutha Chandran Division of Extracorporeal Devices, Department of Medical Device and Engineering, Biomedical Technology Wing, Sree Chitra Tirunal Institute for Medical Sciences and Technology, Trivandrum, Kerala, India

Rashi Chugh Department of Biotechnology, Thapar Institute of Engineering and Technology, Patiala, Punjab, India

Ashish Dwivedi CSIR-Indian Institute of Toxicology Research, Lucknow, India

Subhash Dwivedi University of Alabama, Tuscaloosa, AL, USA

Ranjit Singh Gujjar Division of Crop Improvement, Indian Institute of Sugarcane Research, Lucknow, Uttar Pradesh, India

Shalmali Kamat Department of Microbiology, St Joseph's University, Bengaluru, Karnataka, India

Prachi Kapoor University of Petroleum and Energy Studies, Dehradun, Uttrakhand, India

Malini Kotak Health and Hygiene Segment at ITC Ltd., Bangalore, India

Sacheen Kumar School of Biomedical Sciences, Galgotias University, Greater Noida, Uttar Pradesh, India

Rashmi Mehta Personal Care Products Business, ITC Life Sciences and Technology Centre, Bengaluru, India

Bhartendu Nath Mishra Department of Biotechnology, IET Lucknow, Lucknow, India

Dibyalochan Mohanty Department of Pharmaceutics, Anurag University, Hyderabad, Telangana, India

Renjith P. Nai Division of Thrombosis Research, Department of Applied Biology, Biomedical Technology Wing, Sree Chitra Tirunal Institute for Medical Sciences and Technology, Trivandrum, Kerala, India

Ruchi Pandey Fragrance and Flavour Development Centre, Kannauj, Uttar Pradesh, India

A. B. Pant CSIR-Indian Institute of Toxicology Research, Lucknow, India

Systems Toxicology Group, CSIR-Indian Institute of Toxicology Research, Lucknow, Uttar Pradesh, India

Academy of Scientific and Innovative Research (AcSIR), Ghaziabad, Uttar Pradesh, India

Shiv Poojan bioTox Solutions, Lucknow, India

Ajay Prasad University of Petroleum and Energy Studies, Dehradun, Uttrakhand, India

Akshita Puri PGTD Zoology, RTM Nagpur University, Nagpur, Maharashtra, India

Prabhat Ranjan Division of Cardiovascular Disease, Department of Medicine, The University of Alabama at Birmingham, Birmingham, AL, USA

Rishabh Shah Research & Innovation, Acme Hardesty, Blue Bell, PA, USA

Saniya Shakir Department of Biotechnology, IET Lucknow, Lucknow, India

Saurabh Shrivastav Department of Pediatric Surgery, Dr. Ram Manohar Lohiya Ayurvigyan Institute of Medical Sciences, Lucknow, Uttar Pradesh, India

Shakti Vinay Shukla Fragrance and Flavour Development Centre, Kannauj, Uttar Pradesh, India

Abhishank Singh Growth Pharmaceutical Private Limited, Andhra Pradesh MedTech Zone (AMTZ), Visakhapatnam, Andhra Pradesh, India

Abhishek Singh Growth Pharmaceutical Private Limited, Andhra Pradesh MedTech Zone (AMTZ), Visakhapatnam, Andhra Pradesh, India

Ankita Srivastava Systems Toxicology Group, CSIR-Indian Institute of Toxicology Research, Lucknow, Uttar Pradesh, India

Ankur Kumar Srivastava Systems Toxicology Group, CSIR-Indian Institute of Toxicology Research, Lucknow, Uttar Pradesh, India

Neeraj Kumar Tiwari Babasaheb Bhimrao Ambedkar University (Satellite Centre), Lucknow, Uttar Pradesh, India

Amit Kumar Tripathi School of Biomedical Sciences, Galgotias University, Greater Noida, Uttar Pradesh, India

Anurag Tripathi CSIR-Indian Institute of Toxicology Research, Lucknow, India

Shreya Tripathi School of Biomedical Sciences, Galgotias University, Greater Noida, Uttar Pradesh, India

Atul Kumar Upadhyay Department of Biotechnology, Thapar Institute of Engineering and Technology, Patiala, Punjab, India

Editorial: Hair Throughout History—A Timeline of Its Transformative Journey

1

Ashish Dwivedi, A. B. Pant, Anurag Tripathi, Shiv Poojan, and Malini Kotak

Hair dyeing has a rich history, dating back to ancient times when dyes were derived from plants such as henna, indigo, turmeric, and amla. Synthetic dyes began in the 1860s with the discovery of paraphenylenediamine's reactivity. Eugène Schueller created the first synthetic hair dye in 1907 the first launch of a home color product was "Poly Color," in 1947 (Pointer 2005). Today, hair dyeing is a thriving multibillion-dollar industry, utilizing both plant-derived and synthetic dyes. Throughout history, people have altered their hair color for personal adornment and social status. Ancient Egyptians used henna and dried cow's blood to create red and black shades, while Greeks rinsed their hair with potassium solutions and sun-exposed ointments for lighter hues. In the sixteenth century, Venetian women bleached their hair in the sun. In 1907, Eugène Schueller developed the first synthetic hair colorant, leading to the rise of permanent hair coloring in the 1920s (Pointer 2005). Despite risks like irritant dermatitis and potential carcinogens, hair coloring gained popularity with the introduction of home application products. Over time, hair coloring evolved with changing trends. In the Middle Ages, blonde and red were fashionable, while the Renaissance popularized strawberry blonde. Wigs powdered white were chic in the nineteenth century. The twentieth century saw the advent of chemical coloring treatments, becoming widely popular in the 1920s. Today, hair coloring is a common salon treatment and offers diverse options, reflecting a preference for variety. The future of hair coloring holds continued innovation and alternative options for consumers. Plant-based dyes offer a natural alternative for coloring hair. Henna, derived from the Lawsonia inermis plant, is known

A. Dwivedi (✉) · A. B. Pant · A. Tripathi
CSIR-Indian Institute of Toxicology Research, Lucknow, India

S. Poojan
bioTox Solutions, Lucknow, India

M. Kotak
Health and Hygiene Segment at ITC Ltd., Bangalore, India

A. Dwivedi et al. (eds.), *Hair Care Products*,
https://doi.org/10.1007/978-981-97-6740-3_1

for its orange hue. Its active component, lawsone, binds to keratin, making it semi-permanent to permanent, depending on hair type. Over time, henna builds up from orange to red and then auburn. Indigo, sourced from plants like Indigofera tinctoria, can be combined with henna to create brown to black shades. However, some commercial henna products contain metallic salts that may react unpredictably with hydrogen peroxide used in perms or synthetic hair dyes, leading to unwanted tones and skin irritation. Pure plant-based options are available to avoid such issues. Additionally, in 2018, a sustainable method utilizing blackcurrant fruit skin waste to extract anthocyanin pigments for semi-permanent hair dyes was developed, offering intense blue color with high build-up and stability through multiple washes (Rose et al. 2018).

Synthetic hair dye offers various options for coloring hair, categorized as permanent, demi-permanent, semi-permanent, and temporary. Permanent dyes contain ammonia and require a developer to permanently change hair color by opening the cuticle layer and penetrating the cortex. They provide lasting results and are effective in lightening hair. Demi-permanent dyes utilize alkaline agents other than ammonia and a lower concentration of hydrogen peroxide. They offer coverage for gray hair and a more natural look with less damage compared to permanent dyes. They gradually fade over 20–28 shampoos. Semi-permanent dyes don't contain ammonia or peroxide, making them less damaging to hair (Morel and Christie 2011; Søsted et al. 2002; Gupta 2018). They provide a more natural look with subtle shade variations but cannot lighten hair. Temporary color options include rinses, shampoos, gels, sprays, and foams, often used for special occasions. They offer vibrant colors but do not penetrate the hair shaft, easily washed out with shampoo.

While synthetic dyes provide a range of options, some may contain potentially harmful chemicals like PPD. Temporary options, including plant-based alternatives, offer nonpermanent solutions with less risk. Alternative hair coloring products offer a range of vibrant hues not commonly found in natural hair colors, often referred to as "vivid colors." These include shades like green and fuchsia, with some options available in permanent formulations. Recent innovations include blacklight-reactive dyes that fluoresce under UV light, popular in nightclub settings. These alternative color dyes typically contain only tint and no developer, meaning they create the brightest color when applied to light blond hair. Darker hair shades, from medium brown to black, may require bleaching for optimal color absorption. Fair hair types may achieve more vivid results after bleaching. However, undertones like gold, yellow, and orange in unbleached hair can affect the final color outcome, particularly with pink, blue, and green dyes (Lee et al. 2015). While some alternative colors offer semipermanent results, complete color removal from bleached or pre-lightened hair may take several months.

References

Gupta M (2018) A study of knowledge, attitude and practices regarding hair dye use among general population. Our Dermatol Online 9(1):15–18. https://doi.org/10.7241/ourd.20181.4

Lee S, Chen Y, Lin C, Chen K (2015) Hair dyes resorcinol and lawsone reduce production of melanin in melanoma cells by tyrosinase activity inhibition and decreasing tyrosinase and microphthalmia-associated transcription factor (mitf) expression. Int J Mol Sci 16(1):1495–1508. https://doi.org/10.3390/ijms16011495

Morel O, Christie R (2011) Current trends in the chemistry of permanent hair dyeing. Chem Rev 111(4):2537–2561. https://doi.org/10.1021/cr1000145

Pointer S (2005) The artifice of beauty: a history and practical guide to perfume and cosmetics. The History Press. ISBN: 978-0750938877

Rose PM, Cantrill V, Benohoud M, Tidder A, Rayner CM, Blackburn RS (2018) Application of anthocyanins from blackcurrant (Ribes nigrum L.) fruit waste as renewable hair dyes. J Agric Food Chem 66(26):6790–6798

Søsted H, Agner T, Andersen K, Menné T (2002) 55 cases of allergic reactions to hair dye: a descriptive, consumer complaint-based study. Contact Derm 47(5):299–303. https://doi.org/10.1034/j.1600-0536.2002.470508.x

Anatomy and Physiology Diversity of Hair: A Multidisciplinary Approach to Understanding Structure and Function

2

Shreya Tripathi, Subhash Dwivedi, Saurabh Shrivastav, and Amit Kumar Tripathi

Abstract

This chapter provides a comprehensive exploration of the anatomy and physiology of hair, offering insights into the classification of hair and types of hair follicles. A thorough analysis of the structure of the hair follicle and the hair shaft throws light on the complex makeup of these vital parts. Moving beyond anatomy, this chapter delves into the physiology of hair, with a particular focus on the growth cycle. The hair growth cycle is elucidated through its three distinct stages—Anagen, Catagen, and Telogen. The initiation of the Anagen phase is correlated with the developmental process of the hair follicle, emphasizing the dynamic nature of this cycle. The Catagen phase is explored as a carefully controlled involution process marked by a notable episode of apoptosis, contributing to the regulated progression of the hair growth cycle. Additionally, the maturation of the hair shaft during the Telogen stage is detailed, highlighting the formation of club hairs and their subsequent shedding. Furthermore, this chapter extends its exploration to the secretory glands associated with hair follicles, namely, the sebaceous gland, sweat gland, and apocrine gland. Each gland's role in contributing to the health and characteristics of hair is discussed, with a focus on sebum's impact on hair appearance, the composition of sweat, and the unique features of apocrine gland secretions.

S. Tripathi · A. K. Tripathi (✉)
School of Biomedical Sciences, Galgotias University, Greater Noida, Uttar Pradesh, India

S. Dwivedi
University of Alabama, Tuscaloosa, AL, USA

S. Shrivastav
Department of Pediatric Surgery, Dr. Ram Manohar Lohiya Ayurvigyan Institute of Medical Sciences, Lucknow, Uttar Pradesh, India

A. Dwivedi et al. (eds.), *Hair Care Products*,
https://doi.org/10.1007/978-981-97-6740-3_2

Keywords

Hair · Follicle · Anatomy · Physiology · Shaft

Abbreviations

FGF	Fibroblast growth factor
IGF	Insulin-like growth factor
IRS	Inner root sheath
ORS	Outer root sheath

2.1 Introduction

Hair is a distinctive characteristic present in all mammals, setting them apart from other members of the animal kingdom. In humans, this feature holds particular significance, especially for females, where it is often cherished for its aesthetic value. However, beyond its cosmetic role, the primary functions of hair lie in providing protection to the skin against mechanical insults and contributing to the maintenance of homeothermy (Maderson 2003).

Hair serves as a natural safeguard, acting as a barrier that helps shield the skin from external forces and potential injuries. This protective role becomes especially crucial in environments where the skin is susceptible to abrasions, impacts, or other physical stresses. Additionally, hair contributes to the regulation of body temperature by assisting in the retention of heat close to the body, thereby facilitating homeothermy—the ability to maintain a relatively constant internal temperature despite external fluctuations (Harrison and Davis 1999). The protective functions of eyebrows and eyelashes are crucial in preventing foreign objects from entering the eyes, thereby safeguarding the delicate ocular structures. Simultaneously, scalp hair serves as a multifaceted shield against environmental elements, including sunlight, cold temperatures, and physical damage to the head and neck region. Beyond its protective role, scalp hair contributes to sensory perception, enhancing the skin's responsiveness to tactile stimuli (Branch 2001; Randall and Botchkareva 2009). In addition, hair has a big impact on social and sexual communication, which affects psychological health and overall quality of life. Hair-related conditions like hirsutism and hair loss can have a significant impact on how someone feels about themselves and interacts with others (Branch 2001).

With notable exceptions on particular areas like the palm of the hand, the buccal surface of the lip, the sole of the foot, and some regions of the external genitalia, mammalian skin exhibits a widespread distribution of hair across the body surface. Human hair growth varies tremendously; most body parts have sparser, almost colorless hair, but the scalp, lashes, and eyebrows have thicker, more deeply colored hair that is longer and denser (Randall and Botchkareva 2009; Buffoli et al. 2014). The diversity in human hair extends beyond length and pigmentation, encompassing variations in form such as wavy, spiral, or straight strands. Moreover, variations in hair color are noted, which are determined by the interaction of various melanin subtypes. The two pigments that give hair its overall color are eumelanin, which gives brown to black tones, and pheomelanin, which gives yellow to reddish-brown tones. In addition, features like hair diameter, length, and cross-sectional shape add to the complexity of human hair. This complex interplay of variables emphasizes the remarkable diversity found in the human integumentary system and highlights the intricacy of hair biology and morphology across various body regions (Olabiyi et al. 2008; Ito and Wakamatsu 2011). The primary objective of this chapter is to provide a comprehensive overview of the principal anatomical and physiological characteristics associated with various types of human hair. This investigation seeks to clarify the clinical significance of these variations by examining the complex distributions and structures of hair throughout the human body. Through an integration of data pertaining to the anatomy and physiology of human hair, this chapter aims to provide a sophisticated understanding of the complex interactions between hair properties and their wider clinical implications. By doing this, we hope to further our understanding of the complex nature of human hair and establish the foundation for a deeper understanding of its clinical significance.

2.2 Hair Anatomy

2.2.1 Classification of Hair

Human hair differs in type according to its form, location, and a few essential features. Body posture, embryological development, and the surface of the infundibula—the funnel-shaped portion of the hair follicle—as well as the structural details of the hair follicles, such as density, size of follicular orifices, hair shaft diameter, and volume—are among the classification criteria.

Different body regions have different hair shaft diameters, which typically range from 16 to 42 μm. The forehead has the smallest shaft diameter, measuring 16 μm, while the sural and thigh regions have the largest, measuring 42 and 29 μm, respectively. Given the concentration of hair follicles in this area, forehead skin has the highest average hair follicle density (292 follicles/cm^2). The follicular infundibula, which is thought to be a possible reservoir for substances applied to the skin, also exhibits the largest volume in the sural region (0.18 mm^3/cm^2) and forehead (0.19 mm^3/cm^2). These variances draw attention to the various structural traits of hair in various body parts. The fact that the hair cycle varies in length between body parts only serves to increase the complexity. The hair cycle on the scalp takes 3–4 years to finish, but the brows finish their cycle in about 4 months. This complex interplay of variables highlights how complex human hair is and highlights the need for a thorough understanding when examining its range of characteristics (Buffoli et al. 2014; Otberg et al. 2004) (Table 2.1).

Table 2.1 Basic data of human hair follicles (data from Harkey 1993)

Total count	Scalp hair count	Hair cycle ratios of scalp hair	Duration of hair cycle phase	Hair shaft production rate (scalp)	Hair shaft diameter and length	Hair patterns	Hair shaft pigmentation
Almost 5,000,000	80,000–150,000	Anagen: 85–90%	Anagen: 2–6 years	~0.35 mm/day	Vellus: 0.06 mm; 1–2 mm	Scalp hair	Dark hair: redominance of eumelanin
		Catagen: 1%	Catagen: 2–3 weeks	1 cm/month	Terminal: >0.06 mm; 1–50 cm	Pubic and axillary hair	Blond/red hair: predominance of pheomelanin
		Telogen: 10–15%	Telogen: 3 months			Phalangeal hair	

2.2.2 Types of Hair Follicles

Human hair is categorized in a variety of ways according to a number of unique traits, such as diameter, color, length, texture, and shape. Vellus hair, terminal hair, and intermediate hair are the three main types that are distinguished by this classification.

The parts of the human body that are typically thought to be "hairless" are covered in vellus hair, which is distinguished by its short, fine strands and lack of pigment. These include the eyelids, forehead, and areas of the scalp where normal hair growth is absent. Conversely, terminal hair is distinguished by its texture, length, pigmentation, and large cross-sectional area. It is found mostly in the body's traditionally hairy areas, such as the pubic area, eyebrows, lashes, armpits, scalp, and beard (Wolfram 2003). Between vellus and terminal hair in terms of characteristics, intermediate hair resides in a medium ground. Adults usually have it on their arms and legs, where it exhibits a medium length and shaft size in contrast to the extremes embodied by vellus and terminal hair (Vogt et al. 2007; Wosicka and Cal 2010; Krause and Foitzik 2006).

Three different kinds of follicles produce different types of hair. Men and women both have nonsexual hair follicles situated on the arms and legs that produce intermediate hair. These specific follicles do not respond to hormone changes and remain unaltered after puberty. The axilla, pubic area, and temple region of the scalp are home to ambisexual hair follicles, on the other hand, the characteristics of the hair in the pubic and armpit areas undergo a noticeable change after puberty. The terminal hair changes from being fine, short, and soft at first to being coarse and long. In the temple area, on the other hand, there is an opposite change where the coarse terminal hair initially becomes fine vellus hair. The distinct reshaping of the facial outline is a result of this phenomenon. Male-specific hair follicles linked to male sexual traits are unique to men and are mostly found in certain areas of the body, including the abdomen, chest, ears, nose, and beard area. In response to increased androgen levels, these specific follicles experience notable changes that become apparent after puberty. The formerly fine vellus hair that emerges from these follicles changes dramatically during this developmental stage, becoming thicker and more robust terminal hair. One important part of the secondary sexual traits that males develop throughout their maturation process is represented by this change (Buffoli et al. 2014).

2.3 Structure of the Hair

Hair is a structural component that originates from specialized cells within the epidermis, which is the outermost layer of the skin. Hair is a complex structure comprised two distinct components: the follicle, nestled within the skin, and the hair shaft, which is the visible portion extending beyond the body surface. The follicle serves as the anchor for the hair and is embedded within the skin, housing the root of the hair. This root is responsible for the growth and nourishment of the hair shaft.

Conversely, the portion of the hair that emerges from the skin's surface is known as the hair shaft. It is composed of protein-rich keratin and is the visible, outer layer of the hair. There are several factors that affect the shaft's appearance, texture, and color, including environmental factors and genetics.

2.3.1 Structure of the Hair Shaft

Hair is made up of three different types of cells: the outer cuticle, the central cortex, and possibly the central medulla. Each hair strand is intricately structured. Across one row of elongated cells that overlap one another, the cuticle forms the outermost layer of the cell. There are roughly 45 μm of these cuticle cells, each measuring between 0.5 and 1.0 μm thick. Protecting the internal fibers from the outside environment and holding the hair shaft in place within the follicle are the two main functions of the cuticle. However, the cuticle is susceptible to damage from various sources such as chemicals, heat, light, or mechanical trauma. Consequently, as one moves toward the distal end of the hair shaft, the cuticle may become progressively less intact, leading to fraying and potential disintegration. The cortex, which makes up the bulk of the hair shaft, is an essential structure made up of long, keratinized cells arranged into fibers that are about 100 μm long. A specific chemical cement holds these fibers together. There are tiny air spaces called fusi within the cortex. These gaps contain fluid in the hair root's living segment. But as the hair ages and becomes less hydrated, air takes the place of the fluid in the fusi. The cortical cells also contain pigment granules, and the type and orientation of these pigments determine the color of the hair. Each hair strand's unique strength, flexibility, and aesthetic qualities are enhanced by this complex composition and structure (Wolfram 2003; Publishing 2004) (Fig. 2.1).

Melanin is a predominant pigment within the human hair, skin, and eyes. The intricate process of melanin synthesis occurs within specialized organelles known as melanosomes, situated in the hair bulb and housed within melanocytes. Within this cellular environment, melanin emerges from the transformation of the amino acid tyrosine, catalyzed by the enzyme tyrosinase. Human hair color, ranging from deep black to pristine white, results from variations in the quantity, distribution, and types of pigments present. The foundational composition of hair includes medullar cells, initially loosely arranged but ultimately desiccating and contracting to form vacuoles along the fiber axis. Notably, the prevalence of medullar cells correlates with the diameter of the hair fiber. Fine animal hair predominantly comprises cuticle and cortex cells, while thicker varieties, like those from horses, porcupines, or manes, contain a higher proportion of medullar cells. In the context of human hair, medullar cells constitute a minor percentage of the overall mass and may even be entirely absent in very fine vellus hairs. Their presence can vary from continuous along the central axis to sporadic. Notably, human beard hair exhibits a more intricate medulla, and in some instances, a double medulla configuration may be discerned. This intricate interplay of melanin synthesis and hair cell composition contributes to the diverse spectrum of human hair characteristics (Harkey 1993).

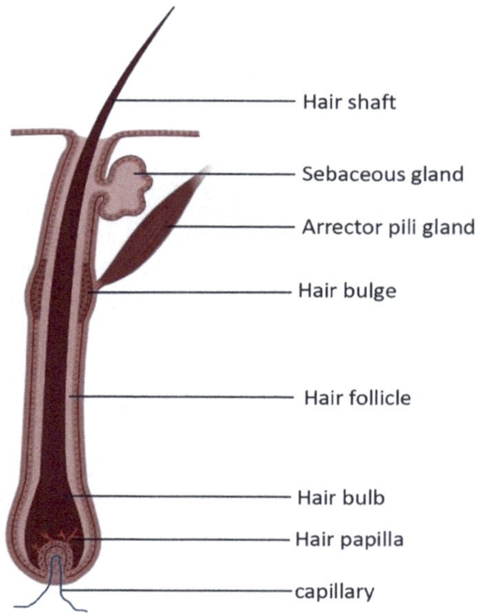

Fig. 2.1 Diagram of an anagen follicle

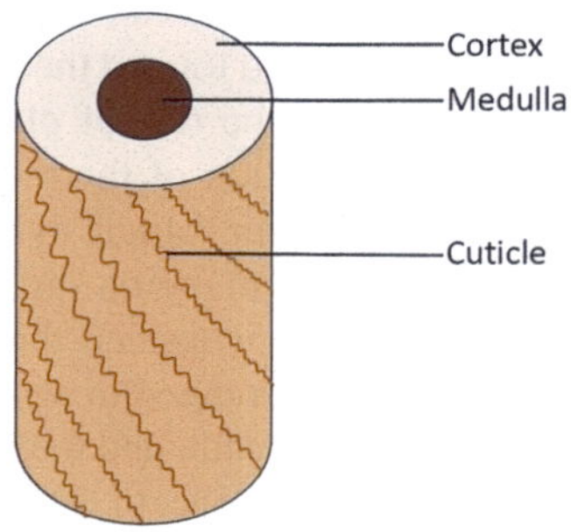

Fig. 2.2 Cross-section of a hair shaft showing the medulla, cortex, and cuticle

2.3.2 The Hair Follicle

The follicle serves as the crucial foundation for hair growth and is fundamentally composed of two discernible components: the upper segment, encompassing the infundibulum and isthmus, and the lower segment, comprising the hair bulb and suprabulbar region (Fig. 2.2). Remarkably, the upper portion maintains a consistent structure, whereas the lower region regenerates repeatedly (Randall and Botchkareva 2009; Buffoli et al. 2014). The infundibulum represents the topmost segment of the hair follicle, spanning from the sebaceous gland's opening to the skin's surface. This funnel-shaped structure is replete with sebum, an oily substance produced by the sebaceous glands. Specifically, the upper region, referred to as the acroinfundibulum, undergoes a process known as "epidermal mode" keratinization. This involves the development of layers such as the stratum granulosum and stratum corneum, mirroring the stratified arrangement observed in the epidermis. The isthmus, situated in the lower segment of the upper part of the hair follicle between the sebaceous gland opening and the insertion point of the arrector pili muscle, marks the initiation of epithelial keratinization. This keratinization, termed "trichilemmal keratinization," is characterized by a notable absence of the granular layer. At this juncture, the epithelial invagination becomes highly permeable to externally applied substances (Blume-Peytavi and Vogt 2011). Notably, only a small number of differentiated corneocytes persist in this region (Buffoli et al. 2014; Harkey 1993; Sperling 1991). Crucially, the bulge area on the isthmus, proximal to the arrector muscle insertion, is believed to harbor hair follicle stem cells (Oshima et al. 2001). Lineage studies have conclusively demonstrated the multipotent nature of bulge cells, with their descendants contributing to the development of the new lower anagen hair follicle (Ito et al. 2004). A defining attribute of stem cells is their slow-cycling behavior, presumed to safeguard their proliferative capacity and minimize potential DNA errors during replication. These cells migrate in a downward trajectory and, upon reaching the hair bulb matrix, undergo proliferation and terminal differentiation, culminating in the formation of the hair shaft and inner root sheath (IRS).

Furthermore, these stem cells exhibit distal migration, participating in the creation of sebaceous glands and responding to wounds by proliferating (Harkey 1993; Oshima et al. 2001; Cotsarelis et al. 1990). It is noteworthy that their migratory pattern is directed toward the lower regions. By presenting this information in a different manner, the risk of plagiarism is mitigated, ensuring a comprehensive yet distinct articulation of the intricate processes associated with the isthmus and its pivotal role in hair follicle dynamics.

The anatomical region situated above the isthmus and below the hair bulb within the hair follicle is termed the suprabulbar region. This segment consists of three distinct layers arranged in an outer-to-inner sequence: the outer root sheath (ORS), the inner root sheath, and the hair shaft.

The outer root sheath (ORS) is an integral component of the hair follicle, spanning from the epidermis at the infundibulum to the hair bulb, exhibiting dynamic cellular transformations along its course. Initially mirroring the characteristics of the epidermis in the infundibulum, the ORS undergoes significant changes as it progresses toward the isthmus level. At this stage, ORS cells embark on a trichilemmal keratinization mode, deviating from their initial epidermal resemblance. Keratinocytes within the ORS play a pivotal role in the formation of the bulge area situated at the base of the isthmus. Notably, at the lower extremity of the hair bulb, the ORS manifests as a single layer of cuboidal cells, undergoing a transition into a multilayered arrangement in the upper hair bulb region. This dynamic alteration in cell organization contributes to the structural complexity of the hair follicle. In certain follicles, an additional layer known as the companion layer emerges, positioned between the outer and inner root sheaths. This distinct single-cell layer assumes a crucial intermediary role within the follicular structure, further highlighting the intricacies of the cellular composition in different regions of the hair follicle (Rothnagel and Roop 1995). Cells within the companion layer display numerous intercellular linkages to the inner root sheath. It is hypothesized that these cells migrate toward the isthmus region, concurrently with the inner root sheath. Ultimately, this migration results in the creation of a separation plane between the inner and outer root sheaths (Buffoli et al. 2014; Harkey 1993; Sperling 1991).

The outer root sheath (ORS) of the hair follicle encompasses various cell types, including melanocytes, Langerhans cells, and Merkel cells. These diverse cellular components play integral roles within the follicle, contributing to its multifunctional nature. Melanocytes, responsible for pigment production, add color to the hair and surrounding skin. Langerhans cells, acting as immunological sentinels, participate in the skin's defense mechanisms. Merkel cells, functioning as sensory receptors, endow the follicle with the ability to perceive stimuli. This intricate collaboration among different cell types within the ORS establishes the hair follicle not only as a structural component but also as a dynamic and responsive organ, showcasing sensory and immunologic functions essential for skin health (Paus and Cotsarelis 1999).

The inner root sheath (IRS) serves a crucial role in anchoring and shaping the hair shaft within the hair follicle. Comprising three distinct layers—Henle's layer, Huxley's layer, and the cuticle layer—the IRS undergoes a process of keratinization that significantly influences the structure and integrity of the hair. The innermost layer, the cuticle of IRS, establishes a formidable connection by interlocking with the cells of the hair cuticle, creating a tightly bound linkage between the hair shaft and the follicle. Remarkably, the inner root sheath undergoes hardening before the emerging hair within it, thereby exerting control over the definitive shape of the hair shaft. This intricate hardening process occurs through abrupt keratinization, a phenomenon observed individually in each of the three layers (Fig. 2.3). While the patterns of keratinization differ at various levels within the IRS, the transformation initiates in Henle's layer, the outermost segment. Specifically, Henle's layer undergoes keratinization first, paving the way for subsequent changes in the remaining layers. Above Henle's layer, at a region known as Adamson's fringe, the Huxley

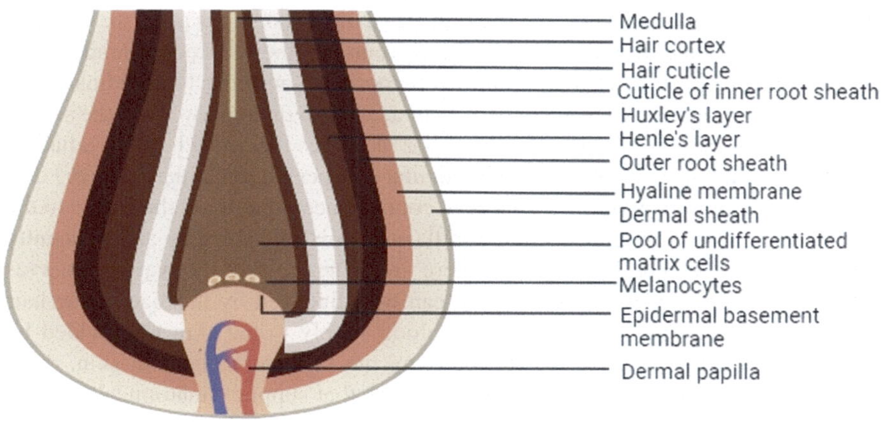

Fig. 2.3 Diagram of proximal hair follicle

layer undergoes keratinization, contributing to the overall structural development of the IRS. The entire process of keratinization in the IRS is instrumental in coating and supporting the hair shaft, extending its influence up to the isthmus level. Notably, the IRS undergoes disintegration at the isthmus level, marking the conclusion of its supportive role in the hair development process.

The hair bulb, which is the active reproductive center of the hair follicle, is the enlarged onion-shaped area found in the lower segment of the hair follicle that includes the follicular papilla as well as the hair matrix. The hair bulb contains the mucopolysaccharide-rich stroma, nerve fibers, capillary loops, and follicular dermal papilla. Essential to hair development, matrix cells are grouped at the base of the follicle and cover the whole follicular papilla. Notably, these matrix cells are the source of the hair shaft and inner root sheath (IRS). The pigment that gives hair its color is produced primarily by melanocytes, which are found among matrix stem cells. As matrix cells go through their differentiation stage, they phagocytose melanin or pheomelanin from the dendritic extensions of melanocytes. The coloration of the hair is consequently determined by both the quantity and the nature of the major pigment that is phagocytosed (Buffoli et al. 2014). The mesenchymal cells that condense into the follicular papilla during the early stages of follicular embryogenesis are essential for the initiation and maintenance of follicular epithelial differentiation. It is important because it identifies the type of follicle and influences the anagen hair bulb size, anagen phase length, and the diameter of the resulting hair shaft (Krause and Foitzik 2006; Sperling 1991; Rogers et al. 2006; Paus et al. 1999).

2.3.3 Neurovascular Network of Hair Follicles

The structure of the nerves supplying our hair follicles is comparable to the skin's nerve network. These nerves comprise autonomic sympathetic nerves as well as sensory (feeling) nerves. They begin in the dermis, or the layer immediately beneath the skin, and proceed from there to the hair follicle via the dermal network, covering the bulb and the outermost layer of the skin. The hair end organ is a unique arrangement of nerves found in some hair follicles surrounding the sebaceous gland, which secretes oil for the skin. This structure is composed of an inner longitudinal layer and an outer circular layer. A secondary, primarily horizontal network in the skin is formed by some of the fibers from this collar of nerves that branch out to adjacent structures. At the meeting point of the reticular (deeper) and papillary (upper) dermis lies this second network. This network contains nerve branches that travel to the upper layers of skin (epidermis and papillary dermis), where they are frequently linked to blood vessels. Interestingly, there are differences in the way that nerves supply individual hair follicles, and this can be influenced by the size of the hair (Iggo and Muir 1969). Additionally, the blood supply to the skin is provided by small arteries called arterioles. After entering the fat layer under the skin, these arterioles ascend into the skin to form a network that supplies anatomical features such as hair follicles. Around the lower third of the hair follicle, these arteries form a dense network connected by cross-shunts. They are more concentrated in the lower portion of the hair follicle (Arck et al. 2001). Variations in the blood flow have been noted at various stages of the hair growth cycle, particularly in the growth phase (anagen), when the number of blood vessels surrounding the hair follicle increases. This increase is correlated with the outer root sheath's (ORS) (Randall and Botchkareva 2009) increased expression of a substance known as vascular endothelial growth factor.

2.4 Hair Physiology: Understanding the Hair Growth Cycle

Every individual hair follicle undergoes a continuous and cyclic progression through three distinct stages, contributing to the dynamic and perpetual nature of the hair growth process. The three stages are Anagen, Catagen, and Telogen. Multiple growth factors and their corresponding receptors play crucial roles in the normal development and cycling of hair follicles. It's essential to note that there isn't a singular growth factor that exerts ultimate control over these intricate processes. Instead, a complex interplay of various growth factors orchestrates the intricate dance of hair follicle development and cycling (Paus and Cotsarelis 1999).

2.4.1 Anagen Stage

The beginning of the anagen phase corresponds to the developmental process of the hair follicle, since the beginning of the formation of the lower hair follicle is caused by the increased proliferation of secondary germ cells found in the bulge area. This initiates a dynamic sequence that recapitulates the complex development of the hair follicle, in which the production of new structures is closely correlated with the robust activity of secondary germ cells present in the bulge (Cotsarelis et al. 1990; Lyle et al. 1998) (Fig. 2.4). The involvement of identical proteins and signaling pathways in both the embryonic folliculogenesis process and the initiation of anagen in adults remains unclear. Nevertheless, it is crucial to acknowledge that interactions between the dermal papilla and the epithelial cells covering the follicles play a pivotal role in both of these biological processes (Botchkarev and Kishimoto 2003; Hardy 1992). Two crucial molecules integral to the intricate processes of hair-follicle development and cycling are insulin-like growth factor 1 (IGF-1) and fibroblast growth factor 7 (FGF-7). These signaling molecules play pivotal roles in regulating various aspects of hair physiology. IGF-1 and FGF-7 are primarily produced by the dermal papilla, which secretes these molecules to transmit vital signals for healthy hair growth and cycling. Regarding their target cells, the IGF-1 and FGF-7 receptors are primarily found in the matrix cells that are located above the dermal papilla. This arrangement of molecules in space highlights their importance in coordinating the communication between the dermal papilla and the surrounding matrix cells, which are essential to the process of hair growth. IGF-1 and FGF-7 regulate cell proliferation, differentiation, and overall follicular activity by binding to their respective receptors on matrix cells (Danilenko et al. 1996).

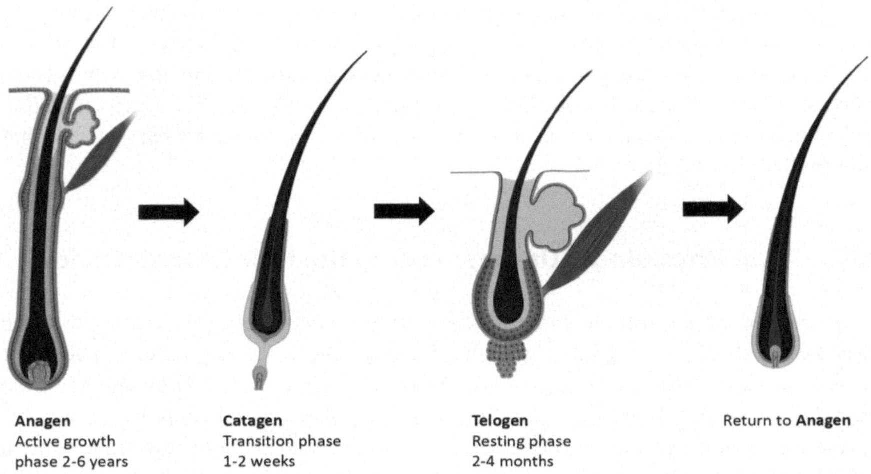

Fig. 2.4 Anagen stage

Hair follicles across various regions of the body exhibit distinct characteristics in terms of hair length, and this phenomenon is closely linked to the duration of the anagen cycle, the growth phase of the hair follicle. To illustrate, the follicles on the scalp maintain an extended anagen stage lasting 2–8 years, resulting in the production of long hairs. In contrast, eyebrow hair follicles have a significantly shorter anagen stage, typically lasting only 2–3 months, leading to the formation of shorter hairs. The termination of the anagen stage is intricately regulated by a key factor known as fibroblast growth factor 5 (FGF5). Interestingly, FGF5 expression initiates within the follicle just before the conclusion of the anagen stage. This regulatory process contributes to the modulation of hair length, ensuring that hair growth is finely tuned and tailored to the specific characteristics of each body region (Rosenquist and Martin 1996). Mice exhibiting a genetic deficiency in fibroblast growth factor 5 (FGF5) manifest a distinctive phenotype known as "angora." This phenotype is characterized by a prolonged anagen stage in the hair growth cycle. As a consequence of this extended anagen phase, the mice display a notable physical trait—their fur is 50% longer than that of mice with typical FGF5 function (Hansen et al. 1997; Hébert et al. 1994; Murillas et al. 1995).

2.4.2 Catagen Stage

During the catagen phase of the hair growth cycle, the hair follicles undergo a carefully controlled involution process. This stage is mostly characterized by a prominent episode of apoptosis, which is the scientific term for programmed cell death that primarily occurs within follicular keratinocytes. This carefully planned series of actions is essential to the natural progression of the hair cycle, contributing to the controlled regression and restructuring of the hair follicle during this specific stage (Lindner et al. 1997). During the catagen stage of the hair growth cycle, follicular melanogenesis, the process responsible for hair pigmentation, undergoes cessation. Additionally, some melanocytes within the hair follicles experience programmed cell death, a biological phenomenon known as apoptosis. As the catagen stage progresses toward its conclusion, a notable transformation occurs in the dermal papilla—the structure at the base of the hair follicle. The dermal papilla condenses and ascends, eventually settling beneath the bulging region of the hair follicle. This intricate sequence of events marks a crucial phase in the dynamic and precisely regulated life cycle of hair (Slominski et al. 1994). A critical interruption happens when the dermal papilla in the catagen stage of the hair growth cycle is unable to reach the bulge region. Loss of hair results from this disruption, which leads to the cessation of the follicle's cycling process, resulting in the loss of hair. This phenomenon is notably observed in individuals, as well as mice, exhibiting mutations in the hairless gene. The inability of the dermal papilla to properly interact with the bulge at this specific stage hinders the normal progression of the hair growth cycle, ultimately leading to hair loss (Panteleyev et al. 1998; Ahmad et al. 1998). The malfunction of this gene disrupts the intricate communication between the dermal papilla and the stem cells, resulting in permanent alopecia (inability of the hair follicle to regenerate and sustain normal hair growth) (Cotsarelis et al. 1990). During the catagen phase of the mouse hair growth cycle, a unique physiological event occurs, which is typified by the deliberate destruction of specific hair follicles via the coordinated action of inflammatory cells. This complex mechanism is suggestive of a planned or coordinated process that results in the intentional removal of specific organ structures. The fact that inflammatory cells are involved at this point highlights a deliberate and controlled aspect of tissue remodeling, which adds to the dynamic and well-regulated nature of the mouse hair follicle life cycle (Eichmüller et al. 1998). For some permanent forms of alopecia, aberrant programmed organ loss may be the cause (Paus et al. 2003).

2.4.3 Telogen Stage

Club hairs are the result of the maturation of the hair shaft during the telogen stage of the hair growth cycle. This club hair is eventually shed from the follicle, commonly occurring during activities like combing or washing. The mechanism behind this shedding remains somewhat ambiguous, including whether it is a passive event that occurs at the same time as the anagen stage, which is when new hair growth starts, or an actively regulated process. Most people naturally lose between 50 and 100 hairs on their scalp each day. It is thought that this shedding phenomenon is a typical aspect of the hair growth cycle. The actual telogen stage lasts for 2–3 months before the scalp follicles revert to the anagen stage, which signals the start of a fresh cycle of hair growth. It is noteworthy that the percentage of follicles in the telogen stage can differ considerably based on the specific area of the body. For example, a proportion of 5–15% of scalp follicles are usually in the telogen stage at any given time, whereas a greater proportion of 40–50% is found in follicles on the trunk of the body (Wilson et al. 2007). Increased percentages of telogen-stage scalp follicles are a factor in increased hair loss. Therefore, drugs that can maintain or reduce the proportion of follicles in this stage have a great deal of potential for treating hair loss (Stenn and Paus 2001).

2.5 Secretory Glands

The apocrine, sebaceous, and sweat glands are the three different glands that are connected to the hair follicle. Since the hair shaft is covered in secretions from these glands, they are extremely important (Hopps 1977). As a result, these glands may function as the source of trace elements present in hair.

2.5.1 Sebaceous Gland

The sebaceous glands are distributed across the entire body surface, with the notable exceptions being the palms, soles, and dorsum of the foot. Positioned just beneath the skin's surface, these glands have ducts that directly connect to the hair follicle tunnel. Consequently, the hair shaft, once permanently formed, receives a coating of sebum just before emerging from the skin. An exception to this pattern is found in beard follicles, where the associated sebaceous gland has a duct that opens directly onto the skin surface. The scalp, forehead, cheeks, and chin house the largest and most abundant sebaceous glands. Sebum, the secretion of these glands, is a waxy substance. The significant volume of sebaceous secretions contributes to the oily appearance of unwashed hair. Composition of sebum exhibits some variation depending on the anatomical region it originates from. However, it generally comprises approximately one-third free fatty acids, one-third combined fatty acids, and one-third unsaponifiable materials such as squalene, cholesterol, and waxes.

2.5.2 Apocrine Gland

Apocrine glands are located in particular areas of the human body, including the eyelids, the perineal region, the external auditory meatus (ear canal), and the axilla (armpits). These glands release their substances by separating the cytoplasm from the secreting cells, which is how they differ from the sweat glands on the scalp in that regard. Apocrine glands also secrete their secretions straight into the hair follicle as opposed to the skin's surface, which is another difference between them and scalp sweat glands. The substance secreted by apocrine glands is an oily, colorless, and odorless liquid. Despite its widespread presence in the body, little is currently known about the precise composition or physiological function of this substance. However, it is established that the development of body odor stems from the bacterial decomposition of the apocrine gland secretions.

2.5.3 Sweat Glands

Merocrine sweat glands, also known as eccrine sweat glands, are widely distributed across almost the entire surface of the body. Though not directly attached to the hair follicles, the ducts from these glands emerge close to where the dermal layer of the skin begins. A coiled segment situated in the dermis and a straight duct opening onto the epidermis make up basic tubular structure of each gland. The fluid secreted by the eccrine sweat glands does not involve the removal of cytoplasm from the secreting cells. The quantity and composition of eccrine sweat can vary significantly based on individual factors and environmental conditions. The primary components of sweat include water, salts such as sodium (Na) and potassium (K), along with urea, amino acids, lactate, and pyruvate (Stowe and Plaa 1968).

2.6 Conclusion

To sum up, this chapter offers a thorough exploration of the complex realm of hair anatomy and physiology. Our knowledge of these essential elements is improved by the thorough examination of hair classification, follicle types, and the structural complexity of the hair shaft and follicle. The investigation delves into the physiology of hair in addition to its anatomy, revealing the active phases of the growth cycle—Anagen, Catagen, and Telogen. The regulated nature of the hair growth process is highlighted by the relationship between the Anagen phase and follicular development, the controlled involution of the Catagen phase marked by apoptosis, and the maturation of hair during Telogen, including the formation and shedding of club hairs. The chapter also discusses the function of secretory glands, including the apocrine, sweat, and sebaceous glands, in determining the type and condition of hair. The whole picture of what influences hair biology includes the effects of sebum on hair appearance, the makeup of sweat, and the unique characteristics of apocrine gland secretions. This chapter, taken as a whole, is a great tool for understanding the complex interactions between the physiology and anatomy of hair. Clarifying these complex mechanisms advances our understanding of hair biology and creates the groundwork for future research into possible uses in disciplines like dermatology and cosmetic science.

References

Ahmad W et al (1998) Alopecia universalis associated with a mutation in the human hairless gene. Science 279(5351):720–724. https://doi.org/10.1126/science.279.5351.720

Arck PC, Handjiski B, Hagen E, Joachim R, Klapp B (2001) Indications for a brain-hair follicle axis: inhibition of keratinocyte proliferation and up-regulation of keratinocyte apoptosis in telogen hair follicles by stress and substance P. FASEB J 15:2536–2538. http://www.fasebj.org/cgi/reprint/00-0699fjev1

Blume-Peytavi U, Vogt A (2011) Human hair follicle: reservoir function and selective targeting. Br J Dermatol 165(Suppl. 2):13–17. https://doi.org/10.1111/j.1365-2133.2011.10572.x

Botchkarev VA, Kishimoto J (2003) Molecular control of epithelial-mesenchymal interactions during hair follicle cycling. J Investig Dermatol Symp Proc 8(1):46–55. https://doi.org/10.1046/j.1523-1747.2003.12171.x

Branch TM (2001) Is alopecia areata an autoimmune disease? Lancet 358:11–13

Buffoli B et al (2014) The human hair: from anatomy to physiology. Int J Dermatol 53(3):331–341. https://doi.org/10.1111/ijd.12362

Cotsarelis G, Sun TT, Lavker RM (1990) Label-retaining cells reside in the bulge area of pilosebaceous unit: implications for follicular stem cells, hair cycle, and skin carcinogenesis. Cell 61(7):1329–1337. https://doi.org/10.1016/0092-8674(90)90696-C

Danilenko M, Ring D, Pierce F (1996) Growth factors and cytokines in insights from and the potentials. Mol Med Today 2(11):460–467

Eichmüller S et al (1998) Clusters of perifollicular macrophages in normal murine skin: physiological degeneration of selected hair follicles by programmed organ deletion. J Histochem Cytochem 46(3):361–370. https://doi.org/10.1177/002215549804600310

Hansen LA et al (1997) Genetically null mice reveal a central role for epidermal growth factor receptor in the differentiation of the hair follicle and normal hair development. Am J Pathol 150(6):1959–1975

Hardy MH (1992) The secret life of the hair follicle. Trends Genet 8(2):55–61. https://doi.org/10.1016/0168-9525(92)90350-D

Harkey MR (1993) Anatomy and physiology of hair. Forensic Sci Int 63(1–3):9–18. https://doi.org/10.1016/0379-0738(93)90255-9

Harrison JL, Davis KD (1999) Cold-evoked pain varies with skin type and cooling rate: a psychophysical study in humans. Pain 83:123–135

Hébert JM, Rosenquist T, Götz J, Martin GR (1994) FGF5 as a regulator of the hair growth cycle: evidence from targeted and spontaneous mutations. Cell 78(6):1017–1025. https://doi.org/10.1016/0092-8674(94)90276-3

Hopps HC (1977) The biologic bases for using hair and nail for analyses of trace elements. Sci Total Environ 7(1):71–89. https://doi.org/10.1016/0048-9697(77)90018-3

Iggo A, Muir AR (1969) The structure and function of a slowly adapting touch corpuscle in hairy skin. J Physiol 200(3):763–796. https://doi.org/10.1113/jphysiol.1969.sp008721

Ito S, Wakamatsu K (2011) Human hair melanins: what we have learned and have not learned from mouse coat color pigmentation. Pigment Cell Melanoma Res 24(1):63–74. https://doi.org/10.1111/j.1755-148X.2010.00755.x

Ito M, Kizawa K, Hamada K, Cotsarelis G (2004) Hair follicle stem cells in the lower bulge form the secondary germ, a biochemically distinct but functionally equivalent progenitor cell population, at the termination of catagen. Differentiation 72(9–10):548–557. https://doi.org/10.1111/j.1432-0436.2004.07209008.x

Krause K, Foitzik K (2006) Biology of the hair follicle: the basics. Semin Cutan Med Surg 25(1):2–10. https://doi.org/10.1016/j.sder.2006.01.002

Lindner G, Botchkarev VA, Botchkareva NV, Ling G, Van Der Veen C, Paus R (1997) Analysis of apoptosis during hair follicle regression (catagen). Am J Pathol 151(6):1601–1617

Lyle S, Christofidou-Solomidou M, Liu Y, Elder DE, Albelda S, Cotsarelis G (1998) The C8/144B monoclonal antibody recognizes cytokeratin 15 and defines the location of human hair follicle stem cells. J Cell Sci 111:3179

Maderson PFA (2003) Mammalian skin evolution: a reevaluation. Exp Dermatol 12(3):233–236. https://doi.org/10.1034/j.1600-0625.2003.00069.x

Murillas R, Larcher F, Conti CJ, Santos M, Ullrich A, Jorcano JL (1995) Expression of a dominant negative mutant of epidermal growth factor receptor in the epidermis of transgenic mice elicits striking alterations in hair follicle development and skin structure. EMBO J 14(21):5216–5223. https://doi.org/10.1002/j.1460-2075.1995.tb00206.x

Olabiyi AO, Akpantah AO, Oyerinde OF, Gbotolorun SC, Eluwa MA, Ekanem TB (2008) The distribution of hair on the phalanges of a sample population of Nigerian Yorubas in relation to sex, age and job type. Niger J Physiol Sci 23(1–2):101–104. https://doi.org/10.4314/njps.v23i1-2.54942

Oshima H, Rochat A, Kedzia C, Kobayashi K, Barrandon Y (2001) Morphogenesis and renewal of hair follicles from adult multipotent stem cells. Cell 104(2):233–245. https://doi.org/10.1016/S0092-8674(01)00208-2
Otberg N, Richter H, Schaefer H, Blume-Peytavi U, Sterry W, Lademann J (2004) Variations of hair follicle size and distribution in different body sites. J Invest Dermatol 122(1):14–19. https://doi.org/10.1046/j.0022-202X.2003.22110.x
Panteleyev AA, Van Der Veen C, Rosenbach T, Muller-Rover S, Sokolov VE, Paus R (1998) Towards defining the pathogenesis of the hairless phenotype. J Invest Dermatol 110(6):902–907. https://doi.org/10.1046/j.1523-1747.1998.00219.x
Paus R, Cotsarelis G (1999) The biology of hair follicles. N Engl J Med 341:491–497
Paus R, Mu S, Botchkarev VA (1999) Chronobiology of the hair follicle: hunting the "hair cycle clock". J Investig Dermatol Symp Proc 4:338–345. https://doi.org/10.1038/sj.jidsp.5640241
Paus R, Ito N, Takigawa M, Ito T (2003) The hair follicle and immune privilege. J Investig Dermatol Symp Proc 8(2):188–194. https://doi.org/10.1046/j.1087-0024.2003.00807.x
Publishing B (2004) Quantitative analysis and classification of AFM images of human hair. J Microsc 215:13–23
Randall VA, Botchkareva NV (2009) The biology of hair growth, 1st edn. William Andrew Inc. https://doi.org/10.1016/B978-0-8155-1572-2.50006-3
Rogers MA, Langbein L, Wunder SP (2006) Human hair keratin-associated proteins (KAPs). Int Rev Cytol 251(6):209. https://doi.org/10.1016/S0074-7696(06)51006-X
Rosenquist TA, Martin GR (1996) Fibroblast growth factor signalling in the hair growth cycle: expression of the fibroblast growth factor receptor and ligand genes in the murine hair follicle. Dev Dyn 205(4):379–386. https://doi.org/10.1002/(SICI)1097-0177(199604)205:4<379::AID-AJA2>3.0.CO;2-F
Rothnagel JA, Roop DR (1995) Hair follicle companiom layer: reacquainting an old friend. J Invest Dermatol 104(5 Suppl):42–43. https://doi.org/10.1038/jid.1995.59
Slominski A, Plonka P, Chakraborty A, Maurer M (1994) Melanogenesis during the anagen-catagen-telogen transformation of the murine hair cycle. J Invest Dermatol 102:862–869. https://doi.org/10.1111/1523-1747.ep12382606
Sperling LC (1991) Hair anatomy for the clinician. J Am Acad Dermatol 25(1):1–17. https://doi.org/10.1016/0190-9622(91)70167-Z
Stenn KS, Paus R (2001) Controls of hair follicle cycling. Physiol Rev 81(1):449–494. https://doi.org/10.1152/physrev.2001.81.1.449
Stowe CM, Plaa GL (1968) Extrarenal excretion of drugs and chemicals. Annu Rev Pharmacol 8:337–356
Vogt A et al (2007) Morphometry of human terminal and vellus hair follicles. Exp Dermatol 16(11):946–950. https://doi.org/10.1111/j.1600-0625.2007.00602.x
Wilson AS, Dodson HI, Janaway RC, Pollard AM, Tobin DJ (2007) Selective biodegradation in hair shafts derived from archaeological, forensic and experimental contexts. Br J Dermatol 157(3):450–457. https://doi.org/10.1111/j.1365-2133.2007.07973.x
Wolfram LJ (2003) Human hair: A unique physicochemical composite. J Am Acad Dermatol 48(6 Suppl):106–114. https://doi.org/10.1067/mjd.2003.276
Wosicka H, Cal K (2010) Targeting to the hair follicles: current status and potential. J Dermatol Sci 57(2):83–89. https://doi.org/10.1016/j.jdermsci.2009.12.005

The Genetics and Molecular Mechanism of Hair Color: From Melanocyte Development to Pigment Synthesis

3

Shreya Tripathi, Sacheen Kumar, Subhash Dwivedi, Saurabh Shrivastav, and Amit Kumar Tripathi

Abstract

Exploring the complex world of hair color, this chapter explains the underlying molecular mechanisms and genetic factors that give rise to the various shades of human hair. The exploration begins with a comprehensive analysis of the molecular processes that control the synthesis of eumelanin and pheomelanin, the pigments that give hair its color, is undertaken. An understanding of the fundamental biochemical pathways involved is provided by the elucidation of the complex processes of melanogenesis, which include the biosynthesis and regulation of each pigment. Subsequently, the emphasis turns to the regulatory factors that impact melanogenesis, specifically highlighting the crucial roles of key players in this complex process. Knowledge about the regulatory processes that transporters like cystinosin and Na/H exchangers orchestrate is presented, providing light on how these mechanisms affect the dynamic equilibrium of pigmentation. The exploration extends into the genetic landscape of hair color, focusing on the melanocortin 1 receptor (MC1R) gene. The function of various MC1R variants in the control of pigmentation and phenotypic expression is investigated, offering a detailed knowledge of the genetic factors influencing individual hair color. The chapter concludes by delving into the field of genome-wide association studies (GWAS), which are an effective means of revealing the intricate genetic foundations of hair color. This chapter provides a comprehensive understanding of

S. Tripathi · S. Kumar · A. K. Tripathi (✉)
School of Biomedical Sciences, Galgotias University, Greater Noida, Uttar Pradesh, India

S. Dwivedi
University of Alabama, Tuscaloosa, AL, USA

S. Shrivastav
Department of Pediatric Surgery, Dr. Ram Manohar Lohiya Ayurvigyan Institute of Medical Sciences, Lucknow, Uttar Pradesh, India

A. Dwivedi et al. (eds.), *Hair Care Products*,
https://doi.org/10.1007/978-981-97-6740-3_3

the genetics and molecular mechanisms underlying the fascinating diversity of human hair color by fusing molecular insights with genetic factors.

Keywords

Hair color · Melanogenesis · Eumelanin · Pheomelanin · Melanocortin 1 receptor (MC1R) · Melanocyte · Melanin biosynthesis · Genome-wide association studies (GWAS) · Genetic variation · Phenotypic expression

Abbreviations

ACTH	Adrenocorticotrophic hormone
cAMP	Cyclic adenosine monophosphate
Co	Cobalt
Cu	Copper
DCT	Dopachrome tautomerase
DHI	Dihydroxyindole
DHICA	Dihydroxyindole-2-carboxylic acid
DOPA	Dihydroxyphenylalanine
GWAS	Genome-wide association study
MC1R	Melanocortin 1 receptor
mRNA	Messenger ribosomal nucleic acid
MSH	Melanocyte stimulating hormone
Ni	Nickel
SNP	Single nucleotide polymorphism
TYRP1	Tyrosinase-related protein 1

3.1 Introduction

Hair color is a highly noticeable trait in people, and it comes in a variety of shades such as black, brown, blond, and red. The color of our hair is something that many people pay a lot of attention to because it has a big impact on how we look. For example, some Asians, particularly females, often choose to change the color of their hair by dyeing it. This is a common trend, and the reason behind it is to make their hair look lighter, which is often seen as more attractive. The decision to dye hair is not just about personal preference; it also has a significant impact on how individuals are perceived in society. People may choose to alter their hair color for various reasons, including cultural preferences, fashion trends, or to enhance their overall appearance. This practice highlights the social and cosmetic importance that hair color holds in our lives. The wide range of hair colors we observe primarily results from variations in the amount and proportion of two main types of pigments: black to brown **eumelanin** and reddish-brown to yellow **pheomelanin**. The diversity in hair pigmentation arises due to differences in the quantities and ratios of these two melanin pigments, giving rise to the various shades of hair color we see in individuals (Ito and Wakamatsu 2011a). The melanocortin 1 receptor (MC1R) gene has been recognized as a key factor influencing the range of human pigmentation. This gene has been pinpointed as a crucial element that contributes to the diversity observed in the general population concerning skin, hair, and eye color. The human MC1R gene has been successfully isolated through cloning techniques and is positioned on the 16th chromosome at the q24.3 locus (Rees 2003).

3.2 Hair Pigmentation

Hair follicles undergo a cyclical pattern of growth that is orchestrated by an enigmatic "biologic clock" situated within the skin itself. This cycle is characterized by three distinct developmental stages: a resting phase known as telogen, a growth phase called anagen, and a regression phase referred to as catagen. The intricate interplay between these stages governs the dynamic nature of hair growth. The process of hair pigmentation is intricately linked to the synthesis of melanin, a pigment responsible for hair color (Slominski et al. 1994; Hardy 1992). This melanin is produced within follicular melanocytes and then transferred to the keratinocytes in the pre-cortical hair matrix. Notably, this pigmentation process is closely tied to the anagen stage of the hair cycle, which is the active growth phase. One of the first indications of catagen induction is the effective shut-off of melanogenesis, or the synthesis of melanin, which is a vital component of this dynamic cycle. The hair cycle's regression is indicated by the catagen phase. The processes involved in the transfer and synthesis of melanin stop during this stage. Significantly, melanogenesis is absent during the telogen, or resting phase. This guarantees that during the times when hair follicles are dormant, the processes involved in hair pigmentation remain dormant (Ortonne and Prota 1993).

3.3 The Melanin Synthesis Pathway

Melanin pigment synthesis is a complex process that occurs primarily in follicular melanocytes within hair follicles. These specialized cells function inside organelles called melanosomes, which are cytoplasmic, membrane-bound structures. The delicate transfer of the synthesized melanin to the surrounding keratinocytes, which together aid in the formation of the hair shaft, is an essential step in this process. Tyrosinase is a multifunctional enzyme that is essential to the synthesis of melanin in melanosomes. By catalyzing the transformation of L-tyrosine into L-dopaquinone, this multifunctional enzyme plays a crucial role. This conversion is the first and most important step in the intricate biochemical process that is known as melanogenesis. The well-planned sequence of events that tyrosinase started emphasizes how important this enzyme is to the process of melanin synthesis. It's critical to realize that L-tyrosine is converted to L-dopaquinone at the beginning of the formation process of melanin, the pigment that gives hair its color. This molecular transformation prepares the way for later phases of melanogenesis, which gives rise to the wide variety of hair colors that people can have. Dopaquinone, identified as an ortho-quinone, exhibits a rapid reactivity with thiol compounds but has a relatively slower reaction with amines. In instances where essential thiol compounds, such as L-cysteine, are absent in melanosomes, dopaquinone undergoes an intramolecular reaction involving the amino group. This reaction leads to the formation of dopachrome through cyclodopa. Subsequently, dopachrome undergoes a spontaneous and gradual transformation, predominantly converting into 5,6-dihydroxyindole (DHI) through a decarboxylative rearrangement. Tyrosinase-related protein-2, or dopachrome tautomerase (DCT), can speed up the conversion of dopachrome to 5,6-dihydroxyindole-2-carboxylic acid (DHICA) (Tsukamoto et al. 1992). The isomerization (tautomerization) of dopachrome to DHICA is catalyzed by DCT. The pigment responsible for the color of black to dark brown hair, eumelanin, is formed through the oxidative polymerization of DHI and DHICA in different ratios. Notably, a redox reaction involving dopaquinone can be used to enhance the final step in eumelanogenesis. The precise function of TYRP1 in human pigmentation is still unknown, but tyrosinase and maybe Tyrp1 in mice may also speed up DHICA oxidation (Boissy et al. 1998). On the other hand, in melanosomes, L-cysteine exclusively produces the isomers 5-*S*- and 2-*S*-cysteinyldopa in a 5:1 ratio when it interacts with dopaquinone at a high enough concentration. Pheomelanin, a reddish-brown pigment, is ultimately produced by the oxidation of these cysteinyldopa isomers, which is dependent on tyrosinase and involves benzothiazine intermediates (Boissy et al. 1998; Ito and Prota 1977). This complex chain of events highlights the intricacy of melanin synthesis in hair follicles and clarifies the molecular processes behind the wide range of hair colors that people can have (Fig. 3.1).

The two main types of melanin pigments, eumelanin and pheomelanin, are known to be complex mixtures or copolymers that make up the majority of natural melanin pigments, including those found in hair. We call this phenomenon "mixed" melanogenesis. More specifically, two different types of melanin—eumelanin and pheomelanin—contribute to the overall pigmentation seen in tissues like hair. Pheomelanin gives red and yellow tones, while eumelanin gives black to dark brown hues. The broad range of colors observed in natural pigmentation is the consequence of the coexistence or combination of these two melanins (Ito and Wakamatsu 2003).

3.4 Ultrastructural Characteristics of Hair Melanogenesis

Many studies have explored the complexities of the ultrastructural and biochemical aspects of melanin pigmentation in hair. We'll take a close look at the results of these scientific investigations and make links between the observed data and the visible representation of hair color (Ortonne and Prota 1993).

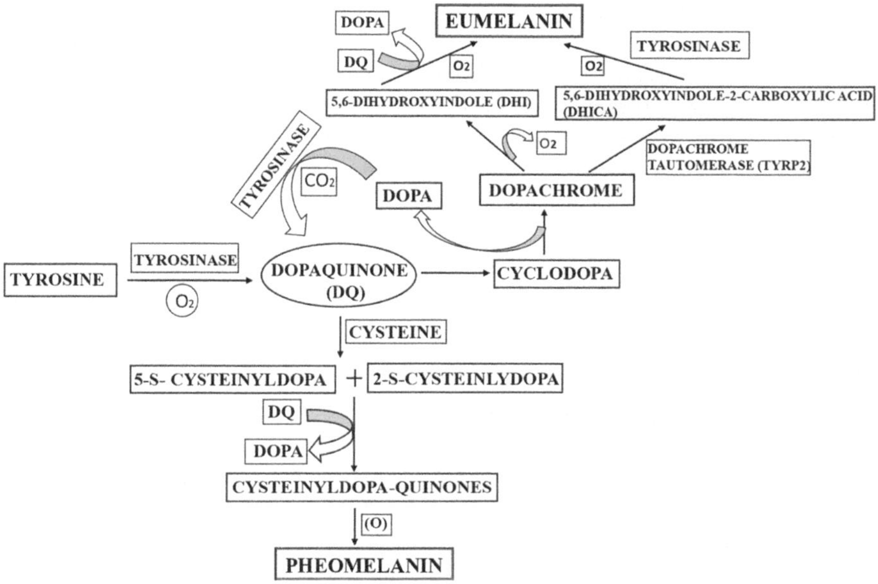

Fig. 3.1 For the production of eumelanin, the activities of tyrosinase, Tyrp1, and Dct/Tyrp2, along with the precursor tyrosine, are involved. In contrast, only tyrosinase, along with the precursors tyrosine and cysteine, is necessary for the production of pheomelanin. *DQ* dopaquinone, *DHI* dihydroxyindole, *DHICA* dihydroxyindole-2-carboxylic acid

3.4.1 Human Red Hair

Melanocytes in people with red hair—which, according to chemical analysis, is classified as pheomelanic—show special structures called melanosomes. These melanosomes are spherical in shape and comprise proteinaceous components as well as microvesicular (vesiculoglobular) matrices. Melanin deposition is sporadic and granular within these structures (Jimbow et al. 1983; Cesarini 1990). Particularly in the areas linked to yellow pigmentation, the melanization process in these organelles closely resembles that seen in pheomelanosomes found in yellow mice and tortoise-shell guinea pigs. Chemical analysis has revealed that the melanogenesis process in people with red hair is a "mixed" type. It is still unknown what exactly these "mosaic" melanosomes are made of—more specifically, whether they are pheomelanic, eumelanic, or a mix of the two. To fully understand the precise makeup and properties of these melanosomes, more investigation is required.

Furthermore, in this particular context, an additional kind of melanocyte has been detected. The production of ellipsoidal melanosomes in the typical form linked to eumelanin is attributed to this second type. The complexity of red hair pigmentation is a result of the diversity of melanocyte types and melanosome structures, underscoring the need for further research and comprehension of the complexities of melanogenesis in people with red hair (Jimbow et al. 1983).

3.4.2 Human Blond Hair

Few electron microscope studies have been conducted on human blond hair, especially since pheomelanosomes were discovered. There are few in-depth studies on human blond hair follicles in the body of current literature. Interestingly, melanin granules in blond hair are smaller and less prevalent than those in dark hair, according to electron microscope analyses. One interesting finding is that melanosomes do not fully melanize even in the dendritic processes of melanocytes in people with blond hair. This suggests that a quantitative decrease in melanosome production and melanization could be the cause of the characteristic light coloration of blond hair (Jimbow et al. 1976; Menon et al. 1980).

3.4.3 Human Black and Brown Hair

When follicular melanocytes are examined in the context of human hair color—more especially, black and brown hair—typical ellipsoidal melanosomes at different stages of melanization are found. These findings are in line with the ultrastructural traits found in the skin of people of Negroid and Caucasian ancestry (Birbeck et al. 1956). Melanosomes at distinct developmental stages are exhibited by follicular melanocytes in black hair, which is consistent with the ultrastructural characteristics of the epidermis of Negroid and Caucasoid people. The melanosomes, responsible for pigment production, are subsequently transferred to neighboring keratinocytes and are singly distributed (Jimbow et al. 1983). Similarly, in brown hair, follicular melanocytes encompass all the developmental stages of eumelanosomes, the melanosomes associated with black and brown pigment production. It is noteworthy that lighter brown hairs exhibit smaller melanosomes, indicating a variation in size among individuals. These characteristics are consistently observed, irrespective of the racial background (Tobin et al. 1999).

3.4.4 Senile Grey and White Hairs

The cells that produce melanin, known as melanocytes, have been found to be either normal or in lower numbers in the melanocytic zone of senile gray hair bulbs (Sadick and Laughlin 2004). Nevertheless, these cells have very few melanosomes—cellular organelles involved in the synthesis of melanin—and very little melanogenic activity. Conversely, dopa-positive melanocytes are absent in senile white hair. Melanocytes are either rare or nonexistent when viewed under an electron microscope, and melanin is not present in the hair's matrix or cortex. Similar observations have been made in white hair from vitiligo macules, indicating parallels in the lack or scarcity of melanocytes and melanin content. Importantly, in senile white hair bulbs, there is a notable absence of immunoreactive tyrosinase antigen, a key enzyme in melanin production (Lloyd et al. 1987). Tyrosinase mRNA or its protein has been found in senile white hairs according to more recent studies. This finding implies that the outer root sheath contains amelanotic melanocytes.

In other words, tyrosinase-related genes or proteins imply that there may still be melanocyte-like cells inside the outer root sheath, even though melanocytes in senile white hair may not exhibit typical melanogenic activity or be easily identified using standard techniques. This complicates our knowledge of the molecular alterations linked to senile gray and white hairs.

3.5 Follicular Melanin Unit: Understanding the Aging Process

The natural evolution of hair color is a notable and common phenomenon that occurs to most people at some point in their lives. During the early years of life, fair-haired people frequently experience a slow change in their hair color. This process is such that these people usually have brown hair by the time they hit middle age. Conversely, some kids who are born with brown hair might notice that as they grow older, their hair becomes lighter and more auburn or sandy in color (Ortonne and Prota 1993). This dynamic process of changing hair color adds an intriguing dimension to the wide range of appearances that people have and is a reflection of the complex interactions between our environment and genetics during our developmental journey (Rodrigues et al. 2017). Graying hair is an extremely common occurrence that affects almost everyone to some extent. Although genetic factors are the primary determinant of the onset of physiologic canities, or the natural graying of hair, other factors probably play a part as well. Graying can happen at any time; it usually happens in the late fourth or early fifth decade of life and is not gender specific. Those with dark hair and those with light hair exhibit distinctly different times of graying. Graying occurs more quickly in dark-haired people than in light-haired people. But it's interesting to note that a greater percentage of people with fair hair eventually turn completely gray. This seeming inconsistency could be clarified by keeping in mind that the early indicators of graying are easier to notice against a dark background than a light one. The precise reason behind senile greying of the hair is still unknown, though several factors have been discussed in scientific discourse. One noteworthy finding is that senile white hair follicles lack the Alpha-MSH (melanocyte-stimulating hormone) binding sites that are normally found in pigmented human scalp hair follicles (Nanninga et al. 1991). This disparity points to a possible connection between the aging process's development of grey hair and the loss of these binding sites. Although some theories suggest that a disruption in immune tolerance is a factor in senile greying, there is little evidence to support this theory (Burch and Jackson 1966). Scientists have studied senile greying of hair in animals, including guinea pigs, rats, rabbits, dogs, sheep, and horses (Hoekstra 2006; Phillips 2008). This has allowed researchers to obtain a better understanding of the phenomenon. With the use of these animal models, researchers hope to gain a better understanding of the underlying mechanisms and possible pathways pertaining to the age-related loss of hair pigmentation (Ortonne and Prota 1993).

According to a number of experimental results, some intermediate metabolites in the melanin pathway have cytotoxic properties. The demise of pigment cells may be caused by the slow deterioration of innate defense mechanisms in melanocytes of aging hair follicles. According to recent research, these processes may play a role in early greying, which is more common in light (BLf) mice. Because of clumping, uneven distribution, and a decrease in melanosomes, the hair in these mice appears pigmented at the tip but shows variable or no pigmentation at the base. The premature death of follicular melanocytes occurs next. Because this phenotype is unique to pigmented mice, it is likely mediated by the inherent cytotoxicity of pigment production (del Marmol and Beermann 1996). A single base change at the mouse brown locus, which codes for the TRP-1 protein, is the cause of this phenomenon. Although the exact role of TRP-1 remains unclear, this observation implies that it is essential for maintaining the stability of the melanosome structure. It is yet unknown if studying senescence and early graying of the hair in humans can benefit from using light mice as a model.

3.6 Regulatory Factors of Melanogenesis

For quite some time, it has been established that the enzymatic activity of tyrosinase, a copper-containing enzyme, plays a pivotal role in the process of melanin formation (Wang and Hebert 2006; Hearing and Jiménez 1987; Lukas and Griffiths 1982). Tyrosinase is a distinctive enzyme with the ability to catalyze three distinct reactions within the biosynthetic pathway of melanin. (1) Hydroxylation of Tyrosine to 3,4-Dihydroxyphenylalanine (DOPA): Tyrosinase facilitates the conversion of tyrosine into 3,4-dihydroxyphenylalanine (DOPA) through hydroxylation. (2) Oxidation of DOPA to DOPAquinone: The enzyme is involved in the oxidation of DOPA to DOPAquinone, marking a crucial step in the melanin biosynthetic pathway. (3) Oxidation of 5,6-Dihydroxyindole (DHI) to Indole-quinone: Tyrosinase also participates in the oxidation of 5,6-dihydroxyindole (DHI) to form indole-quinone.

The hydroxylation of tyrosine to DOPA is thought to be the most biologically significant of these reactions since the following steps in the sequence can happen on their own at physiological pH levels. New research suggests that there is active regulatory control over a number of other reactions in the pathway leading to the biosynthesis of melanin. But tyrosinase activity stands out as the most crucial, serving as the process's rate-limiting step overall. There are two main kinds of melanin that result from this process: Eumelanins: The metabolites of DOPAchrome are the source of these melanins. Phaeomelanins: Derived from cysteinylDOPA metabolites (Prota et al. 1998). The original theory that indole-quinone was a homogeneous polymer made up of eumelanins has been refuted. According to new discoveries, melanins are complex polymers that exhibit heterogeneity. The subunits of melanins are made up of all cyclized derivatives of DOPAquinone, possibly even DOPAquinone itself. Phaeomelanin synthesis seems to be dependent on sulfhydryl abundance rather than eumelanin synthesis. Glutamate or cysteine is likely to bind quantitatively to DOPAquinone as soon as it is generated in the presence of these sulfhydryls (Prota 1993; Jara et al. 1988). Thiols, or compounds containing sulfhydryl, are known to interact with these enzymes, altering their function and, as a result, affecting the synthesis of melanin as a whole (Prota 1980). Since the melanosome is an intracellular organelle that is membrane-bound, the regulatory mechanisms connected to the melanosomal membrane may have some effect on the concentration of melanogenic substrates within the melanosome. Notably, the concentration of substances involved in melanogenesis may be influenced by other factors within the intracellular environment of the melanosome, even though the uptake of tyrosine through this membrane is unregulated (Jara et al. 1990). Other essential regulatory processes take place after tyrosinase's enzymatic activity in the synthesis of eumelanin. One significant step is the use of catalytic processes to direct the synthesis to produce 5,6-dihydroxyindole-2-carboxylic acid (DHICA) instead of 5,6-dihydroxyindole (DHI). DHI can be produced spontaneously from DOPAchrome without the need for a specific enzymatic intervention (Pawelek 1991). This catalytic activity was first known as DOPAchrome conversion factor (Korner and Pawelek 1980), but it has since been renamed as DOPAchrome oxidoreductase (Barber et al. 1984; Leonard et al. 1988), DOPAchrome isomerase (Pawelek 1990), and finally DOPAchrometautomerase (Aroca et al. 1989). These changes in terminology reflect a growing comprehension of the role of the enzyme in directing DOPAchrome conversion to the intended DHICA product. Beyond the initial catalysis by tyrosinase, this enzymatic activity is critical to the complex regulatory network governing eumelanin synthesis. When it is noted that some divalent transition metal cations, such as Co_2, Cu_2, or Ni_2, can produce a comparable catalytic effect, the discussion of the catalytic activity in relation to DOPAchrome becomes more complex (Palumbo et al. 1987, 1988). A new level of complexity is added by this, since it raises the question of whether metal-induced activity or enzymatic action has more biological significance. It is noteworthy to acknowledge that there is no absolute necessity for these two mechanisms to be mutually exclusive. Currently, there is disagreement regarding the definitive assessment of the biological relevance of either metal- or enzyme-induced activity. Acknowledging the potential involvement of both mechanisms is imperative, and comprehending their mutual influence is critical to gaining a thorough understanding of melanin formation. This complexity is further enhanced by the distinction between melanins derived from DHICA and melanins derived from DHI. Although precise structural or functional differences between the two are unclear, their observable traits are remarkable. While DHICA-melanin appears yellowish-brown and finely dispersed, DHI-melanin is black and flocculent (Aroca et al. 1989). There must be notable differences in the underlying mechanisms or characteristics that control the synthesis of these melanins if there are clear differences in appearance and texture. To clarify the specifics of these variations and their significance within the framework of melanin biology, more study is necessary. Melanosomes contain DOPAchrometautomerase, which interacts with tyrosinase and possibly other proteins in the melanosomal membrane (Hearing et al. 1982; Kwon 1993). there has been a recent finding that suggests peroxidase can use DHI as a substrate (d'Ischia et al. 1991; Hearing and Tsukamoto

2017). scientists are currently looking into how peroxidase might be involved in melanogenesis.

3.7 Regulation of Melanogenesis by Various Transporters

There is presently a great deal of interest in the control of mixed melanogenesis, the process that gives human hair its variety of pigmentations (Ito and Wakamatsu 2011b; Ancans et al. 2001; Smith et al. 2004; Cheli et al. 2009). Tyrosinase, an essential enzyme in the synthesis of melanin, has been shown to exhibit progressive inhibition of its activity as pH drops. The production of lighter pigments is the hallmark of the pheomelanic phenotype, which is the result of this inhibition (Fig. 3.2).

The amount of cysteine in melanosomes, the cellular organelles that produce melanin, is another important factor regulating mixed melanogenesis (Chintala et al. 2005; Del Marmol et al. 1996). It is suggested that the relationship between melanosome cysteine levels and pH plays a major role in determining the different

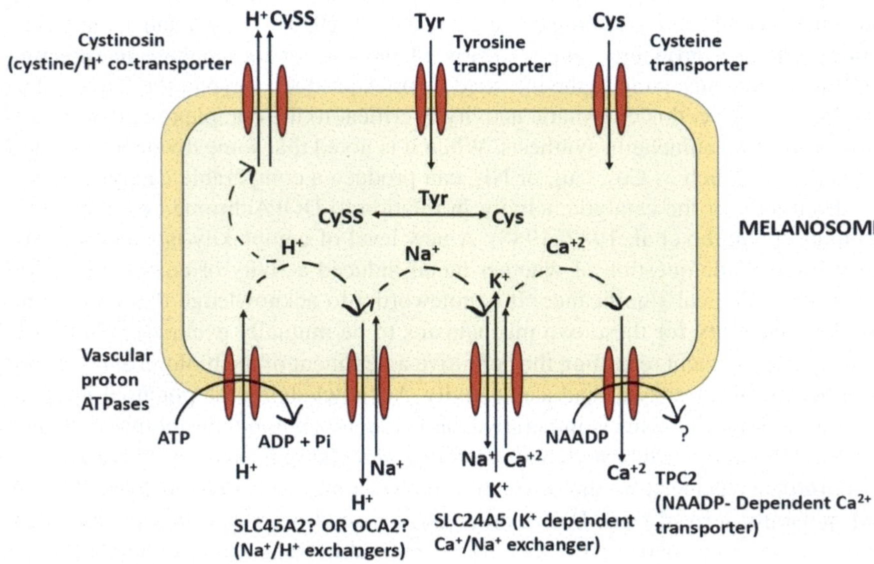

Fig. 3.2 The regulation of melanogenesis involves the active transport of precursors, tyrosine and cysteine, into melanosomes. Cysteine and its oxidized form, cystine, undergo redox exchange. Cystinosin (CTNS) facilitates the pumping out of cystine, regulating proton levels through vacuolar proton ATPases, Na^+/H^+ exchangers (SLC24A5 and TPC2). The overall effect is a neutral pH, high tyrosine, and low cysteine, promoting pro-eumelanogenicconditions (dark hair). Polymorphisms in ion transporters may make melanosomes acidic, suppressing eumelanogenesis (light hair). MC1R polymorphisms can down-regulate tyrosinase and transporters, shifting to pheomelanogenesis. The diagram illustrates these regulatory pathways (Smith et al. 2004; Walker and Gunn 2010; Potterf et al. 1999; Lamason et al. 2005). *CTNS* cystinosin, *MC1R* melanocortin 1 receptor, *CySS* dimer of cys

hair phenotypes that people experience, which can range from red to light to dark hair. The "dark" hair phenotype, which is characterized by black to dark brown hair, is essentially linked to full activity of ion transporters and cystinosin (the cystine transporter), as well as high levels of signaling through the melanocortin-1 receptor (MC1R). This results in neutral melanosomes that lack cysteine, which in turn causes significant amounts of eumelanin—the pigment that gives colors their darkness—to be produced. On the other hand, the "light" hair phenotype, which includes light brown to blond hair, is defined by increased MC1R signaling (Walker and Gunn 2010), decreased ion transporter activity, and preserved cystinosin activity (Kalatzis et al. 2001). As a result, lower amounts of eumelanin are produced by acidic melanosomes that are also cysteine-deficient. Numerous pigmentary genes are down-regulated in the "red" hair phenotype, which is linked to low levels of MC1R signaling. In these circumstances, melanosomes produce low to medium levels of pheomelanin (as well as some eumelanin) and become acidic but rich in cysteine (Walker and Gunn 2010; Le Pape et al. 2009).

3.8 Genetics of Hair Color

The melanocortin 1 receptor (MC1R), a crucial gene linked to the regulation of pigmentation, was identified as the extension locus by Cone and associates in 1993 (Robbins et al. 1993). This locus was previously associated with a number of mouse mutations, where recessive mutations produced yellow or pheomelanin-dominant hair as opposed to the brown/black eumelanic wild type (Rees 2003; Silvers 1979). On the other hand, because of elevated eumelanin synthesis, dominant gain-of-function mutations resulted in a black coloring. The seven-pass transmembrane receptor is encoded by the MC1R gene. When this receptor is activated, it modifies the ratio of eumelanin to pheomelanin by sending signals via cyclic AMP (cAMP) (Silvers 1979). Activation encourages the synthesis of brown or black melanin at the expense of yellow or red pheomelanin, acting as a sort of pigmentary switch. α-melanocyte stimulating hormone (αMSH) is the endogenous ligand for MC1R in mice, though related peptides such as adrenocorticotrophic hormone (ACTH) also demonstrate receptor activity. Furthermore, mice have a physiological antagonist called AGOUTI that is made in the skin. AGOUTI counteracts the effects of αMSH (Rees 2003) by acting in a paracrine way. The interaction between agonists and antagonists at the MC1R demonstrates the intricacy of the regulatory mechanisms controlling the synthesis of melanin and accounts for the variation in coat color seen in mice.

After successfully cloning (Chhajlani et al. 1993; Mountjoy et al. 1992) the human Melanocortin 1 Receptor (MC1R) gene, its genomic location was determined to be 16q24.3 (Gantz et al. 1994). Cone's prediction, which used red hair as an example to illustrate how variations in the MC1R gene might be linked to human pigmentary differences, has proven to be remarkably insightful (Robbins et al. 1993). Variations in the MC1R gene show a variety of traits that contribute to individual differences in nature and coloration. Determining the genetic basis of pigmentary variations in the human population requires an understanding of the spectrum of MC1R mutations. Humans with MC1R genes encode a protein with roughly 317 amino acids (Healy et al. 2000) that is similar to many other G-coupled receptors in structure. Unlike a number of other genes, MC1R is intron-free, though reports have suggested that it contains an extra exon 5′ (Tan et al. 1999; Harding et al. 2000) which may or may not be physiologically significant. In addition to MC1R, the melanocortin family contains four (Mountjoy et al. 1992; Cone et al. 1993) other well-known receptors: MC2R (ACTH receptor), MC3R, and MC4R, which are mostly found in the central nervous system and are important for energy homeostasis; and MC5R, which is widely expressed but has a notable impact on the function of mouse exocrine glands, particularly those found in the skin sebaceous glands (Hatta et al. 2001; Chen et al. 1997). The different receptors respond to ligands in somewhat similar ways, but the activity characteristics of MC2R seem different from those of the others (Schiöth et al. 1996, 1997; Schioth et al. 1995), and there seem to be differences between mouse and human receptors as well. Another cleavage product of pro-opiomelanocortin (POMC), ACTH, is also active in humans at the MC1R. Early human MC1R sequencing revealed sequence diversity was widespread and that certain alterations appeared to be connected to individuals with pale skin and red hair (Valverde et al. 1995). Additional data regarding the function of this signaling pathway in human pigmentary physiology is cited before going into greater detail about these sequence variants (Fig. 3.3).

The seven-segmented receptor that the MC1R gene encodes resides on cell membranes and interacts with G-proteins. A sequence of events within the cell is triggered when the MC1R receptor binds to specific stimulating molecules, also

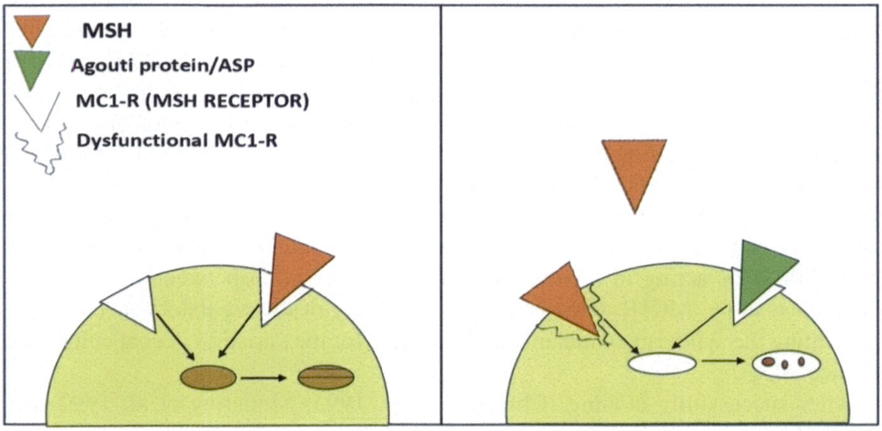

Fig. 3.3 Regulation of the melanogenesis in mammals. Melanocortin 1 receptor (MC1-R) dysfunction or the binding of agouti signaling protein (ASP) to MC1-R causes pheomelanogenesis, whereas the binding of α melanocyte-stimulating hormone (MSH) to MC1R increases eumelanogenesis. The latter process has only been studied in mice, not in people. *MC1-R* melanocortin 1 receptor, *ASP* agouti signaling protein, *MSH* α melanocyte-stimulating hormone (Schaffer and Bolognia 2001)

known as agonists, such as α-melanocyte-stimulating hormone (α-MSH) or adrenocorticotropic hormone (ACTH). Adenylyl cyclase activation, which results in the synthesis of cyclic AMP (cAMP) (Mountjoy et al. 1992), is one example of this. The transcription factor family (cAMP responsive-element-binding protein) CREB is then phosphorylated by elevated cAMP levels (D'Orazio et al. 2006; Barsh et al. 2000), activating a number of genes, including the one encoding the microphthalmia transcription factor (MITF) (Levy et al. 2006). The expression of enzymes and other components involved in the synthesis of pigments and cell differentiation depends on MITF. An antagonistic protein against MC1R is encoded by the agouti gene (Barsh et al. 2000). The agouti gene encodes this antagonist, which is in charge of giving mice's fur its distinctive color patterns. This gene can be mutated to cause color changes in fur, such as the black fur of the C57BL6 mouse strain (Rana et al. 1999). MC1R plays a special role in human hair pigmentation. At least 30 distinct variants of the gene have been found, and the gene exists in multiple forms known as alleles (Kanetsky et al. 2002). In the MC1R protein, some of these variants cause a single amino acid to change. Red hair is associated with specific changes, such as R151C, R160W, and D294H. Not only do these variations result in red hair, but they also typically have fair skin, have trouble tan, and are prone to freckle. According to functional studies, these variants may result in MC1R receptors that are less potent. In other words, these variants may struggle to bind to their activating molecules or activate adenylyl cyclase properly (Healy et al. 2001; Ringholm et al. 2004). It's also possible that having two variant alleles could have a cumulative effect (Naysmith et al. 2004; Lin and Fisher 2007).

3.9 Phenotypic Expression of Different MC1R Variants

Red hair color is strongly correlated with three variants of the MC1R gene (R151C, R160W, and D294H), indicating the gene's important role in determining this trait (Dessinioti et al. 2011). An individual with two copies of the variant gene will exhibit red hair color, which is an autosomal recessive trait (Valverde et al. 1995; Healy et al. 2001; Sulem et al. 2007; Han et al. 2006; Duffy et al. 2004). Variations in the MC1R gene have been linked to sun sensitivity in addition to hair color, which increases the risk of skin cancer in non-redheads for both melanoma and nonmelanoma types (Stratigos et al. 2006). D84E, R151C, R160W, and D294H are variant alleles that have been found to be substantially associated with red hair and fair skin in a study on Australian twins. Certain MC1R variants, particularly D294H, were linked to fair skin and red hair in an Australian population. Similarly, red hair and light skin type were strongly linked to variants R151C, R160W, and D294H in an Irish population. People with two of these three variants were found to have red hair on a regular basis. Certain variations in the MC1R gene, specifically the V60L, V92M, and R163Q alleles, known as Weak Red Hair Color (RHC) alleles, were found to have a combined odds ratio (OR) of 5 (with a 95% confidence interval of 3–11) (Duffy et al. 2004; Sturm et al. 2003). This indicates an increased likelihood of having red hair when carrying these alleles. Some MC1R gene variants, most notably the D294H variant, have been associated with red hair and fair skin in members of the British population (Valverde et al. 1995) as well as in Australian twins (monozygotic and dizygotic) (Box et al. 1997). Variants such as R151C, R160W, and D294H were significantly associated with lighter skin types and red hair in an Irish population. It's interesting to note that people who carried two of these three variants regularly had red hair. It's important to note that although MC1R variants were present in a large number of people with dark hair or skin tones, there was no conclusive proof linking these variants to a particular darker phenotype (Smith et al. 1998). Distinct variants in the MC1R gene—namely R151C, R160Q, and D294H—have been statistically linked to the expression of red hair. Another allele, Val60Leu, is associated with blonde or light brown hair, while any deviation from the consensus allele correlates with lighter skin tones (Box et al. 1997; Smith et al. 1998). Additionally, the minor alleles R142H and I155T may function as recessive alleles contributing to red hair color when combined with the three more common R alleles (Sturm et al. 2001; Davies et al. 2012).

3.10 Genome-Wide Association Studies of Hair Color

Hair follicles are intricate structures with a complex three-dimensional cell arrangement, comprising multiple cell lineages that undergo cycles during the growth and regression phases of the hair shaft. The primary component of the hair follicle is made up of keratinocytes, while the base houses specialized dermal papilla cells. These dermal papilla cells interact with melanocytes within the follicular epithelium, imparting pigment to the developing hair shaft. The dermal papilla plays a crucial role by producing signals that regulate the activity of hair follicle melanocytes, influencing pigment production (Pavan and Sturm 2019). The genetic factors influencing skin color are predicted to be less diverse than those influencing hair color because of this complex process. Twin studies indicate that 97% of heritability of hair color is present. Hair color is highly heritable. Hysi et al. (2018) conducted a genome-wide association meta-analysis in a large-scale study involving almost 300,000 European individuals from the 23andMe and UK Biobank cohorts. More than 120 loci, 107 of which were recently discovered, were substantially linked to hair color, reflecting the polygenic nature of this trait. Numerous important loci have been discovered in the investigation of genetic factors influencing hair color, including MC1R (rs1805007), IRF4 (rs12203592), and OCA2/HERC2 (rs12913832). Furthermore, genes linked to the development of melanocytes have been identified, including FGF5 (rs7681907), TWIST2 (rs11684254), MITF (rs9823839), HPS5 (rs201668750), and EDNRB (rs1279403). After a thorough analysis that included Gene Ontology entries for every gene in the set, a robust association with pigmentation processes and melanin biosynthesis and metabolic pathways was found. This thorough analysis shed light on the heritable variation in hair color, attributing red hair to 34.6% of the variability, blond hair to 24.8%, and black hair to 26.1%. Interestingly, the study found that women preferred lighter hair shades, indicating a sex bias in hair color.

A comprehensive genome-wide association study (GWAS) was conducted using a substantial cohort of over 343,000 individuals from the UK Biobank, leading to the discovery of a more extensive set of genes related to hair color (Morgan et al. 2018). This research, particularly focused on the red hair phenotype, highlighted the substantial contribution of MC1R polymorphisms, explaining 73% of the heritability of red hair traits. The study also uncovered eight additional associations for red hair at a genome-wide significant level, employing statistical fine mapping to identify likely causal single nucleotide polymorphisms (SNPs). Notably, it was discovered that variations in the POMC and OCA2/HERC2 loci reduce the chance of having red hair. The gene responsible for producing the α-MSH ligand for MC1R is likely to have its expression enhanced by the SNP rs76645354, which is situated 2 kb upstream of POMC. The significance of SNPs surrounding the ASIP locus for the red hair phenotype was also demonstrated by the study. rs6059655*G/A functions as an expression quantitative trait locus (eQTL) for ASIP in the skin. Red hair was linked to the TSPAN10 gene, which is highly expressed in melanocytes, via a lead SNP in strong linkage disequilibrium with rs6420484*A/G Y177C. Previous research has connected this allele to the hue saturation of eye color. Furthermore, it was discovered that the PKHD1 gene has epistatic interactions. More than 200 genetic variations have been linked to hair color, ranging from light brown to blond, and from black to dark. In total, 73% of the heritability of blond hair was explained by SNPs, which is equivalent to 163 different genes. A polygenic phenotype score confirmed that hair color is a complex trait that is influenced by several genes. Seven loci that were previously linked to variations in hair shape were also found to influence the difference between blond and non-blond hair color. This implies that hair color is influenced by genes that affect hair texture and may be expressed by keratinocytes (Hysi et al. 2018).

3.11 Conclusion

In conclusion, a complex interaction between genetic, biochemical, and structural factors governs the intricate processes that determine hair pigmentation. The vast range of hair colors that people have is mainly ascribed to the synthesis of two primary melanin pigment types: pheomelanin, which contributes to red and yellow tones, and eumelanin, which gives black to dark brown hues. The 16th chromosome contains the melanocortin 1 receptor (MC1R) gene, which is essential for controlling human pigmentation. It affects the synthesis of melanin and gives rise to the variety of colors seen in skin, hair, and eyes. The process of hair pigmentation is closely linked to the active growth phase by the telogen, anagen, and catagen phases of hair follicle growth, which form a cyclical pattern. Tyrosinase is an essential enzyme in the pathway leading to the synthesis of melanin, as it converts L-tyrosine into precursors of melanin. The range of hair colors seen in the human population is produced by the oxidative polymerization of these precursors, which leads to the synthesis of eumelanin and pheomelanin. Different hair colors ultrastructural features of hair melanogenesis show unique patterns in melanosome and melanocyte types. Research on different hair types—red, blond, black, brown, and gray/white—offers insights into the structural differences connected to distinct pigmentations. The cytotoxic characteristics of intermediate metabolites in the melanin pathway, oxidative stress, and genetic factors are some of the complex mechanisms that underlie the aging process and cause senile greying of hair. Enzymatic activityTyrosinase, which functions as a rate-limiting step, and different transporters that regulate the transport of precursors into melanosomes are among the regulatory factors of melanogenesis. A hair phenotype is determined by the balance between eumelanogenesis and pheomelanogenesis, which is influenced by the pH, cysteine levels, and ion transporter activity. Different skin tones and hair colors are correlated with variations in the genetics of hair color, specifically in the MC1R gene. Red hair, fair skin, and sun sensitivity are linked to specific alleles of the MC1R gene, demonstrating the gene's important function in phenotypic expression. All things considered, the rich interplay of genetics, biochemical reactions, and environmental factors is reflected in the complexity of hair pigmentation, which adds to the amazing diversity of human appearances.

Acknowledgments Critical reading and suggestions to improve the chapter by Prof. Amit Kumar Tripathi are highly appreciated.

References

Ancans J, Tobin DJ, Hoogduijn MJ, Smit NP, Wakamatsu K, Thody AJ (2001) Melanosomal pH controls rate of melanogenesis, eumelanin/phaeomelanin ratio and melanosome maturation in melanocytes and melanoma cells. Exp Cell Res 268(1):26–35

Aroca P, Garcia-Borron JC, Solano F, Lozano JA (1989) Regulation of mammalian melanogenesis I: partial purification and characterization of a dopachrome converting factor: dopachrome tautomerase. Biochim Biophys Acta 1035:266

Barber JI, Townsend D, Olds DP, King RA (1984) Dopachrome oxidoreductase: a new enzyme in the pigment pathway. J Invest Dermatol 83(2):145–149. https://doi.org/10.1111/1523-1747.ep12263381

Barsh G, Gunn T, He L, Schlossman S, Duke-Cohan J (2000) Biochemical and genetic studies of pigment-type switching. Pigment Cell Res 13(Suppl. 8):48–53

Birbeck MSC, Mercer EH, Barnicot NA (1956) The structure and formation of pigment granules in human hair. Exp Cell Res 10(2):505–514

Boissy RE, Sakai C, Zhao H, Kobayashi T, Hearing VJ (1998) Human tyrosinase related protein-1 (TRP-1) does not function as a DHICA oxidase activity in contrast to murine TRP-1. Exp Dermatol 7(4):198–204

Box NF, Wyeth JR, O'Gorman LE, Martin NG, Sturm RA (1997) Characterization of melanocyte stimulating hormone receptor variant alleles in twins with red hair. Hum Mol Genet 6(11):1891–1897

Burch PR, Jackson D (1966) The greying of hair and the loss of permanent teeth considered in relation to an autoimmune theory of aging. J Gerontol 21(4):522–528

Cesarini JP (1990) Chapter 8: Hair melanin and hair color. In: Hair and hair diseases. Springer

Cheli Y, Luciani F, Khaled M, Beuret L, Bille K, Gounon P et al (2009) αMSH and cyclic AMP elevating agents control melanosome pH through a protein kinase A-independent mechanism. J Biol Chem 284(28):18699–18706

Chen W, Kelly MA, Opitz-Araya X, Thomas RE, Low MJ, Cone RD (1997) Exocrine gland dysfunction in MC5-R-deficient mice: evidence for coordinated regulation of exocrine gland function by melanocortin peptides. Cell 91(6):789–798
Chhajlani V, Muceniece R, Wikberg JES (1993) Molecular cloning of a novel human melanocortin receptor. Biochem Biophys Res Commun 195:866–873
Chintala S, Li W, Lamoreux ML, Ito S, Wakamatsu K, Sviderskaya EV et al (2005) Slc7a11 gene controls production of pheomelanin pigment and proliferation of cultured cells. Proc Natl Acad Sci USA 102(31):10964–10969
Cone RD, Mountjoy KG, Robbins LS, Nadeau JH, Johnson KR, Roselli-Rehfuss L et al (1993) Cloning and functional characterization of a family of receptors for the melanotropic peptides. Ann N Y Acad Sci 680(1):342–363
d'Ischia M, Napolitano A, Prota G (1991) Peroxidase as an alternative to tyrosinase in the oxidative polymerization of 5,6-dihydroxyindoles to melanin(s). Biochim Biophys Acta 1073(2):423–430
D'Orazio JA, Nobuhisa T, Cui R, Arya M, Spry M, Wakamatsu K et al (2006) Topical drug rescue strategy and skin protection based on the role of Mc1r in UV-induced tanning. Nature 443(7109):340–344
Davies JR, Randerson-Moor J, Kukalizch K, Harland M, Kumar R, Madhusudan S et al (2012) Inherited variants in the MC1R gene and survival from cutaneous melanoma: a BioGenoMEL study. Pigment Cell Melanoma Res 25:384
del Marmol V, Beermann F (1996) Tyrosinase and related proteins in mammalian pigmentation. Cancer Res 381:25–28
Del Marmol V, Ito S, Bouchard B, Libert A, Wakamatsu K, Ghanem G et al (1996) Cysteine deprivation promotes eumelanogenesis in human melanoma cells. J Invest Dermatol 107(5):698–702
Dessinioti C, Antoniou C, Katsambas A, Stratigos AJ (2011) Melanocortin 1 receptor variants: functional role and pigmentary associations. Photochem Photobiol 87(5):978–987
Duffy DL, Box NF, Chen W, Palmer JS, Montgomery GW, James MR et al (2004) Interactive effects of MC1R and OCA2 on melanoma risk phenotypes. Hum Mol Genet 13(4):447–461
Gantz I, Yamada T, Tashiro T, Konda Y, Shimoto Y, Miwa H et al (1994) Mapping of the gene encoding the melanocortin-1 (α-melanocyte stimulating hormone) receptor (mc1r) to human chromosome 16q24.3 by fluorescence in situ hybridization. Genomics 19:394–395
Han J, Kraft P, Colditz GA, Wong J, Hunter DJ (2006) Melanocortin 1 receptor variants and skin cancer risk. Int J Cancer 119(8):1976–1984
Harding RM, Healy E, Ray AJ, Ellis NS, Flanagan N, Todd C et al (2000) Evidence for variable selective pressures at MC1R. Am J Hum Genet 66:1351–1361
Hardy MH (1992) The secret life of the hair follicle. Trends Genet 8(2):55–61
Hatta N, Dixon C, Ray AJ, Phillips SR, Cunliffe WJ, Dale M et al (2001) Expression, candidate gene, and population studies of the melanocortin 5 receptor. J Invest Dermatol 116(4):564–570
Healy E, Birch-Machin M, Rees JL (2000) The human melanocortin-1 receptor gene. In: The melanocortin receptors, vol 9. Humana Press, pp 341–359
Healy E, Jordan SA, Budd PS, Suffolk R, Rees JL, Jackson IJ (2001) Functional variation of MC1R alleles from red-haired individuals. Hum Mol Genet 10(21):2397–2402
Hearing VJ, Jiménez M (1987) Mammalian tyrosinase—the critical regulatory control point in melanocyte pigmentation. Int J Biochem 19(12):1141–1147
Hearing J, Tsukamoto K (2017) Enzymatic control of pigmentation in mammal. FASEB J 5(14):2902–2909
Hearing VJ, Korner AM, Pawelek JM (1982) New regulators of melanogenesis are associated with purified tyrosinase isozymes. J Invest Dermatol 79(1):16–18. https://doi.org/10.1111/1523-1747.ep12510422
Hoekstra HE (2006) Genetics, development and evolution of adaptive pigmentation in vertebrates. Heredity (Edinb) 97(3):222–234
Hysi PG, Valdes AM, Liu F, Furlotte NA, Evans DM, Bataille V et al (2018) Genome-wide association meta-analysis of individuals of European ancestry identifies new loci explaining a sub-

stantial fraction of hair color variation and heritability. Nat Genet 50(5):652–656. https://doi.org/10.1038/s41588-018-0100-5
Ito S, Prota G (1977) A facile one-step synthesis of cysteinyldopas using mushroom tyrosinase. Experientia 33(8):1118–1119
Ito S, Wakamatsu K (2003) Quantitative analysis of eumelanin and pheomelanin in humans, mice, and other animals: a comparative review. Pigment Cell Res 16(5):523–531
Ito S, Wakamatsu K (2011a) Diversity of human hair pigmentation as studied by chemical analysis of eumelanin and pheomelanin. J Eur Acad Dermatol Venereol 25(12):1369–1380
Ito S, Wakamatsu K (2011b) Human hair melanins: what we have learned and have not learned from mouse coat color pigmentation. Pigment Cell Melanoma Res 24(1):63–74
Jara JR, Aroca P, Solano F, Martinez JH, Lozano JA (1988) The role of sulfhydryl compounds in mammalian melanogenesis: the effect of cysteine and glutathione upon tyrosinase and the intermediates of the pathway. Biochim Biophys Acta 967:296–303
Jara JR, Martinez-Liarte JH, Solano F, Peñafiel R (1990) Transport of L-tyrosine by B16/F10 melanoma cells: the effect of the intracellular content of other amino acids. J Cell Sci 97(3):479–485
Jimbow K, Quevedo WC, Fitzpatrick TB, Szabo G (1976) Some aspects of melanin biology: 1950–1975. J Invest Dermatol 67(1):72–89. https://doi.org/10.1111/1523-1747.ep12512500
Jimbow K, Ishida O, Ito S, Hori Y, Witkop CJ, King RA (1983) Combined chemical and electron microscopic studies of pheomelanosomes in human red hair. J Invest Dermatol 81(6):506–511
Kalatzis V, Cherqui S, Antignac C, Gasnier B (2001) Cystinosin, the protein defective in cystinosis, is a H+-driven lysosomal cystine transporter. EMBO J 20(21):5940–5949
Kanetsky PA, Swoyer J, Panossian S, Holmes R, Guerry DP, Rebbeck TR (2002) A polymorphism in the agouti signaling protein gene is associated with human pigmentation. Am J Hum Genet 70:1–9
Korner AM, Pawelek J (1980) Dopachrome conversion: a possible control point in melanin biosynthesis. J Invest Dermatol 75(2):192–195
Kwon BS (1993) Pigmentation genes: the tyrosinase gene family and the pmel 17 gene family. J Invest Dermatol 100(2 Suppl):S134–S140. https://doi.org/10.1038/jid.1993.2
Lamason RL, Mohideen MAPK, Mest JR, Wong AC, Norton HL, Aros MC et al (2005) Genetics: SLC24A5, a putative cation exchanger, affects pigmentation in zebrafish and humans. Science 310(5755):1782–1786
Le Pape E, Passeron T, Giubellino A, Valencia JC, Wolber R, Hearing VJ (2009) Microarray analysis sheds light on the dedifferentiating role of agouti signal protein in murine melanocytes via the Mc1r. Proc Natl Acad Sci USA 106(6):1802–1807
Leonard LJ, Townsend D, King RA (1988) Function of dopachrome oxidoreductase and metal ions in dopachrome conversion in the eumelanin pathway. Biochemistry 27(16):6156–6159
Levy C, Khaled M, Fisher DE (2006) MITF: master regulator of melanocyte development and melanoma oncogene. Trends Mol Med 12(9):406–414
Lin JY, Fisher DE (2007) Melanocyte biology and skin pigmentation. Nature 445(7130):843–850
Lloyd T, Garry FL, Manders EK, Marks JG (1987) The effect of age and hair colour on human hairbulb tyrosinase activity. Br J Dermatol 116(4):485–489
Lukas SE, Griffiths RR (1982) Science 217:1161–1163
Menon IA, Persad S, Haberman HF, Kurian CJ (1980) A comparative study of the physical and chemical properties of melanins isolated from human black and red hair. J Invest Dermatol 80(3):202–206. https://doi.org/10.1111/1523-1747.ep12534045
Morgan MD, Pairo-Castineira E, Rawlik K, Canela-Xandri O, Rees J, Sims D et al (2018) Genome-wide study of hair colour in UK Biobank explains most of the SNP heritability. Nat Commun 9(1):1–10. https://doi.org/10.1038/s41467-018-07691-z
Mountjoy KG, Robbins LS, Mortrud MT, Cone RD (1992) The cloning of a family of genes that encode the melanocortin receptors. Science 257(5074):1248–1251
Nanninga PB, Ghanem GE, Lejeune FJ, Bos JD, Westerhof W (1991) Evidence for alpha-MSH binding sites on human scalp hair follicles: preliminary results. Pigment Cell Res 4(4):193–198

Naysmith L, Waterston K, Ha T, Flanagan N, Bisset Y, Ray A et al (2004) Quantitative measures of the effect of the melanocortin 1 receptor on human pigmentary status. J Invest Dermatol 122(2):423–428
Ortonne JP, Prota G (1993) Hair melanins and hair color: ultrastructural and biochemical aspects. J Invest Dermatol 101(1 Suppl):82S–89S
Palumbo A, d'Ischia M, Misuraca G, Prota G (1987) Effect of metal ions on the rearrangement of dopachrome. Biochim Biophys Acta 925(2):203–209
Palumbo A, d'Ischia M, Misuraca G, Prota G, Schultz TM (1988) Structural modifications in biosynthetic melanins induced by metal ions. Biochim Biophys Acta 964(2):193–199
Pavan WJ, Sturm RA (2019) The genetics of human skin and hair pigmentation. Annu Rev Genomics Hum Genet 20:41–72
Pawelek JM (1990) Dopachrome conversion factor functions as an isomerase. Biochem Biophys Res Commun 166(3):1328–1333
Pawelek JM (1991) After dopachrome? Pigment Cell Res 4:53–62
Phillips PC (2008) Epistasis—the essential role of gene interactions in the structure and evolution of genetic systems. Nat Rev Genet 9(11):855–867
Potterf SB, Virador V, Wakamatsu K, Furumura M, Santis C, Ito S et al (1999) Cysteine transport in melanosomes from murine melanocytes. Pigment Cell Res 12(1):4–12
Prota G (1980) Cysteine and glutathione in mammalian pigmentation. In: Natural sulfur compounds. Springer, pp 391–392
Prota G (1993) Regulatory mechanisms of melanogenesis: beyond the tyrosinase concept. J Invest Dermatol 100(2 Suppl):S156–S161. https://doi.org/10.1038/jid.1993.69
Prota G, D'Ischia M, Napolitano A (1998) The chemistry of melanins and related metabolites. Pigment Syst 8(4):97–106
Rana BK, Hewett-Emmett D, Jin L, Chang BHJ, Sambuughin N, Lin M et al (1999) High polymorphism at the human melanocortin 1 receptor locus. Genetics 151(4):1547–1557
Rees JL (2003) Genetics of hair and skin color. Annu Rev Genet 37:67–90
Ringholm A, Klovins J, Rudzish R, Phillips S, Rees JL, Schiöth HB (2004) Pharmacological characterization of loss of function mutations of the human melanocortin 1 receptor that are associated with red hair. J Invest Dermatol 123(5):917–923
Robbins LS, Nadeau JH, Johnson KR, Kelly MA, Roselli-Rehfuss L, Baack E et al (1993) Pigmentation phenotypes of variant extension locus alleles result from point mutations that alter MSH receptor function. Cell 72(6):827–834
Rodrigues M, Ezzedine K, Hamzavi I, Pandya AG, Harris JE (2017) New discoveries in the pathogenesis and classification of vitiligo. J Am Acad Dermatol 77(1):1–13. https://doi.org/10.1016/j.jaad.2016.10.048
Sadick NS, Laughlin SA (2004) Effective epilation of white and blond hair using combined radiofrequency and optical energy. J Cosmet Laser Ther 6(1):27–31
Schaffer JV, Bolognia JL (2001) The melanocortin-1 receptor. Arch Dermatol 137:1447
Schioth HB, Muceniece R, Wikberg JE, Chhajlani V (1995) Characterisation of melanocortin receptor subtypes by radioligand binding analysis. Eur J Pharmacol Mol Pharmacol Sect 288(3):3311–3317
Schiöth HB, Muceniece R, Wikberg JES (1996) Characterisation of the melanocortin 4 receptor by radioligand binding. Pharmacol Toxicol 79(3):161–165
Schiöth HB, Muceniece R, Larsson M, Mutulis F, Szardenings M, Prusis P et al (1997) Binding of cyclic and linear MSH core peptides to the melanocortin receptor subtypes. Eur J Pharmacol 319(2–3):369–373
Silvers WK (1979) The coat colors of mice. Springer
Slominski A, Paus R, Plonka P, Chakraborty A, Maurer M, Pruski D et al (1994) Melanogenesis during the anagen-catagen-telogen transformation of the murine hair cycle. J Invest Dermatol 102(6):862–869
Smith R, Healy E, Siddiqui S, Flanagan N, Steijlen PM, Rosdahl I et al (1998) Melanocortin 1 receptor variants in an Irish population. J Invest Dermatol 111(1):119–122

Smith DR, Spaulding DT, Glenn HM, Fuller BB (2004) The relationship between Na+/H+ exchanger expression and tyrosinase activity in human melanocytes. Exp Cell Res 298(2):521–534

Stratigos AJ, Dimisianos G, Nikolaou V, Poulou M, Sypsa V, Stefanaki I et al (2006) Melanocortin receptor-1 gene polymorphisms and the risk of cutaneous melanoma in a low-risk southern european population. J Invest Dermatol 126(8):1842–1849

Sturm RA, Teasdale RD, Box NF (2001) Human pigmentation genes: identification, structure and consequences of polymorphic variation. Gene 277(1–2):49–62

Sturm RA, Duffy DL, Box NF, Chen W, Smit DJ, Brown DL et al (2003) The role of melanocortin-1 receptor polymorphism in skin cancer risk phenotypes. Pigment Cell Res 16(3):266–272

Sulem P, Gudbjartsson DF, Stacey SN, Helgason A, Rafnar T, Magnusson KP et al (2007) Genetic determinants of hair, eye and skin pigmentation in Europeans. Nat Genet 39(12):1443–1452

Tan CP, Kulju McKee K, Weinberg DH, MacNeil T, Palyha OC, Feighner SD et al (1999) Molecular analysis of a new splice variant of the human melanocortin-1 receptor. FEBS Lett 451(2):137–141

Tobin DJ, Slominski A, Botchkarev V, Paus R (1999) The fate of hair follicle melanocytes during the hair growth cycle. J Investig Dermatol Symp Proc 4(3):323–332

Tsukamoto K, Jackson IJ, Urabe K, Montague PM, Hearing VJ (1992) A second tyrosinase-related protein, TRP-2, is a melanogenic enzyme termed DOPAchrome tautomerase. EMBO J 11(2):519–526

Valverde P, Healy E, Jackson I, Rees JL, Thody AJ (1995) Variants of the melanocyte–stimulating hormone receptor gene are associated with red hair and fair skin in humans. Nat Genet 11(3):328–330

Walker WP, Gunn TM (2010) Shades of meaning: the pigment-type switching system as a tool for discovery. Pigment Cell Melanoma Res 23(4):485–495

Wang N, Hebert DN (2006) Tyrosinase maturation through the mammalian secretory pathway: bringing color to life. Pigment Cell Res 19(1):3–18

Raw Materials in Hair Care Industry and Regulations

4

Rishabh Shah

Abstract

Current hair care market is dynamic and competitive as consumers demand varied functionalities like shine, softness, hair repair, color protection, etc. from their hair care products in addition to cleansing. Personal care industries that formulate these hair care products use a range of raw materials and ingredients to meet the evolving needs of consumers. This chapter will explore the latest trends in hair care industry and highlight the opportunities and challenges for brands and manufacturers. The chapter will also focus on innovation opportunities through raw materials and ingredients, regulations, and sustainability initiatives being deployed in hair care raw materials industry.

Keywords

Hair care raw materials · Sustainable · Hair care trends · Innovation

4.1 Trends and Raw Ingredients in Hair Care Industry

The global beauty and personal care market is constantly evolving, driven by changing consumer preferences, emerging technologies, and environmental concerns. Let's explore the latest trends in hair care, styling, and color, and how they reflect the needs and desires of today's consumers. We will also highlight the opportunities and challenges for brands and manufacturers in this dynamic sector.

R. Shah (✉)
Research & Innovation, Acme Hardesty, Blue Bell, PA, USA
e-mail: rshah@acme-hardesty.com

A. Dwivedi et al. (eds.), *Hair Care Products*,
https://doi.org/10.1007/978-981-97-6740-3_4

4.1.1 Peace-of-Mind Hair Care

One of the key drivers of hair care innovation in 2023 is the increasing demand for products that enhance the emotional wellbeing of consumers, as well as their physical appearance. "According to Mintel, 69% of US consumers agree that using beauty products makes them feel good, while 58% of UK consumers say that beauty products help them cope with stress." These findings suggest that consumers are looking for more than just functional benefits from their hair care products; they also want to experience positive emotions, such as happiness, relaxation, and confidence.

To meet this demand, brands are integrating mood-enhancing elements into their hair care products, such as aromatherapy, color therapy, and sound therapy. For example, Aveda launched its Botanical Repair range in 2023, which features a blend of plant-based ingredients and essential oils that repair and strengthen the hair, while also stimulating the senses with soothing scents. Similarly, L'Oréal introduced its Elvive Dream Lengths Wonder Water in 2023, which claims to transform the hair in 8 s with a liquid formula that contains lamellar technology and a floral fragrance.

Another way that brands are elevating the hair care experience is by offering multifunctional products with a purpose that addresses multiple hair concerns in one step, saving time and money for consumers. These products also cater to the growing awareness of the environmental impact of hair care, as they reduce the amount of packaging and water usage. "For instance, **Garnier** launched its **Ultimate Blends Shampoo Bars** in 2023, which are vegan, biodegradable, and plastic-free solid shampoos that last up to two months and come in four variants for different hair types." Likewise, "**Dove** released its **Hair Therapy** range in 2023, which combines the benefits of a shampoo, conditioner, and treatment in one product, and claims to nourish the hair from the inside out."

4.1.2 "Skinification" of Hair Care

Another major trend in hair care is the "skinification" of hair care, which refers to the application of skincare principles and ingredients to scalp and hair care. This trend is driven by the growing recognition of the link between scalp health and hair quality, as well as the desire for more personalized and effective solutions. "According to Mintel, 54% of US consumers agree that scalp health is important for healthy hair, while 46% of Chinese consumers are interested in scalp care products that are customized to their scalp condition."

To tap into this trend, brands are embracing proven skincare ingredients, such as hyaluronic acid, niacinamide, and ceramides, and incorporating them into their hair care products, to provide hydration, anti-inflammation, and barrier protection for the scalp and hair. For example, Kérastase launched its Genesis range in 2023, which is designed to prevent hair fall and strengthen the hair fiber and features a serum that contains a high concentration of hyaluronic acid, edelweiss native cells, and ginger root. Similarly, Pantene introduced its Miracles range in 2023, which is formulated with pro-vitamin B5, lipids, and antioxidants, and includes a scalp care shampoo that contains niacinamide and panthenol.

Another way that brands are adopting the "skinification" approach is by using innovative technology and delivery systems to enhance the efficacy and convenience of hair care products. These include devices, apps, and sensors that analyze the scalp and hair condition, and recommend personalized products and routines. For instance, Schwarzkopf launched its Salon Lab system in 2023, which consists of a handheld device that measures hair quality, a smartphone app that creates a digital hair profile, and a customized shampoo that is mixed on the spot. Likewise, L'Oréal released its Perso device in 2023, which is a smart beauty device that creates personalized skincare, makeup, and hair care products, based on the user's skin type, environment, and preferences.

4.1.3 Personalization of Hair Care

The third trend that we will discuss in this chapter is the personalization of hair care, which is closely related to the "skinification" trend, but goes beyond the scalp and hair condition, and considers the individual preferences, needs, and goals of consumers. "According to Mintel, 64% of US consumers agree that everyone's hair is different and needs a unique approach, while 57% of UK consumers are interested in hair care products that are tailored to their hair type, style, or color."

To cater to this trend, brands are offering customized products that are created based on the consumer's input, such as a quiz, a consultation, or a DNA test. These products allow consumers to choose the ingredients, fragrance, color, and packaging of their hair care products, and provide personalized advice and guidance. For example, **Function of Beauty** is a direct-to-consumer brand that offers customized hair care products, based on a hair quiz that assesses the hair type, goals, and preferences of the consumer. Similarly, **Prose** is a personalized hair care brand that uses consultation, an algorithm, and a team of chemists to create customized hair care products and provides a hair care plan and progress tracking for the consumer.

Another way that brands are enhancing the value of their products is by showcasing their exceptional performance, quality, and sustainability throughout the customer journey. These include providing transparent information about the ingredients, sourcing, and manufacturing of their products, as well as offering guarantees, trials, and subscriptions. For instance, "Olaplex is a premium hair care brand that claims to repair, protect, and strengthen the hair, and uses patented technology and scientific research to back up its claims." Likewise, Love Beauty and Planet is a sustainable haircare brand that uses natural ingredients, recycled packaging, and carbon offsetting, and supports social and environmental causes.

In summary, these trends reflect the changing needs and desires of today's consumers, who are looking for more than just functional benefits from their hair care products; they also want to experience positive emotions, personalized solutions, and superior results. Brands and manufacturers that can tap into these trends and deliver on these expectations will be able to differentiate themselves in the competitive and dynamic hair care market.

4.2 Innovation Through Ingredients

Consumers are increasingly aware of the health of the hair and scalp and are constantly looking for products that provide protection, nourishment, and applicable for different hair types. Raw ingredients manufacturers are making a conscious effort to comply with consumer growing needs while delivering effectiveness and enhanced functionality. Over the years, we have seen a paradigm shift in product innovation within hair care segmentation.

4.2.1 Actives

Actives in hair care segmentation are growing rapidly, as the global and regional innovators are accentuating claim-driven innovation with purpose. Below are examples of trending actives with focus on hair care.

Rosemary extract is known for its anti-inflammatory properties, promoting hair growth (Frutarom.com).

Maxiliss is a blended active comprising Glycerin, Stevia Rebaudiana Leaf/Stem Extract, Papyrus Stem Extract, and Hydrolyzed Rice Protein is an adaptogenic natural active ingredient with a focus on hair structure (Novachem.com).

4.2.2 Cationic Agents

Cationic conditioners such as **Merquat™ polymers** are developed and manufactured by Lubrizol. They are used as conditioners and sensory modifiers for hair and scalp conditioning leaving a soft, silky feel. These cationic polymers are differentiated based on the degree of substitution, molecular weight, and charge density. For example, Merquat™ 100 Polymer is a high charge density, Merquat™ 2001 is ampholytic terpolymer and Merquat® 2003 PR is polymer specializing for use in color protection products (Lubrizol.com/personalcare).

4.2.3 Emollient

Emollients are ingredients that moisturize, soften, and nourish the hair. They are categorized into Butters, Esters, Hydrocarbons, Organosilicons, and Natural Oils. They offer a cascading sensory modification, solvency, moisturization, wet/dry combability, and foam modification for surfactant-driven formulas. To select the right emollient, one may need to consider the hair type, region-specific development, and core objectives. Let us review a few emollient chemistries and their benefits.

Cupuacu butter is a triglyceride with balance composition of saturated and unsaturated fatty acids, providing the butter low melting point and the appearance of a soft solid that melts quickly upon contact with skin. These triglycerides contain sterol that helps replenish skin's lipid barrier, minimizes dryness, and imparts hydration (citrooleogroup.com).

Triethyl citrate (TEC) is a triethyl ester of citric acid. It is produced via esterification of citric acid, where ethyl 25 groups from the ethanol replace all three of the carboxyl groups in the citric acid. It is widely used as a solvent, diluent, and fixative for fragrance. TEC has a high refractive index imparting shine on hair, helps transform foam profile from copious to dense-looking foam (mamtapolycoats.com). There are several options available, and one can pick the right ester by evaluating polarity, refractive index, co-efficient of friction, and molecular weight properties.

Castor oil is a natural oil touted to be the "liquid gold" in the hair care market segment. Over the years, castor oil and its derivatives have gained significant popularity. Castor oil is a vegetable oil, primarily cultivated in India, South America, and Africa. Ricinoleic fatty acid is one of the primary components. Derivatives such as castor oil #1, pale pressed castor oil, dehydrated castor oil, sulfated castor oil (red turkey oil), blown castor oil, jamaican black castor oil are few examples. These products help promote hair growth (per empirical evidence), nourish, and condition the scalp and hair, and help align cuticles imparting high shine.

Major producers such as Gokul Overseas, Jayant Agro, and Adani are market leaders and supply ethical, sustainable castor oil solutions to global regions, suppliers, and customers.

Silicones are a family of synthetic polymers used widely across hair care segments since the last 80+ years. Primary benefits include conditioning, shine, wet/dry hair manageability, heat protection, color retention, frizz reduction, and volume. Silicones are further classified into water soluble and water insoluble. Three major categories of silicones are:

(a) Dimethicone: Linear silicone fluid offered in different viscosities and molecular weights. These products form a protective layer on the hair and reduce friction, water loss, and protection from heat styling (dow.com).

(b) Amodimethicone: These are modified silicone fluid with amino groups attached to the silicone backbone. Binds to the damaged areas of the hair and improves the hair fiber strength, elasticity, color retention, and protection (dow.com).

(c) Cyclomethicone: These are cyclic silicone fluids that are volatile and have a tendency to evaporate quickly from the hair-delivering actives, other ingredients to improve substantiation to the hair, and help impart a smooth and silky feel without the buildup (dow.com).

4.2.4 Fatty Acids and Fatty Alcohols

Fatty acids and fatty alcohols are the building blocks for various commercially sold products globally. They are also used as an intermediate for derivatives in space of esters, surfactants, thickeners, etc.

Fatty acids: They are classified into short, medium, and long-chain fatty acids. KLK Oleo's Palmera fatty acids offer a wide range of functionality (klkoleo.com/palmera-fatty-acids):
- (a) Caproic acid (C6)—used as a perfume agent.
- (b) Caprylic acid (C8) and capric acid (C10) are derived into an ester–medium chain triglyceride used predominantly as a sensory modifier in hair care formulations.
- (c) Lauric acid (C12) is used as an intermediate in the production of esters, fatty alcohols, and fatty acid surfactants such as isethionates and sarcosinate. Lauric acid offers a foam boost.
- (d) Myristic acid (C14) helps promote foaming.
- (e) Palmitic acid (C16) is used as a slip agent and a sensory modifier.
- (f) Stearic acid (C18) is predominantly used as a structuring and stability aid. It acts as a binder for powder-based products.
- (g) Oleic acid (C18:1) is unsaturated fatty acid, supplied in liquid. It is used predominantly in hair color application.

Fatty alcohols: Like fatty acid, fatty alcohols are classified into three categories. Predominantly, they are used as an intermediate in the production of esters, surfactants, actives, etc. (klkoleo.com/palmera-fatty-acids).
- (a) Cetyl alcohol (C16)—used as a stabilizer and an opacifier agent.
- (b) Ceto-stearyl alcohol (C16–18)—used as a stabilizing agent and a viscosity modifier.

4.2.5 Fragrance

Fragrance/perfume have gained significant popularity in recent years. Fragrance is a consumer's signature and is driven on perception of choice. The fragrance in the formulas provides a unique opportunity for consumers to venture on a nostalgic experience. Global fragrance houses have successfully developed solutions that are region specific. Rise in multifunctional benefits have driven innovation and acceptance to the next level. IFF has a product portfolio that comprises clean labeled, natural with inclusions to probiotics and botanical extracts (iff.com/fragrance).

Below is the list of leading fragrance houses:

- Firmenich (firmeniche.com)
- Givaudan (Givaudan.com)
- Symrise (Symrise.com)
- Takasago (takasago.com)

- Mane SA (mane.com)
- Robertet (robertet.com)
- Sensient (sensient.com)
- Hasegawa (thasegawa.com)
- Huabao (hbglobal.com)

4.2.6 Humectants

Humectant such as glycerine helps retain moisture in the hair. It also provides solvency and acts as a dispersing aide for powdered ingredients. KLK Palm Oleo is one of the world's largest standalone oleochemical producers. Glycerine is manufactured from sustainable palm oil. Glycerine is a natural, biodegradable, and renewable product that conforms to USP, EP, and BP standards (klkoleo.com). Glycerine is traditionally used in multitude of different formulations within hair care segments and has been a go-to for industry experts globally.

Propanediol is another humectant used as a primary ingredient or together with glycerine in hair care formulations. Propanediol gained popularity, owing to changes in regulatory landscape and requirements to move from synthetic to sustainable, naturally derived option. Benefits such as hydration, improvement of hair elasticity, detangling, and wet and dry combing properties, while reducing static. Covation Bio PDO is one of the leading manufacturers of pure, petroleum-free diol, and Zemea® Propanediol (covationbiopdo.com/zemea). They currently offer standard, USP-FCC and USP-NF grades catering to a wide range of formulations.

4.2.7 Rheology Modifier and Thickeners

Rheological modifiers and thickeners are critical components of hair care formulations used to improve texture, flow property, stability, and performance. Sensory modification such as the spread ability, application, and substantiation on hair and scalp. These products also help improve the foaming and rinsing properties of shampoos and conditioners (personal-care.basf.com).

Carbopol® Ultrez 20 polymer is a hydrophobically modified cross-linked acrylic acid polymer that provides viscosity modification and suspension stability (Lubrizol.com).

Carboxymethyl cellulose is a cellulose derivative that is used as rheology modifier and emulsion stabilizer. Viscosity modification, degree of substitution, and molecular weight will impact the stability and sensory modification of the products developed. Aqualon® Carboxymethyl Cellulose manufactured by Ashland is market leader in offering CMC grades applicable for leave-on and rinse-off products (Ashland.com).

Clay-based additives such as **Bentonite Clay** and **Kaolin Clay** are popular clay-based additives used for their ability to detoxify and promote cleansing properties. Clay helps with sebum control, while promoting moisturizing and softening properties, enabling variety of texture modifications, such as gels, creams, lotions, and pomades. BYK is a global leader and manufacturer in clay-based additives classified into water based and oil-based thickeners catering to multitude of product formulations (byk.com).

4.2.8 Surfactants

Surfactants are surface-active agents found in shampoo, conditioners, and other specialty hair care products. They act as a cleansing agent, foam modifier that helps remove dirt, dead cells, residues, and excessive sebum from scalp and hair. Surfactants are primarily categorized into:

Anionic surfactant: Provides detergency, foaming, and cleaning (e.g., Sodium Lauryl Sulfate, Sodium C12–14 Olefin Sulfonate).
Amphoteric surfactant: Reduces irritation potential, acts as a foam booster, and helps enhance the active content level (e.g., Cocamidopropyl Betaine, Sodium Cocoamphoacetate).
Nonionic surfactant: The Alkanoamides are viscosity modifiers and improve quality of the lather (e.g., Cocamide MEA, Cocamide MIPA). Glucosides are another chemistry under the nonionic category that has gained popularity to create safe, mild chassis with improved sustainability.

The surfactant category is going through a paradigm shift from synthetic to natural due to regulatory changes. Some of the major manufacturers of surfactants are leading the change:

- BASF SE (basf.com)
- The Dow Chemical Company (dow.com)
- Akzo Nobel N.V. (akzonobel.com)
- Clariant AG (Clariant.com)
- Hunstman Corporation (huntsman.com)
- Kao Corporation (kao.com)
- Solvay (solvay.com)
- Stepan Company (stepan.com)
- Evonik Industries AG (corporate.evonik.com)
- Croda International Plc (croda.com)

4.2.9 Preservatives

Preservatives are key components of hair care formulation. Inclusion of preservatives in the product helps prevent the growth of bacteria, fungi, mold, and yeast and improves shelf stability. Preservatives, similar to surfactants, are scrutinized by the regulatory bodies and have gone through changes regarding the approvals, and use limits with a focus on safety and well-being in a commercially sold product. Preservatives are present in a product through indirect and direct addition. Preservatives are classified into the following categories:

Traditional: Parabens, DMD Hydantoin, MIT, CIT, and PHMB are a few examples. Due to changes in regulatory requirements, cosmetic product manufacturers are looking to phase out of these options.

Free-off: Organic acids and blends, phenoxyethanol, and blends are touted as alternatives to traditional. Globally, the rise of free-off solutions has enabled product manufacturers to find alternative solutions as well as meeting safety and well-being criteria for consumers.

Nature identical and natural derived: Preservative solutions developed in the lab that are perceived from the alternative markets (cross over from food, pharma, nutra markets). Nature's identical solutions provide robust protection against known microbes. Furthermore, naturally derived solutions are gaining popularity due to rise in indie brands on a global scale. Marketers of indie brands are targeting a specific set of population with improved transparency and clarity of the claim messaging and benefits.

Some of the major manufacturers of preservatives that are leading the change in innovation and safety:

- Sharon Personal Care (sharonpc.com)
- Ashland (Ashland.com)
- Lonza Group Ltd. (lonza.com)
- Clariant AG (Clariant.com)
- Symrise AG (Symrise.com)

4.3 Global Regulation

Globally, the hair care regulatory and sustainability landscape is evolving and going through a paradigm shift. Requirements are normalizing and standards across hair care segmentation are becoming diverse, and inclusive. Ethical, sustainable, ingredient transparency steered by conscious and informed consumer demand on the rise. Year over Year, regulatory bodies and sustainability establishments are embracing the changing times and ensuring a steady pace of product innovation while prioritizing health and well-being of consumers and improving environmental footprint.

The global hair care market is expected to grow to \$148 billion by 2030 at a CAGR of 5.8% during the forecasted period, as reported by Fortune Business Insights. Increase in consumer demand coupled with enhanced safety and well-being, global regulatory bodies are implementing stricter regulations on chemicals used in hair care chassis.

4.3.1 North America

The FDA regulates hair care products to ensure safety and effectiveness (fda.gov). FDA provides safety and resource guidance on hair dyes, shampoo, conditioners, and hair treatment products, for consumers, salon professionals, and manufacturers. For instance, the FDA has set limits on the % VOCs that can be present in hair care products (Hair sprays are required to reduce the current 55% limit to 45% in 2027; Dry shampoos have a new 50% limit that would lower to 45% in 2027). Below are examples of regulatory acts operated in North America with global impact.

- Fair Packaging and Labeling Act (ftc.gov). The FLPA is designed to facilitate value comparisons and prevent deceptive packaging and labeling of commercial products.
- Prop 65; officially known as the Safe Drinking Water and Toxic Enforcement Act of 1986, to prevent contamination of chemicals known to cause cancer, birth defects. They are monitored by the state of California and have established safe harbor limits (oehha.co.gov).
- The Modernization of Cosmetic Regulations Act expands the FDA's authority over the personal care products and facilities (MoCRA). The establishment of MoCRA will see widespread impact and heightened regulatory requirements.

4.3.2 Europe

The Cosmetic Products Regulation (CPR) and Registration, Evaluation, Authorization and Restriction of Chemicals (REACH) regulate hair care products to ensure safety and effectiveness (cometicseurope.eu). Both the regulations provide guidance and have established substance restrictions, safety assessment, labeling and documentation requirements for cosmetic products.

4.3.3 Japan

Cosmetic products are regulated by the Ministry of Health, Labor, and Welfare (MHLW) under the Pharmaceutical and Medical Devices (PMD) Act. Per MHLW, Health Authority in Japan approves quasi-drug applications, accepts cosmetics notifications, monitors their adverse effect reports, and audits the participating companies.

Cosmetics in Japan are classified into two primary categories:

- General cosmetics (General categories)
- Quasi-drugs (Treatment, claims-driven product)

Regulatory requirements in Japan may vary depending on the requirements and the type of claims one may market. Follow MHLW guidelines on how to market cosmetic products and registration requirements.

4.3.4 China

Cosmetics Supervision and Administration Regulations have divided cosmetics into special cosmetics and ordinary cosmetics. Per CIRS group, under special cosmetics, products are regulated per function of anti-whitening, sunscreen, hair dye, new claim-driven efficacy, whereas ordinary cosmetics are general categories (including toothpaste).

Follow regulatory guidelines as established with respect to the National Medical Products Administration (NMPA), CFDA, and State Administration of Market Regulation (SAMR).

4.4 Sustainability Platforms and Programs

1. The **Roundtable on Sustainable Palm Oil (RSPO)** is a global, nonprofit organization that aims to promote the production and procurement of sustainable palm oil. The RSPO was established in 2004 with the support of the World Wildlife Fund (WWF) in response to growing concerns about the environmental and social impacts of palm oil production (ethicalconsumer.org). The organization brings together stakeholders from across the palm oil industry to develop and implement global standards for producing and sourcing certified sustainable palm oil (rspo.org). The RSPO's certification program guarantees that palm oil production meets environmental and social principles and criteria. The RSPO has over 4000 members, including NGOs, institutions, purchasers, manufacturers, and the biggest palm oil producers and traders (ethicalconsumer.org). The organization's standards embody its objectives and goals (rspo.org).

   Types of Certified Sustainable Palm Oil

   (a) Identity Preserved—Traceable to a single plantation.
   (b) Segregated—Separated from uncertified oil but may be mixed from several certified mills.
   (c) Mass Balance (MB)—May be mixed with uncertified oil, but the refinery can only sell the same amount of oil as it purchased.
2. The No-deforestation, No-peat, and No-exploitation (NDPE) is a commitment made by companies to ensure that their palm oil supply chains are free from deforestation, peatland development, and exploitation (ndpe-irf.net). The NDPE Implementation Reporting Framework (NDPE IRF) is a reporting tool designed to help companies systematically understand and track progress in delivering NDPE commitments in their palm oil supply chains (ndpe-irf.net). The NDPE commitments are most seen in reference to agricultural commodity production used by downstream companies, traders, mills, and growers (proforest.net).

NDPE is considered the most powerful private instrument for breaking the link between deforestation and oil palm (econusantara.org). These commitments help to ensure land is managed properly. Valuable forests are saved, peatland is protected and preserved, and local people are treated fairly and given safe places to work and live (sustainablepalmoilchoice.eu). NDPE is the next step in sustainability after RSPO.

3. **Ecovadis** is a global cloud-based SaaS platform that provides comprehensive corporate social responsibility (CSR) assessment services (ecovadis.com). The platform evaluates the sustainability of a company's nonfinancial management systems, including environmental impact, labor practices and human rights, ethics, and sustainable procurement (ecovadis.com). The EcoVadis Rating covers a broad range of nonfinancial management systems including Environmental, Labor & Human Rights, Ethics and Sustainable Procurement impacts (ecovadis.com). The assessment is evidence-based and results in an easy-to-read scorecard with a score ranging from 0 to 100 and a medal (bronze, silver, gold) if applicable (ecovadis.com). EcoVadis helps companies manage value chain risk and compliance, build performance for net-zero and sustainability targets, and drive value and impact (ecovadis.com).
4. **SEDEX**, or the Supplier Ethical Data Exchange, is a platform for sharing ethical supply chain data. It is the largest collaborative platform of its kind and is used by over 75,000 members worldwide (sedex.com). SEDEX's mission is to empower sustainable supply chains by providing a platform to store, analyze, share, and report on sustainability practices (sedex.com). The platform helps businesses manage their supply chain sustainability from one place, proactively identify and manage risks, and protect their business by staying ahead of current and upcoming legislation (sedex.com). SEDEX also offers SMETA audits to gain visibility into working conditions and environmental performance in the supply chain (sedex.com). In addition to these services, SEDEX provides consulting services to help businesses maximize the impact of their sustainability efforts (sedex.com). Finally, SEDEX offers exclusive member training programs to improve businesses' sustainability operations (sedex.com).
5. The **Castor Success program** is a sustainable castor code that aims to improve the livelihood of castor farmers and contracted workers by supporting them to optimize yield and reduce environmental impact (castorsuccess.org). The program is developed, owned, and managed by Indian Castor industry stakeholders and brings smallholders into the sustainability fold. The program invests in highly experienced field staff and field offices to assist the producers to be compliant with the code (castorsuccess.org). The Castor Success program offers two codes: the Global SuCCESS Farming Code and the Success Supply Chain Code V1.1. The former focuses on sustainable farming practices, while the latter focuses on sustainable supply chains (castorsuccess.org). The Castor Success program is a step toward enabling sustainable castor farming and sustainable castor oil and derivatives supply chain. The program focuses on five pillars (castorsuccess.org):
   (a) Using Good agricultural practices to increase yield and farmer income.

(b) Enabling better health and safety practices, and respecting human rights.
(c) Efficiently using water resources and maintaining soil fertility.
(d) Driving adoption of good waste management practices.
(e) Develop and manage pilot projects for supply chain certification standard for SuCCESS certified castor arriving from farm.

6. There are several other programs that are market driven to make innovative solutions in hair care to address circular beauty through clean label, natural alternatives to synthetically derived ingredients, use of upcycled products to minimize waste creation. Furthermore, efforts are being made to address water conservation through innovation promoting the water-less concept and higher active ingredient delivery. To improve and promote health, wellbeing, and inclusivity, community-led programs have been initiated by a few select manufacturers to empower local communities economically and socially.

   Below are a few examples of product certifications that are gaining popularity on a global scale:
   (a) COSMOS: Standardization of Organic and Natural Cosmetics
   (b) Cruelty-free (Crueltyfreeinternational.org)
   (c) Ecocert (Ecocert.com)
   (d) EU Ecolabel (environment.ec.europa.eu)
   (e) Halal (isahalal.com)
   (f) ISO 16128 (iso.org)
   (g) Non-GMO Project (nongmoproject.org)
   (h) USDA Organic (usda.gov)
   (i) Vegan (vegan.org)

Hair Care Products, i.e., Oils, Dyes, Lotions, Serum, etc.

5

Saniya Shakir, Samreen Abbas,
and Bhartendu Nath Mishra

Abstract

This chapter provides a comprehensive overview of the diverse realm of hair care products, encompassing essential categories such as shampoos, conditioners, hair oils, dyes, lotions, and serums. It explores their roles in cleansing, conditioning, styling, and addressing various hair concerns. From traditional remedies to modern formulations, it examines the efficacy and application techniques of popular choices like argan and coconut oil. Additionally, it delves into the vibrant spectrum of hair dyes, discussing types, color selection, and potential damage. Furthermore, it elucidates the science behind lotions and serums, guiding readers in selecting suitable treatments for specific hair types and concerns.

Keywords

Hair care products · Shampoos · Conditioners · Hair oils · Hair dyes · Lotions · Serums · Cleansing · Conditioning · Styling · Hair concerns · Argan oil · Coconut oil · Hair dye types · Color selection · Hair damage · Targeted treatments · Hair loss · Thinning · Frizz · Application techniques · Ingredients · Self-expression

5.1 Importance of Healthy Hair

Various functions such as protection from heat, facilitating perspiration, evaporation of pheromones, showing sensory activities, and in some cases, it also provides camouflage due to consider. Hair is a differentiating characteristic of mammals that serves as a vital organ; the use of exogenous and endogenous substances has a major impact on appearance, and it also acts as an indicator of healthy well-being,

S. Shakir · S. Abbas · B. N. Mishra (✉)
Department of Biotechnology, IET Lucknow, Lucknow, India

A. Dwivedi et al. (eds.), *Hair Care Products*,
https://doi.org/10.1007/978-981-97-6740-3_5

youthfulness, and vitality. Along with the history of civilizations, so does the history of hair care.

Nowadays, everyone is concerned with their physical appearance and hair is a valuable addition to this. Hair that is not maintained can be frizzy, have split ends, and be extremely dry, which is neither healthy physically nor good for one's self-confidence. With time, the thickness of hair is reduced due to many internal and external factors. Complete or partial baldness is becoming a common problem not only with age and sex but also becoming prevalent in the younger generation these days. For the hydration of the scalp and nourishment of hair, the natural sources of skin moisturization play a significant role as well, and it is also accelerated by the effect of exogenously applied substances.

In order to promote well-being and overall health, there is a need to bring changes in hair care routine and most importantly in the diet as intake of essential micro and macronutrients. This overcomes problems like hair fall, dryness, split ends, frizzy hair, and many other unhealthy conditions. With the proper maintenance and care, it is possible to have healthy, vibrant hair at all times. Furthermore, your overall confidence will be significantly improved and allow you to live a more fulfilling life (Alessandrini and Piraccini 2016).

Overall the role of hair care products, oils, and serum are very significant for the maintenance of good and healthy hair and perfect daily life routine. In this chapter, there is an in-depth discussion of the different hair care products, their composition, their effect on hair and scalp, their advantages, and their regulation. There are various regulatory norms that need to be followed and fulfilled for the market release of any new hair care products which will be further discussed.

5.2 Hair Care Products

Hair has certain physical as well as chemical behavior. It is an organic structure having various morphological components together as a unit. The mammalian hair shaft has three main regions that consist of cuticle, cortex, and medulla. Hair products, when applied, go to the scalp and most of the time, into the hair shaft. The hair follicle possesses a unique structural feature that enables the modulation in anatomy and physiology of hair using a range of hair preparations that are available for consumers.

There are different hair care preparations (Fig. 5.1) that are in the form of cosmetics or medical formulations that have the aim to provide hair care or work for hair problem treatment, respectively (Alshehrei 2024). These products have attained immense popularity over time because of their application, heterogeneity available in preparations, easy accessibility, and scope of manipulation. These products are natural as well as synthetic, the advantages and disadvantages of both are listed in Table 5.1. The availability of natural hair care products in the form of herbal formulations is very prevalent in the market and has various uses as hair tonics, hair

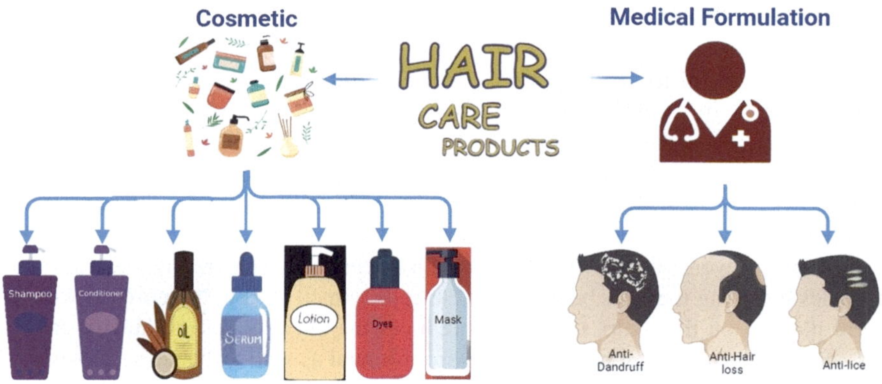

Fig. 5.1 The different types of hair care products

Table 5.1 Table showing advantages and disadvantages of natural and synthetic hair care products

Hair care products	Natural	Synthetic
Component	Phyto-chemical (plants derived)	Chemical compound
Toxicity	No	Yes
Efficacy	Low	High
Long lasting	No	Yes
Eco friendly	Yes	No
Local irritation	Yes	No
Ease of application	Difficult	Easy

conditioners, hair-cleansing agents, hair growth promoters antidandruff agents for the treatment of dandruff, alopecia, and lice infection.

These product formulations are intended for hair care and embellishment purposes of strengthening our hair follicles and enhancing hair growth by hair cleansing and care, hair growth stimulation, and hair color modification (Jose et al. 2024). Depending on the different conditions of hair, there are a variety of products available in the market. So the scope of selecting the product is a personal choice of the customer or if medically recommended. The cosmetic industry is a global, ever-changing, growing sector. Over the past decades, industry innovation has been immense, resulting in a wide range of new products and an increase in sales. In 2020 alone, the global cosmetic market was valued at USD 341.1 billion and is expected to hit USD 560.50 billion by 2030, with a compound annual growth rate of 5.1% from 2021 to 2030 (Jose et al. 2024).

The use of hair care cosmetics should be regulated in order to make them safe and effective to the consumer. However, manufacturers across the globe are conducting safety analysis to ensure the safety, nonirritating potential of these products and that the product should not cause any allergic reaction.

5.2.1 Hair Oil

The use of hair oil has been a routine practice followed since ages; however, its exact effect on the scalp and hair is still a topic that is obscure. Hair oils come in different varieties. It has a major role in protecting hair from damage as oils, when applied deeper into the scalp, cause a reduction in the amount of water that is absorbed by the hair, which lowers swelling. Due to this hydraulic fatigue is lowered which is by repeated swelling and drying and is a major cause of hair damage.

Oil also prevents insoluble salts and substances from penetrating deep into the follicles by filling the gaps between cuticle cells, which go into the pores as well as form plaques on the scalp. Oiling also reduces the dryness that ultimately causes less dandruff and also protects the scalp from fungal as well as bacterial growth.

The oiling process not only enhances lubrication of the hair shaft but also brings luster to the hair as well as prevents breakage, provides nourishment, helps in regeneration, and improves hair density. The damaging effects of sun rays can also be avoided due to oils acting as a shield in preventing loss of hydration (Sattar et al. 2024).

With the practice of oiling comes the habit of combing from roots to tips and this also increases blood circulation in the scalp. Understanding the practices and necessity of the cosmeceutical industry captures this opportunity. This belief has been exploited by the cosmeceutical industry, with numerous personal hair care products. There are different oils like coconut oil, almond oil, castor oil, argan oil, olive oil, amla oil, fenugreek oil, sesame oil, linseed oil, and mustard oil, which have been used from primitive times due to their antibacterial, antifungal, and emollient action and lubricant properties. These oils also have more distinct functions, which are shown in Fig. 5.2. As per the "Indian Cosmetic Market Outlook 2018" accessed on the date December 21, 2020, the share of hair-care products is huge in the cosmetic industry, with hair oils dominating the market.

5.2.2 Advantages of Using Essential Oils for Hair Oiling

Essential oils are low molecular weight compounds that are slightly soluble in water. These are volatile in nature, generally oils or lipid-like textures with a strong fragrance. The content may vary from 20 to 60 or approx. 100, and it is extracted from different parts of plants like citrus fruits, peels, leaves, grass, herbs, flowers, bark, seeds, and roots.

Apart from commonly formulated hair oils, the use of essential oils is prevalent nowadays as they have an effect on the cellular function of the skin. It also has anti-inflammatory, antibacterial, antioxidant as well as antifungal benefits on the scalp. Due to these properties, the overall density of the hair shaft increases, and also the entire hair bulb system is strengthened.

Due to the above beneficial properties, natural source precedence, and low toxicity, essential oils are added to cosmetics products that are available on the market. These EOs formulated products are promoted by the company, which promises to

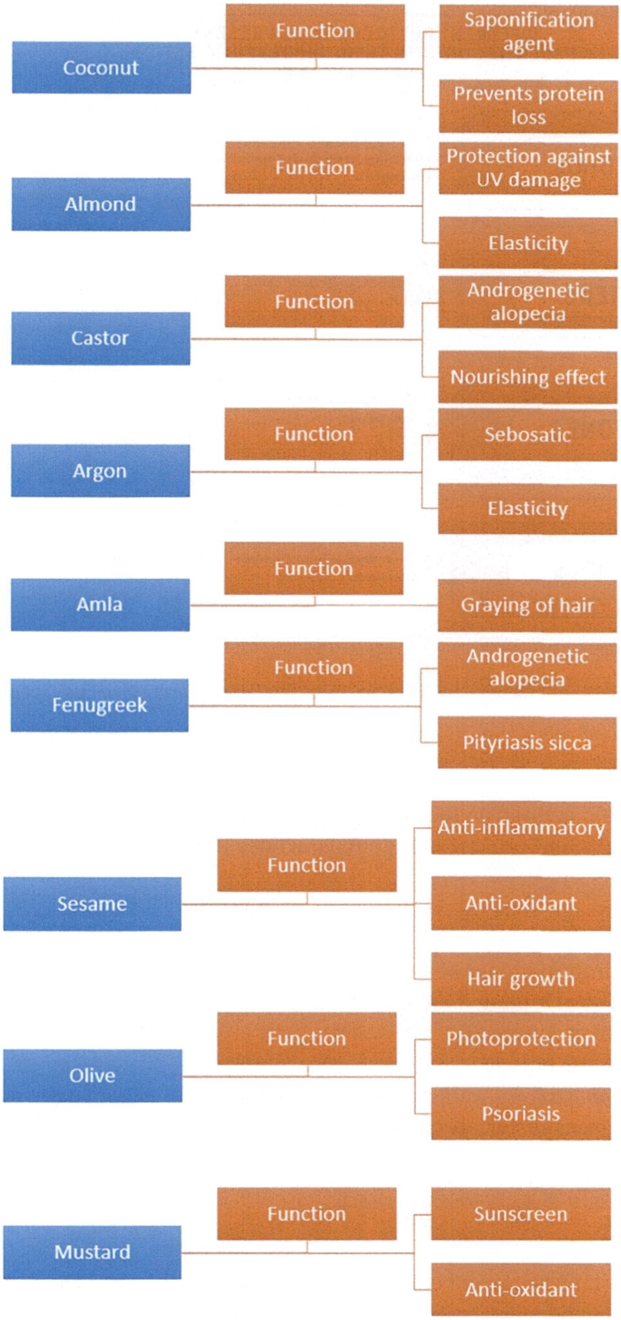

Fig. 5.2 Different hair oils and its functions

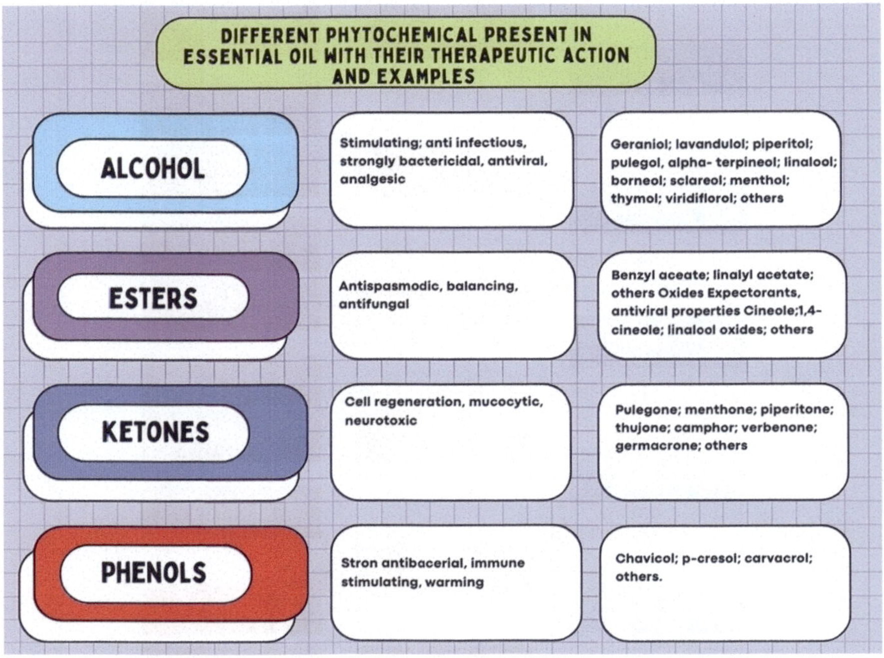

Fig. 5.3 Different phytochemical present in essential oil with their therapeutic action and examples

have therapeutic actions on the scalp and hair shafts (Abelan et al. 2022). Essential oils contain different organic compounds, whose therapeutic functions are listed in Fig. 5.3.

Listed below are some of the functions of hair oil:

- It helps in improving compatibility.
- The 18-MEA (18-methyl eicosanoid acid) of the hair is maintained.
- Restoring hydrophobicity.
- It also seals the cuticle.
- Oiling avoids frizz as well as friction by neutralizing negative net charge.
- It also enhances luster, softness, and manageability.

5.2.3 Dyes

The history of hair dyes can be traced back thousands of years. This was commonly seen in Egyptian people as early as 5000 years BC when Rameses II used henna for coloring hair in red. Hair dyes are classified into temporary, semi-permanent, demi-permanent, and permanent on the basis of color resistance. The durability of these products varies differently. Temporary dyeing products deposit on the cuticles of hair, but semipermanent go deep into the cortex, due to which the color remains intact up to seven to eight washes. Color precursors into demi-permanent and permanent dyes penetrate deeply due to their small molecular size. The development of the final shade is by interaction with oxidizing agents. For longer retention of color, due to the oxidation systems, the molecules are diffused deeply into the cortex. Dyes and color precursors have differences in chromophore groups, photostability, water solubility, and hair fiber affinity. There are many well-known skin hair dyes active, like para-phenylenediamine and para-toluenediamine, which are also skin sensitizers. Various studies explain how the diffusion of dye molecules to the inner hair fiber is established. In this path, molecules are permeated into the interarticular regions and pass through regions of non-keratinized endocuticles and then into the intracellular cement. Before being incorporated into the matrix, it migrates to keratinized regions and, eventually, reaches the microfibrils (Yun and Ahn 2023).

Using hair dye can cause both long-term and short-term health issues. On application, instant immune responses are seen in some individuals, which include, skin irritations, allergy, dermatitis, respiratory sensitization, and hair loss. Long-term practice of dyeing may result in reproductive toxicity, bladder cancer, disruption of thyroid hormone synthesis, and breast cancer. In the case of pregnant ladies, the maternal hair dyeing process can induce childhood tumors. As the awareness regarding the harmful effects of synthetic hair dyes is spreading, there is an increasing demand for environmentally friendly and safer dyes.

Biological dyes from natural products have many advantages, such as they are not toxic, less allergenic, and environmentally friendly. Natural dyes show natural anti-oxidant property, which consists of flavonoids, catechins, ascorbic acid, etc. For the treatment of scalp destruction, premature graying, and head lice, herbal hair colors are also used. Herbal hair colors are highly pigmented with anthocyanins, carotene (golden), and lutein (yellow) (Mandal and Venkatramani 2023). With the advantages of sustainability and health benefits, there is still a need to look into the following challenges:

- Complex extraction/purification process.
- Highly susceptible to environmental, UV, metal ions, and temperature.
- The lower intensity of coloration especially, on unbleached hairs.
- Color reproducibility is poor on human hair and testing models like wool and yak hair.
- Lack of sufficient toxicological data.

5.2.4 Lotion and Serum

Hair serum is a styling product that coats the surface of your hair. It's made with a base of silicone, a rubber-like substance that sits on top of your strands. There are different types of hair serums for different hair goals. Depending on the product's formula, a hair serum might reduce frizz, add shine, or straighten your hair. Some formulas may also protect against various forms of damage.

5.2.5 Shampoos

Shampoos help prevent hair damage by cleaning the scalp and shaft. The active ingredients that are added to shampoo's formulation are useful in the treatment of many diseases. Whatever the condition of the hair, like dermatitis, psoriasis, seborrhea, or alopecia, while treating the scalp, the hair must be shiny, soft, presentable, and compatible (Gubitosa et al. 2019). The composition of shampoo consists of 10–30 ingredients, but there are products available that have as few as four ingredients. The products are divided into (1) cleansing agents; (2) additives that contribute to the comfort and stability of the product; (3) conditioning agents to enhance disentangling, softness, gloss, and softness; and (4) special care ingredients to treat problems like dandruff and greasy hair.

The difficulty arises in untangling strands and frizz effects by using harsh shampoos. A condition named attrition, which can be the main cause of frizziness, can be reduced by the proper formulation of cleansing products. If the formulas for shampoo are not accurate, there is an aggravation in fiber attrition. There are many allergens present in shampoo such as formaldehyde-releasing preservatives, cocamidopropyl betaine, methylchloroisothiazolinone, propylene glycol, Vitamin E (tocopherol), benzophenones, parabens (Fig. 5.4).

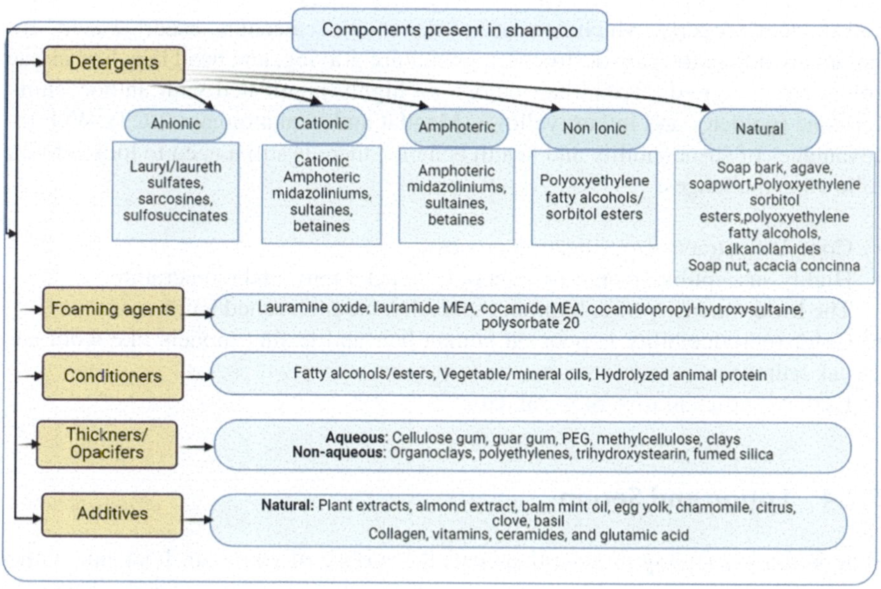

Fig. 5.4 Different components present in shampoo

5.2.6 Conditioners

The use of conditioner decreases friction, improves combability, and detangles the hair. When conditioners are applied, they add a positive charge and neutralize the electrical negative charge. It also reduces the hydrophilicity of fiber by lubricating the cuticle. Conditioners contain substances that lubricate hair, which are divided into five major groups: waxes, polymers, oils, cationic molecules, and hydrolyzed amino acids. Silicone is one of the most active and used agents in conditioners (Hordinsky et al. 2022). Conditioners are of various types depending on their way of application like:

Leave-on conditioners are directly applied to dried and washed hair and stay on till the next shampoo. Their function is to smoothen down the cuticle and safeguard the hair from weathering. They are easy to use and well-known. They contain thickeners, water, lipids, and cationic detergents. Their availability is in a thick and thin formulation state, but their impact is for less time.

Leave-in conditioners have been used for the maintenance of curly or unmanageable hair preceding styling or blow drying. It is also used on chemically damaged hair. Their main role is to confer shine and trap moisture because they contain silicone, vegetable oils, petroleum, and mineral oils. These are not reasonable for fine hair.

Deep conditioners are popularly used in parlors. This formulation can have an oil or cream base and is applied for 20–30 min, with or without using steam, and afterward, it is rinsed off. These contain agents like proteins or quaternary ammonium salts. The application smoothens the cuticle more and imparts better shine and gloss.

Anti-hair loss conditioners are especially for split or frayed ends that need deep conditioning. It helps in the reduction of hair fall, brittleness, and damage.

The functions of the conditioners are:

- Improve combability
- Mimetize the hair's natural lipid outer layer: 18-MEA
- Restore hydrophobicity
- Seal the cuticle
- Avoid or minimize frizz, and friction: Neutralize the negatively charged net
- Enhance shine, smoothness, and manageability

5.3 Regulation of Hair Care Products

A regulation is the most suitable legal vehicle because it establishes precise and comprehensive guidelines that prevent Member States from adopting different versions of the regulations. Furthermore, a regulation ensures that all of the community implements the law at the same time. The purpose of this regulation is to reduce administrative burden and ambiguities by consolidating terminology and simplifying procedures.

In addition, it reinforces specific components of the cosmetics regulatory system, like in-market supervision, to guarantee a high degree of human health protection. This regulation ensures a high degree of human health protection while completely harmonizing Community regulations to create an internal market for cosmetic products.

Because the cosmetics industry is a highly creative, dynamic, and fast-paced sector, regulations are necessary to guarantee the quality and safety of cosmetic products and prevent negative effects on consumer health. It is extremely difficult for a worldwide industry to offer the same product on all markets, nevertheless, because regulatory frameworks vary widely across markets and nations (Ferreira et al. 2022).

International Cooperation on Cosmetics Regulation (ICCR) is a voluntary group of cosmetic regulatory authorities from Brazil, Canada, China Taipei, the European Union (EU), Japan, the Republic of Korea, and the United States of America (USA) that meets annually to discuss various topics related to cosmetic safety and regulation (e.g., alternatives to animal testing, nanotechnology, microbiological limits). The ICCR was established in 2007 and serves as the regulatory framework for the regulation of cosmetic products (Milstein et al. n.d.). Other examples include the Organization for Economic Co-operation and Development (OECD) and the International Organization for Standardization (ISO).

5.4 Definition with Reference to Cosmetics

This regulation lays out guidelines that any cosmetic product placed on the market must follow in order to maintain both a high standard of human health protection and the smooth operation of the domestic market (Union P 2009). There are different definitions of cosmetic products by country or region, as shown in Fig. 5.5.

The following terms shall be used for the purposes of this Regulation:

1. "**Cosmetic product**" is defined as any formulation or combination that is meant to come into contact with external areas of the body of humans (like external genital organs, epidermis, lips, hair system) or with the oral cavity, including the teeth and mucous membranes. The sole or primary purpose of cleaning, perfuming, altering, protecting, maintaining, or reducing body odors.
2. "**Substance**" refers to any chemical element and its compounds in their natural state or as obtained through any manufacturing process, including any additive required to maintain the substance's stability and any impurities resulting from the process employed; however, solvents that can be separated without compromising the substance's stability or composition are not included in this definition.
3. "**Mixture**" is a form of solution or combination made up of two or more substances.

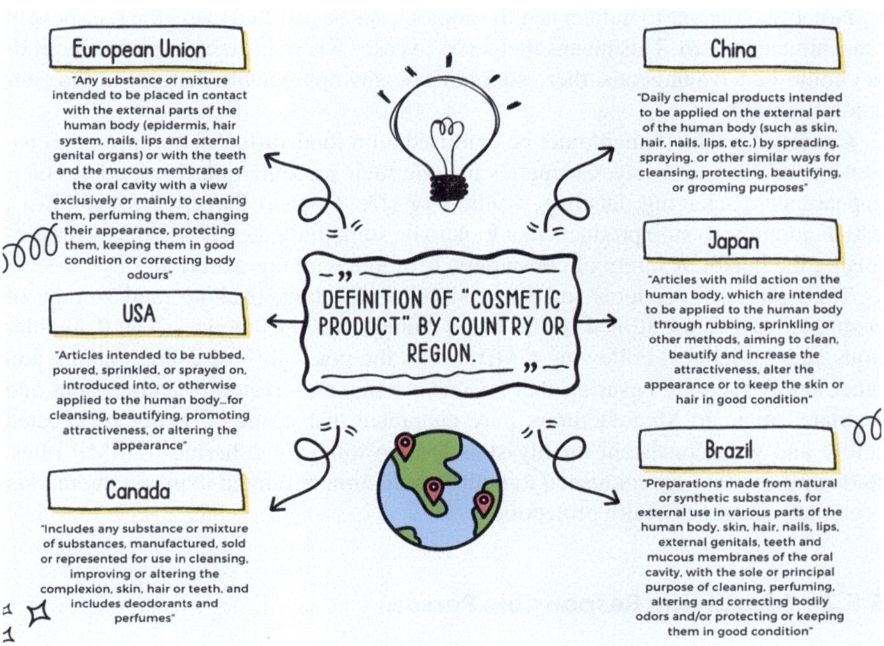

Fig. 5.5 Different definitions of cosmetic products by country or region

4. "**Manufacturer**" refers to anyone, natural or legal, who designs, manufactures, or commissions the design of a cosmetic product and then markets the product under his name or brand.
5. "**Distributor**": Any natural or legal supply chain participant that places a cosmetic product on the Community market, aside from the maker or importer, is referred to as a "distributor."
6. An "**end user**" is a person who uses a cosmetic product, be they a professional or a consumer.
7. "**Making available on the market**" refers to the provision of cosmetic products for distribution, use, or consumption on the Community market as part of a commercial activity, whether at no cost or in exchange for payment.
8. "**Placing on the market**" refers to a cosmetic product's initial availability on the Community market.

5.5 Roles of Regulatory Commission

5.5.1 Safety of the Product

To ensure the safety of cosmetic products under normal or reasonably anticipated conditions of use, the Commission's task is to establish the categories of items that are relevant for the implementation of regulations.

Notably, a danger to human health should never be justified using the risk-benefit reasoning approach. This means that, even in cases where utilizing a cosmetic product could be advantageous, there shouldn't be any appreciable dangers to the consumers' health.

Cosmetic products should not be confused with foodstuffs that compromise customers' health and safety. Cosmetics include their presentation, form, odor, color, appearance, packaging, labeling, volume, or size. Council Directive 87/357/EEC, which attempts to stop products that look to be something they're not from compromising the health or safety of consumers, is in line with this criterion.

The optimal procedures for the production, labeling, packing, and storage of cosmetic items are outlined in the Good Manufacturing Practices (GMP) regulations and guidelines. Following GMP lowers the possibility of contamination and other safety risks by ensuring that cosmetic items are created in a consistent and regulated manner. Manufacturers may guarantee that cosmetic items are created safely and with consistent quality standards in mind by adhering to GMP rules. Reducing the dangers connected to faulty production or tainted items on the market promotes consumer health protection.

5.5.2 Identifying Responsible Person

To ensure clear responsibilities and accountability, each cosmetic product should be linked to a responsible person who is established within the Community. This responsible person will have the obligation to ensure that the cosmetic product meets all legal requirements and is safe for consumers when used as intended. It is necessary to determine under which conditions a distributor can be considered responsible for a cosmetic product.

According to Regulation (EC) No 1223/2009 on cosmetic products, all legal or natural persons in the wholesale trade, as well as retailers selling directly to consumers, are covered by reference to the distributor. However, the obligations of the distributor should be adapted based on their respective role and part of the activity.

This means that depending on their specific involvement in the distribution chain, distributors may have different responsibilities and obligations regarding ensuring compliance with safety requirements for cosmetic products. The regulation takes into account these varying roles and activities in order to ensure clear responsibilities and accountability throughout the supply chain.

5.5.3 Ensuring the Traceability of a Cosmetic Product

Effective market surveillance requires ensuring the traceability of cosmetic items along the whole supply chain. Authorities in charge of market surveillance can quickly track down business owners who distribute cosmetics thanks to a strong traceability system.

Authorities can swiftly identify and track down the accountable parties in the event of any problems or non-compliance with cosmetic products by implementing a transparent and effective traceability system. In order to safeguard the health and safety of consumers, this improves market surveillance efforts and makes prompt interventions possible.

Additionally, a strong traceability system aids in locating and resolving any possible dangers or problems connected to certain cosmetic items. Better batch recalls, product quality monitoring, and the detection of illegal or fake goods on the market are all made possible by it.

In the cosmetics sector, transparency, accountability, and safety are all greatly enhanced by a robust traceability system.

5.5.4 Regulating Information of the Product

Cosmetic product importers and producers must establish and maintain a Product Information File (PIF) in accordance with Regulation (EC) No 1223/2009. The PIF includes all the information that is required to understand the cosmetic product, such as its ingredients, manufacturing process, safety evaluation, and proof that it complies with applicable laws. To ensure effective market surveillance and regulatory oversight, the competent authority of each Member State within the European Union should have access to the PIF. The manufacturer or importer must make the PIF readily accessible at one single address within the Community (EU), allowing easy access for inspections and other regulatory purposes.

Authorities find evaluating regulatory compliance simpler and carrying out market monitoring operations when PIFs are housed in one central location. Additionally, it makes it easier for importers and manufacturers to communicate with the appropriate authorities as needed. This criterion helps guarantee that, before being made available to customers, cosmetic items put on the market are thoroughly evaluated for their safety and adhere to all applicable laws.

5.5.5 Regulation of Ingredients

Specific limitations on the use of certain compounds that have been determined to be potentially dangerous or unsafe for use in cosmetics are established by Regulation (EC) No 1223/2009 on cosmetic products. These limitations could be on concentration, prohibitions on particular drugs, or demands for particular directions or warnings. Regulatory bodies must perform sampling and analysis to ensure that the regulations are being followed in order to properly enforce these limits. In order to conduct testing, representative samples of cosmetic items from the market are chosen through the sampling process. During sampling and analysis, it's critical to adhere to defined techniques and guidelines to guarantee reproducibility and standardization.

After the Scientific Committee for Consumer Safety (SCCS) completes its risk evaluation of these compounds, hair colorants should be included in the discussion of safety concerns, which are currently limited to skin colorants. This would enable a thorough evaluation of hair colorants' safety and guarantee that they fulfill all requirements before being approved for usage. According to Regulation (EC) No 1272/2008 of the European Parliament and of the Council of December 16, 2008, on classification, labeling, and packaging of substances and mixtures, given the hazardous properties of substances classified as carcinogenic, mutagenic, or toxic for reproduction (CMR), categories 1A, 1B, and 2.

When it is appropriate, the SCCS should provide comments on the safety of using nanomaterials in cosmetics. The whole information provided by the accountable party needs to serve as the foundation for these opinions. Prohibited compounds should only be allowed in trace quantities if they are technologically necessary with proper production processes and if the product is safe. This will help to assure product safety.

5.5.6 Labeling of the Products

Transparency is required for the components used in cosmetic products. A cosmetic product's packaging should list the components in order to attain this level of transparency. In cases when it is not feasible to list the components on the package, the consumer should nonetheless have access to this information by having it enclosed. The Commission should create a dictionary of common ingredient names to promote consistency in labeling and make it easier to identify compounds in cosmetics. It is not intended for this dictionary to serve as an exhaustive list of all the ingredients found in cosmetics.

Cosmetic items should have clear, concise labels that provide information to consumers about how long they are durable. It is crucial to understand the date of minimal durability or the optimal time to use a cosmetic product, as customers have a right to know how long it will last to fulfill its intended purpose and stay safe. When a cosmetic product has a minimum durability of more than 30 months, the user should be made aware of how long they can use it without risk after opening. However, this condition shouldn't be applied to things that are single-use or for whom the idea of durability after opening is irrelevant.

The SCCS has identified a number of compounds that are likely to trigger allergic responses. Thus, it will be required to limit their usage or apply certain guidelines when using them. The existence of these drugs should be noted in the ingredient list, and consumers' attention should be drawn to their presence in order to guarantee that they are properly informed. Customers should be better able to diagnose contact allergies thanks to this knowledge, which will also help them avoid using cosmetics that they can't tolerate.

Other restrictive measures, like a ban or a concentration restriction, should be taken into consideration for chemicals that are likely to cause allergies in a significant percentage of the population. To help them make well-informed product decisions, customers should have the option to request specific product-related information from the responsible party in addition to the information that is labeled. Regulatory bodies and industry players can improve consumer safety and knowledge, encourage well-informed decision-making, and help to establish trust in the cosmetics goods market by putting these transparency measures into place.

5.5.7 Regulating Nonclinical Safety Studies

To guarantee comparability and high quality, it is crucial that the outcomes of nonclinical safety investigations for cosmetic products adhere to applicable Community regulations.

According to Regulation (EC) No 1223/2009, nonclinical safety studies must be carried out in order to evaluate the safety of cosmetic goods. Tests for skin irritation, ocular irritation, sensitization, and other pertinent criteria are commonly included in these investigations.

These studies must meet certain conditions specified in different community legislation in order to guarantee their dependability and quality. As an example:

1. **Good Laboratory Practice (GLP)**: Research on nonclinical safety should adhere to GLP guidelines as outlined in Directive 2004/10/EC or other pertinent laws. GLP makes sure that protocols and documentation guidelines are strictly followed and that the studies are carried out in an organized manner.
2. **Test procedures**: The nonclinical safety research should employ tested procedures that have been approved by international organizations like the Organisation for Economic Co-operation and Development (OECD) or by experts in the field. These techniques must adhere to strict standards for relevance, accuracy, and dependability, as well as be in line with the state of science.
3. **Animal testing**: Under Directive 2010/63/EU on the protection of animals used for scientific purposes, the use of animals in cosmetic product testing is highly restricted. To reduce the amount of animal testing, alternative testing techniques such as in vitro experiments or computer models should be taken into account whenever feasible.

Manufacturers can guarantee the validity and comparability of their results during nonclinical safety investigations by adhering to pertinent Community legislation. Reducing the risks associated with dangerous cosmetics on the market promotes consumer protection and assists regulatory bodies in making well-informed judgments concerning product safety.

5.5.8 Regulating Animal Testing

Customers, who are becoming more concerned about animal welfare, have been debating and worrying about the matter of animal experimentation in the cosmetics sector. With many nations implementing bans on such testing, the campaign to replace animal testing in cosmetics with other alternatives has gained traction. The disparities between national marketing bans and laws prohibiting animal testing have created difficulties and impacted global trade. For example, regulations in one nation that forbids animal testing on cosmetics may apply to those tested in another, creating obstacles to market entry.

The treaty's appendix, the Protocol on the Protection and Welfare of Animals, emphasizes how crucial it is to take animal welfare into account while implementing laws, especially those pertaining to the internal market. A shift is underway from using animals in cosmetic ingredient testing to nonanimal alternatives. Organizations such as the European Centre for the Validation of Alternative Methods (ECVAM), which follow the validation guidelines of the Organization for Economic Cooperation and Development (OECD), have evaluated and certified these alternative techniques as scientifically validated. The cosmetics sector may go closer to doing away with animal testing while maintaining the quality and safety of its products by progressively switching to nonanimal testing techniques and harmonizing legislation globally. This change not only aligns with consumer preferences but also spurs advancements in morally sound and efficient testing techniques. The Commission set deadlines that ran out on March 11, 2009, to forbid the commercialization of cosmetic goods, their final formulation, substances, or mixtures of ingredients that have been subjected to animal testing, as well as all existing animal testing. However, considering the results of studies on reproductive toxicity, repeated-dose toxicity, and toxicokinetics, it is fair to set March 11 as the final date for banning the marketing of cosmetic goods that use those tests.

References

Abelan US, de Oliveira AC, Cacoci ÉSP, Martins TEA, Giacon VM, Velasco MVR, de Castro Lima CRR (2022) Potential use of essential oils in cosmetic and dermatological hair products: a review. J Cosmet Dermatol 21(4):1407–1418

Alessandrini A, Piraccini BM (2016) Essential of hair care cosmetics. Cosmetics 3(4):34

Alshehrei FM (2024) Microbiological quality assessment of skin and body care cosmetics by using challenge test. Saudi J Biol Sci 31:103965

Ferreira M, Matos A, Couras A, Marto J, Ribeiro H (2022) Overview of cosmetic regulatory frameworks around the world. Cosmetics 9(4):72

Gubitosa J, Rizzi V, Fini P, Cosma P (2019) Hair care cosmetics: from traditional shampoo to solid clay and herbal shampoo, a review. Cosmetics 6(1):13

Hordinsky M, Chu S, Caramori APA, Donovan JC (2022) Hair physiology and grooming. In: Cosmetic dermatology*:* products and procedures. John Wiley & Sons, pp 299–308

Jose J, Shenoy S, Shetty S, Zulfa S, Haseena T (2024) Nanomaterials in cosmetics. In: Handbook of nanomaterials, vol 2. Elsevier, pp 497–522

Mandal S, Venkatramani J (2023) A review of plant-based natural dyes in leather application with a special focus on color fastness characteristics. Environ Sci Pollut Res 30(17):48769–48777

Milstein SR, Jurkiewicz M, Ciarlo G, Taborska I, Teshima R, Roesler C, Breslawec H, Ansell J, Renner G, Kasai Y et al (n.d.) Report for the International Cooperation on Cosmetics Regulation

Sattar M, Saeed F, Afzaal M, Rasheed A, Asif A, Sharif S, Hussain M, Asad Ur Rehman H, Raza MA, Munir H et al (2024) An overview of the nutritional and therapeutic properties of amaranth. Int J Food Prop 27(1):263–272

Union P (2009) Regulation (EC) No 1223/2009 of the European Parliament and of the Council. Official Journal of the European Union L, 342, 59

Yun K, Ahn C (2023) Effect of surfactant type on the dyeability and color resistance of semipermanent basic hair dye. Fash Text 10(1):4

Regulatory Requirements for the Efficacy/Safety Assessment of Cosmetics/Nano-Cosmetics Products: Opportunities and Challenges for New Approach and Methodologies

6

Amrutha Chandran, Renjith P. Nair, and Anugya Bhatt

Abstract

Ensuring the safety and efficacy of cosmetics, particularly those containing nanomaterials (nano-cosmetics), requires robust regulatory frameworks. This chapter explores the current landscape of safety assessment for these products, highlighting the opportunities and challenges associated with adopting new approach methodologies (NAMs). The growing emphasis on ethical and animal-free testing methods necessitates the exploration of NAMs. However, challenges remain in fully integrating NAMs into regulatory frameworks. The need for robust validation, data interpretation, and addressing knowledge gaps regarding the unique properties of nanomaterials are crucial considerations.

The chapter explores the potential of NAMs to revolutionize cosmetic safety assessment while acknowledging the need for ongoing research and regulatory adaptations to ensure consumer protection in the ever-evolving field of nano-cosmetics.

A. Chandran
Division of Extracorporeal Devices, Department of Medical Device and Engineering, Biomedical Technology Wing, Sree Chitra Tirunal Institute for Medical Sciences and Technology, Trivandrum, Kerala, India
e-mail: amrutha@sctimst.ac.in

R. P. Nair · A. Bhatt (✉)
Division of Thrombosis Research, Department of Applied Biology, Biomedical Technology Wing, Sree Chitra Tirunal Institute for Medical Sciences and Technology, Trivandrum, Kerala, India
e-mail: renjith@sctimst.ac.in; anugyabhatt@sctimst.ac.in

A. Dwivedi et al. (eds.), *Hair Care Products*,
https://doi.org/10.1007/978-981-97-6740-3_6

Keywords

Cosmetic safety assessment · Nano-cosmetics · New approach methodologies (NAMs) · Regulatory frameworks · Omics technologies · Data validation · Nanomaterial properties · Consumer protection

6.1 Introduction

The size of the world market for cosmetics was estimated at $380.2 billion in 2019 and is expected to increase to $463.5 billion by 2027 at a compound annual growth rate (CAGR) of 5.3% (Cosmetics Market 2021). Over the past few decades, the industry has seen great innovation, which has resulted in a wide range of new products and a boost in sales (Cosmetics Market 2021).

The most crucial elements in the development of the cosmetic sector are the efficacy, safety, regulatory environment, and marketing of cosmetic products. The safety of cosmetic products is overseen by numerous regulatory authorities from across the world, each with its own laws and regulations. Distinct countries have different laws governing cosmetics, such as nomenclature, labeling, and the safety of colorants. The focus of this chapter is to briefly discuss different aspects under regulation to ensure safety and efficacy of cosmetics in the United States, Europe, and India along with the global challenges and opportunities for new approaches under regulation.

6.2 United States of America (USA)

The main regulatory body in charge of upholding the laws controlling the marketing of cosmetics in the United States is the Food and Drug Administration (FDA). The FDA's main organization in charge of planning, organizing, and managing important matters pertaining to food and cosmetics is the Center for Food Safety and Applied Nutrition (CFSAN) (Ferreira et al. 2022). The FDA enforces laws approved by Congress and, when authorized by Congress, issues regulations in order to carry out its statutory power. Regulations can create binding commitments and have legal authority.

The two most important laws controlling cosmetics sold in the United States are the Fair Packaging and Labeling Act (FPLA) and the Federal Food, Drug, and Cosmetic Act (FD&C Act). The FDA is able to regulate cosmetics based on these laws (Rampal and Kanti 2020). The rules that govern cosmetics in the United States and many other countries range significantly in a number of ways, such as how cosmetics are defined legally, how color additives and other substances are used, and registration requirements.

For instance, in Europe, some products are controlled as drugs in the United States are controlled as cosmetics. One such example is sunscreen. Additionally, there are distinctions between substances that are forbidden and regulated, particularly color additives. In United States, cosmetic manufacturers may be required to register their organizations and declare their products and ingredients with the government and this registration is voluntary but strongly advised (FD&C Act 1938a; Katz et al. 2022).

6.2.1 Definition of Cosmetics Under US Law

The definition of cosmetics according to the Federal Food, Drug & Cosmetic Act (FD&C Act) is "articles intended to be rubbed, poured, sprinkled, or sprayed on, introduced into, or otherwise applied to the human body, for cleansing, beautifying, promoting attractiveness, or altering the appearance." Products including moisturizers for the skin, lipsticks, nail polishes, eye and face makeup preparations, shampoos, permanent waves, hair dyes, toothpastes, deodorants, and any substance meant to be used as an ingredient in a cosmetic product are all included in the definition (FD&C Act 1938a, b).

"Soap" products are generally exempt from the cosmetic rules and definition of cosmetics under the FD&C Act. The Consumer Product Safety Commission (CPSC) is in charge of regulating soap products, which are typically outside the FDA's jurisdiction. "Alkali salts of fatty acids" are the main ingredient required for a product to qualify as "soap," and they are the only ones that have the ability to clean. A product is classified as a cosmetic rather than a soap if it contains synthetic detergents. Body cleaning must be the main purpose of the soap's formulation. If a product is meant to moisturize skin, improve one's scent, or eliminate body odor, it is considered a cosmetic (Paye et al. 2005).

Some cosmetics goods may also be regulated as drugs in the United States; this can happen when a product has two claimed purposes and contains components that do two distinct functions. For example, shampoo is a cosmetic because its purpose is to clean hair. Anti-dandruff therapy meets the criteria for drugs category because its primary goal is to treat dandruff. An antidandruff shampoo is therefore a drug as well as a cosmetic. Other examples of cosmetic/drug combinations include toothpaste that contains fluoride, deodorants that serve as antiperspirants, moisturizers, and cosmetics that claim to protect against sun damage. For legal purposes, a product can be either a pharmaceutical or a cosmetic, or both. However, the term "cosmeceutical" has no legal meaning (Ruhela et al. 2018).

6.2.2 Major Cosmetics Laws and Regulations

6.2.2.1 Federal Food, Drug, and Cosmetic Act (FD&C Act)

These statutes were passed by the US Congress in 1938 and gave the US Food and Drug Administration (USFDA) the power to monitor the safety of cosmetics, drugs, food, and medical devices. The different provisions for cosmetics are regulated under Chapter VI of the statute. Except for color additives, the act does not mandate that cosmetic products and ingredients receive FDA approval prior to being put on the market (FD&C Act 1938a).

The FDA has the power to take enforcement action against businesses that violate the law and against items that are already on the market but don't follow the law. When creating a cosmetic product, a company can generally use any ingredient—the only exceptions being color additives and substances that are restricted or illegal.

Provided that the following requirements must be adhered to:

- The product is appropriately labeled.
- If the substance is used, the cosmetic product won't be adulterated or mislabeled, which would be against FDA-enforced laws.
- The ingredients and the finished cosmetic are safe under the listed or customary conditions of use.

Additionally, the Act forbids the importation of "adulterated" or "misbranded" cosmetics into interstate commerce as well as their receipt within (FD&C Act, Section 601 and 602, respectively). Table 6.1 summarizes sections and specifics of FD&C Act Chapter VI: Cosmetics.

6.2.2.2 Fair Packaging and Labelling Act

A consumer protection act known as the Fair Packaging and Labelling Act (FPLA) of 1966 gives the FDA and FTC authority to create rules governing product packaging and labeling. The act requires product makers to provide accurate and comprehensive information on their labels and packaging (Fair Packaging and Labeling Act 1966).

6.2.2.3 Microbead Free Waters Act, 2015

On December 18, 2015, the Microbead-Free Waters Act of 2015 was passed by Congress, amending the Federal Food, Drug and Cosmetic Act (FD&C Act). Microbead-Free Waters Act of 2015 prohibits the manufacturing, packaging, and distribution of rinse-off cosmetics containing plastic microbeads. The term "plastic microbead" is defined by law as any solid plastic particle that is:

- 5 millimeters or less in size.
- Intended to be used to exfoliate or cleanse the body or any part of the body.
- This new rule also covers non-prescription (commonly known as "over-the-counter" or "OTC") medications, including toothpaste, and cosmetics (The Microbead-Free Waters Act 2015; McDevitt et al. 2017).

Table 6.1 FD&C Act Chapter VI: Cosmetics, subchapters, and section numbers

Title	FD&C Act section number	U.S. code section number
Adulterated cosmetics	601	361
Misbranded cosmetics	602	362
Regulations making exemptions	603	363

6.2.3 Title 21 of the Code of Federal Regulations (21 CFR)

Subchapter G of Title 21 of the Code of Federal Regulations contains the regulations pertaining to cosmetics in the United States. These regulations may have legally enforceable obligations. Table 6.2 is a compilation of the main CFR requirements.

6.2.4 Safety Assessment for Cosmetics

Currently, medicines, biologics, and medical devices are subject to FDA safety and effectiveness requirements; cosmetics are not. Manufacturers are not required to test their products for safety under the FDA, or are they required to utilize preclinical (i.e., animal or in vitro alternative studies) or human clinical safety tests to support the safety of cosmetic items. This is because neither the FD&C Act nor the FDA specify the types of tests that must be used. Cosmetic product manufacturers and marketers are exempt from submitting test findings for safety substantiation to the FDA for premarket approval (Ferreira et al. 2022).

According to the Federal Register dated March 3, 1975 (40 FR 8912 at 8916), a product's safety can be sufficiently demonstrated by:

(a) Relying on toxicological test results for individual chemicals that have already been published, as well as on product formulations that are comparable to the cosmetic in terms of composition.

Table 6.2 Title 21 of the Code of Federal Regulations (21 CFR) for cosmetics

21 CFR Part No.	Title
1	General Enforcement Regulations
2	General Administrative Rulings and Decisions
20	Public Information
250	Requirements for Drugs and Cosmetics—Hexachlorophene
700 Subpart A	Cosmetics—General Provisions
700 Subpart B	Requirements for Specific Cosmetic Products
701 Subpart A	Cosmetic Labeling—General Provisions
701 Subpart B	Package Form
701 Subpart C	Labeling of Specific Ingredients
710	Voluntary Registration of Cosmetic Product Establishments
720	Voluntary Filing of Cosmetic Product Ingredient and Cosmetic Raw Material Composition Statements
740	Cosmetic Product Warning Statements

(b) Execution of any additional toxicological and other tests that are necessary in light of the data and knowledge already available.

To ensure the safety of the final cosmetic product, toxicological testing involving the entire formulation will still be required, even though there may be sufficient data regarding the safety of each ingredient in a cosmetic product. According to the exclusivity wording in 21 CFR 740.10, the product could be considered misbranded and face regulatory action if the label omits one of the following statements, (a): "Warning—The safety of this product has not been determined."

6.2.5 Prohibited and Restricted Ingredients in Cosmetics

The FD&C act does not impose any prior approval for the ingredients used in the cosmetics before they are introduced into the market, with the sole exemption for color additives, which are subject to premarket approval.

FDA has regulations that specifically prohibit or restrict the use of some ingredients as they will make the cosmetic product harmful if incorporated. These are given in Table 6.3 (Code of Federal Regulations 2024a; Rodriguez-Perez 2022)

Table 6.3 Cosmetic ingredients prohibited or restricted in the United States

Ingredients	21 CFR Part No.	Reason for prohibition
Bithionol	700.11	May cause photo contact sensitization
Chlorofluorocarbon propellants	700.23	The application of chlorofluorocarbon propellants in aerosol cosmetics meant for home usage
Chloroform	700.18	Causes cancer in animals and may have negative health effects on people
Halogenated salicylamides (di-, tri-, metabromsalan and tetrachlorosalicylanilide)	700.15	May cause serious skin disorders
Hexachlorophene	250.250	Possibility of toxicity and skin penetration in humans
Mercury compounds	700.13	May result in skin irritation, allergic reactions, or neurotoxic issues
Methylene chloride	700.19	Causes cancer in animals and is likely to be harmful to human health
Prohibited cattle materials	700.27	To ward off "mad cow disease," often known as bovine spongiform encephalopathy, or BSE
Vinyl chloride	700.14	Aerosol product ingredient: causes cancer and other health issues
Zirconium-containing complexes	700.16	Animal lung toxicity as well as the development of granulomas in human skin

6.2.6 Colour Additives

According to U.S. legislation [Federal Food, Drug, and Cosmetic Act (FD&C Act), Sec. 721; 21 U.S.C. 379e], color additives must pass a rigorous clearance process. Failure to comply with US color additive requirements renders a cosmetic adulterated, except for coal-tar hair dyes (FD&C Act 2024a).

According to the 1960 Colour Additive Amendments to the FD&C Act, color additives used in food, drugs, medical devices, and cosmetics must first receive FDA approval for their intended use. As part of this process, color chemistry and toxicology staff experts at the FDA must evaluate the color additive's chemistry and safety. A cosmetic is deemed adulterated and is subject to regulatory action if it contains an unlisted color additive (i.e., one that the FDA has not approved for its intended purpose) (Paye et al. 2005).

6.2.6.1 Categorization of Color Additives

The permitted color additives are divided into two primary categories under the FD&C Act Section 721(c) [21 U.S.C. 379e(c)] and the color additive rules [21 CFR Parts 70 and 80]: those subject to certification (sometimes known as "certifiable") and those exempt from certification. The rules also provide other classes, such as straight colours, and lakes.

(a) **Colours subject to certification**. These color additives, also referred to as "coal-tar dyes" or "synthetic-organic colours," are mostly sourced from petroleum. With the exception of coal-tar hair dyes, these colors cannot be used unless the FDA certifies that the batch in question has successfully completed analysis of its composition and purity in FDA-owned facilities.

The names of these authorized colors often consist of three parts. The names also contain a color, a number, and the prefix FD&C, D&C, or External D&C. "FD&C Yellow No. 5" is a prime illustration. In cosmetic ingredient declarations, certified colors can simply be specified by colour and number without a prefix (such as "Yellow 5") (FD&C Act 2024b).

(b) **Colours exempt from certification**. Most of these colorants come from mineral, plant, or animal sources. The rules for batch certification do not apply to them. However, because they continue to be regarded as artificial colors, they must adhere to the regulations' criteria for identification, specifications, uses, and labeling when they are used in cosmetics or other FDA-regulated items [21 CFR 73].

(c) **Straight color**. "Straight colour" refers to any color additive listed in 21 CFR 73, 74, and 81 [21 CFR 70.3(j)].

(d) **Lake**. A lake is a straight color that has been spread over a surface by a process other than simple mixing, such as adsorption, co-precipitation, or chemical combination [21 CFR 70.3(l)].

To ensure that one is only utilizing color additives for their allowed applications, it is crucial to review both the regulations themselves (21 CFR 82, Subparts B and C) and the summary of color additives listed for use in the United States in Foods, Drugs, Cosmetics, and Medical Devices (FD&C Act 2024b).

6.2.7 Safety of Nanomaterials in Cosmetic Products

Because of their distinctive features, nanomaterials are used in several FDA-regulated products, providing potential benefits to the new product. Due to their nanoscale size, such materials may have different chemical, physical, and biological characteristics when compared to other materials of larger size. Changes in magnetic characteristics, electrical or optical activity, enhanced structural integrity, or altered chemical or biological activity are a few examples of such variations. The performance, quality, safety, and/or efficacy, if applicable, of a product that integrates the nanomaterial may be impacted by these new or altered characteristics (US FDA 2014).

The use of these materials in cosmetic goods may raise concerns about the product's fitness for its intended application, according to the FDA's guidelines on the safety of nanomaterials in cosmetics. Nanomaterials can have distinct chemical, physical, and biological properties when compared to larger-scale particles with the same chemical composition.

Data requirements and testing procedures should be assessed to address any special features and functions of the nanomaterials used in cosmetic products as well as the ongoing concerns about the applicability of conventional safety testing procedures to products involving nanotechnology. This is true for any cosmetic product that has novel or modified properties (US FDA 2014).

The FDA advises that the safety assessment for cosmetic items employing nanomaterials should take into account a number of crucial factors, as shown in Fig. 6.1.

6.2.8 Labeling Requirements

Cosmetics sold in the United States must adhere to the FDA's labeling rules, which were published in compliance with the FD&C Act and the FP&L Act. Labeling includes all written, printed, and graphic content attached to or on a product, including labels (FD&C Act 2024c). Both the inside and any exterior container or wrapper must have the label statements mandated by the FD&C Act. The labeling specifications are outlined in 21 CFR Sections 701 and 740 (Fair Packaging and Labeling Act 1966).

Cosmetics that are misbranded and that include label statements that are untrue or deceptive or that are otherwise not labeled in line with these rules may be subject to regulatory action. Label requirements under the FD&C Act must be displayed on both the inside and exterior of any container or wrapper that may be present. The label of the exterior container or wrapper is the only place where

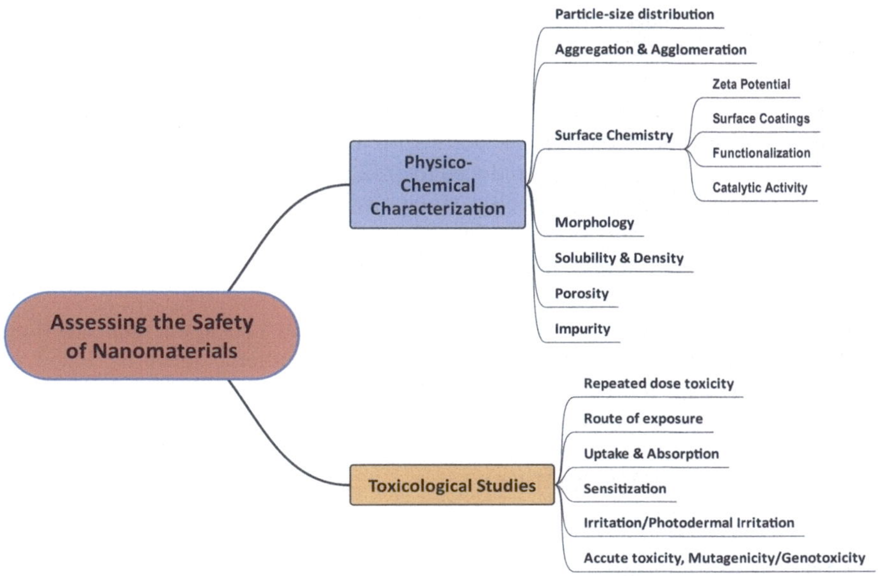

Fig. 6.1 FDA recommended safety assessment considerations for nanomaterials in cosmetics

FPLA criteria must be stated. The principal display panel is the part of the label that is most likely to be displayed or examined under standard conditions of display for sale. It must contain the product name, a description of the intended use of the product, and an indication of the total quantity of cosmetics in the package in terms of weight, measure, numerical count, or a combination of weight and measure (21 CFR 701.10).

The declaration must, as required by law, be readable, placed in the lower portion of the panel approximately parallel to the base that the package rests on, and use a font size that is appropriate for the size of the container (Code of Federal Regulations 2024b; US FDA 2024).

6.2.9 Declaration of Ingredients

Cosmetics created or sold for consumer retail use for personal care must have an ingredient declaration as per 21 CFR 701.3. Products like skin cleansing or emollient creams used by people at their places of job, or hair preparations or makeup products used by professionals on clients at their establishments, are exempt from this requirement as long as they are not also sold to customers at professional establishments or workplaces for consumption at home (US FDA 2024).

To ensure that the consumer is likely to read the ingredient declarations at the time of purchase, they must be "conspicuous" and "promlinent" in their placement on any information panel of the outer container and not less than a certain size in relation to the size and shape of the product package (Paye et al. 2005).

The regulations specify the following specific requirements for how cosmetic ingredients must be presented on the label:

- **Font size**: According to 21 CFR 701.3(b), the lettering for ingredients must be at least 1/16 of an inch tall. For packages with a labeling surface less than 12 sq. inches, the minimum height is reduced to 1/32 of an inch (21 CFR 701.3(p)).
- **Off-package labeling**: In specific situations, ingredient information can be placed outside the primary packaging. This is allowed if the cosmetics are held in organized trays or racks, not in a folding carton, and the total labeling surface is under 12 square inches (21 CFR 701.3(i)).
- **Ingredient order and nomenclature:** As per the Fair Packaging and Labelling Act (FPLA) labeling requirements, cosmetic ingredients must be listed in descending order of predominance using standardized names established by law (21 CFR 701.3(c) and 701.30). Alternatively, grouping of ingredients with declaration based on specific regulations is also permitted (21 CFR 701.3(f); Fair Packaging and Labeling Act 1966).

6.2.10 Label Warnings

Cosmetics that could harm customers if they are misused must have the proper label warnings and usage instructions. The declarations must be obvious and prominent. Some cosmetics are required by law to have warnings or cautions on the label (21 CFR 740). Examples of products that demand such disclosures include children's bubble bath products, feminine deodorant sprays, and cosmetics in self-pressurized containers (aerosol products) (US FDA 2024).

The FDA encourages cosmetic producers to conduct necessary toxicological or physico-chemical or in vitro tests required to substantiate the safety of their products, even though the FD&C Act does not require cosmetic manufacturers or marketers to test their products for safety. If a cosmetic's safety is not sufficiently supported, the product may be deemed misbranded and be the target of regulatory action unless the label includes the following wording: Warning: The safety of the product is not determined (Code of Federal Regulations 2024c).

6.2.11 Tamper-Resistant Packaging

A product is considered tamper resistant if it has a clear sign or barrier that shows if someone has opened it. Examples include aerosol containers and tape seals. In the United States, regulations require certain personal care products, like douches, vaginal tablets, mouthwashes, and breath fresheners, to be sold in tamper-resistant containers. As per Section 21 CFR 700.25, this indicator should be easily noticeable. It can be designed in a specific way (like a breakaway cap or blister pack), have a clear visual element (logo, image), or both. The tamper-resistant feature can be on the inner container, outer container, or both. The packaging must clearly explain this tamper-resistant feature to consumers. If the tamper-evident feature is damaged or missing, this explanation should still be clear and not be affected (US FDA 2024).

6.2.12 Voluntary Cosmetic Registration Program

The FD&C Act does not demand that cosmetic companies provide information on product adverse reaction reports or register production facilities or formulations with the FDA. Manufacturers, packers, and distributors of cosmetic items that are sold commercially in the US may use the FDA's Voluntary Cosmetic Registration Program (VCRP) as a reporting method. The VCRP program has three parts.

Part I of the Cosmetic Establishment Registration Program (CERP) calls for the registration of cosmetic production facilities with the FDA (see 21 CFR 710).

Part II of the Cosmetic Product Ingredient Statements (CPIS) regulation stipulates that the FDA must be notified of cosmetic formulations and cosmetic raw material composition statements (21 CFR 720). The FDA has designated 13 product category codes (PCC) in this regulation, 21 CFR 720.4, for acceptable cosmetic uses. In the early 1990s, semiquantitative raw material disclosures were discontinued and removed from the VCRP database.

Part III, "Product Experience Reports (PER)," requires manufacturers to submit "reportable" adverse events (defined in 21 CFR 700.3(q)) associated with the use of cosmetic products on an annual basis to the FDA (21 CFR 730) (Paye et al. 2005).

6.3 European Union (EU)

The primary legislative framework for finished cosmetic products introduced to the EU market is Regulation (EC) No. 1223/2009, which took effect on July 11, 2013, replacing the EU Cosmetics Directive (Directive 76/768/EC). This regulation unifies and streamlines the laws governing cosmetics across all EU member states. Its objectives include standardizing terminology, simplifying procedures, and strengthening the regulatory framework to ensure a high level of human health protection. The legislation applies to suppliers of cosmetic ingredients, as well as manufacturers and importers of cosmetic products (EU 2009). Table 6.4 illustrates the major annexes to the regulation.

6.3.1 Recent Amendments

Regulation (EU) 2022/1531, which amends Annexes II, III, and V of the Cosmetics Regulation (EC) 1223/2009 concerning the use of particular compounds found to be carcinogenic, mutagenic, or hazardous for reproduction (CMR substances), was published by the EU Commission on September 15, 2022. Substances listed in Part 3 of Annex VI to Regulation (EC) No. 1272/2008 (CLP) as CMR categories 1A, 1B, or 2 are not allowed to be used in cosmetic goods, as per Article 15 of the Cosmetics

Table 6.4 Annexes of EU Cosmetic Regulation 1223/2009

Annexes	Title
I	Cosmetic product safety report
II	Prohibited substance list
III	List of ingredients that cosmetic goods cannot contain, with the exception of the specified limitations
IV	A list of colorants that are permitted in cosmetics
V	A list of preservatives that are acceptable in cosmetic goods
VI	List of UV filters allowed in cosmetic products
VII	Symbols used on packaging/container
VIII	List of validated alternative methods to animal testing
IX—Part A	Repealed directive with its successive amendments
IX—Part B	A list of deadlines for incorporating into national legislation and its implementation
X	Correlation table

Regulation. Articles 15(1) and 15(2) of the Cosmetics Regulation, on the other hand, offer an exception, permitting the use of CMR compounds in cosmetic goods subject to specific limitations.

Fourteen new compounds have been added to Annex II of the Cosmetics Regulation, which enumerates substances prohibited in cosmetic products. Sodium *N*-(hydroxymethyl) glycinate (CAS No. 70161-44-3), previously listed in Annex II as entry 1669 and classified under specific conditions as Carcinogenic category 1B and Mutagenic category 2 according to the (EU) 2020/1182 amendment to CLP, has been removed from the Annex (EU Regulation 2022).

6.3.2 Registration, Evaluation, Authorization, and Restriction of Chemicals (REACH) Regulation (EC No. 1907/2006)

The EU regulation (EC) No. 1907/2006 on the registration, evaluation, and authorization of chemicals is known as REACH. On June first, 2007, it became effective. REACH has an impact on both finished cosmetics goods and chemical raw ingredients. Most cosmetic products fall under the REACH regulation's definition of chemical preparations (mixtures). Each chemical substance or ingredient must be pre registered with the European Chemicals Agency (ECHA) if its annual tonnage exceeds 1 tonne. Companies outside the EU can appoint a REACH-only representative to submit pre-registrations or registrations on their behalf. EU cosmetics importers and manufacturers must ensure that all ingredients, whether used in finished products or as components in cosmetics, are correctly registered by their suppliers to avoid needing to perform their own REACH registrations. Additionally, the use of chemicals of very high concern in cosmetic products may require EU manufacturers and importers to seek prior authorization (EU 2009; Williams et al. 2009).

6.3.3 Definition of a Cosmetic Product

According to the European Commission Regulation 1223/2009 [Article 1], a cosmetic product is any substance or preparation intended for contact with various parts of the human body (such as the epidermis, hair system, nails, lips, and external genital organs) or with the teeth and the mucous membranes of the oral cavity, primarily for the purposes of cleaning, perfuming, changing appearance, correcting body odors, or protecting and maintaining them in good condition. This definition outlines the target sites for cosmetic product applications and their permissible functions. Therefore, in the European Union, items like soap, shampoo, permanent waves, hair dyes, toothpaste, deodorants, skin creams, lotions, fragrances, lipsticks, nail polishes, and makeup are categorized as cosmetic products. Additionally, products like suntanning preparations, antiperspirants, and antidandruff shampoos are also considered cosmetics in the EU, even though this classification may differ in other regions of the world (EU 2009).

6.3.4 Safety Requirements for Cosmetics

According to the regulation, a cosmetic product sold within the Community must not pose a risk to health when used under typical or foreseeable conditions. This includes considering the product's presentation, labeling, usage, and disposal instructions, and any other information provided by the manufacturer, their authorized agent, or any other person responsible for marketing the product. It is the duty of the manufacturer, their authorized agent, or any responsible party to ensure the safety of cosmetics for consumer use.

A qualified safety assessor, holding a relevant diploma in pharmacy, toxicology, dermatology, medicine, or a related field, must sign off on the safety assessment of the cosmetic product. This assessor takes responsibility for ensuring the product's safety when used as intended or under reasonably foreseeable conditions.

6.3.5 Material Safety Data Sheet (MSDS) and REACH

All compounds used in cosmetic products, including those specifically utilized in cosmetic powders, must adhere to REACH regulations as distinct chemicals. Standard studies on acute inhalation toxicity are requested for substances produced in quantities greater than 10 tonnes annually and for substances that are expected to be inhalable, or potentially breathed in, based on their volatility (Steiling et al. 2018). The material safety data sheet will provide information on these tests and any classification under the GHS/CLP (Globally Harmonized System), which was implemented in the EU in the form of Regulation (EC) No. 1272/2008 on the classification, labeling, and packaging of chemicals and mixtures (MSDS). In evaluating the safety of the cosmetic ingredient's constituents, each distinct danger classification must be considered (Pauwels and Rogiers 2004).

6.3.6 The Directive on Animals Used for Scientific Procedures (EC/2010/63)

According to the EU's Directive 2010/63, animal testing is prohibited if an alternative method is available, making this directive unique globally in its regulations on the use of animals in scientific procedures. The directive states that "Member States shall ensure that a procedure is not carried out if another method or testing strategy for obtaining the result sought, not entailing the use of a live animal, is recognised under the legislation of the Union, without prejudice to national legislation prohibiting certain types of methods" (First Article, 13(1)).

This provision clarifies for regulatory bodies which alternative techniques are considered "reasonably and practically available," specifically those approved by law.

However, Article 4(1) further mandates that "Member States shall ensure that, wherever possible, a scientifically satisfactory method or testing strategy not entailing the use of live animals shall be used instead of a procedure." This provides a broader prohibition on animal testing when alternatives are available (Pauwels and Rogiers 2004; Taylor and Rego 2020).

These special provisions demonstrate to the rest of the world the EU's value for avoiding animal testing and the significance of creating substitute methods.

6.3.7 Cosmetic Products and Exposure Levels in Use

The safety assessment of cosmetics is based on how the product will be used. This mode of use determines the amount of substances ingested, absorbed through the mucous membrane or skin, or inhaled, etc. Therefore, it is necessary to determine the extent and mode of consumer exposure prior to conducting any safety evaluation or risk assessment of a final product (Pistollato et al. 2021; Filaire et al. 2022). The major factors to be considered while assessing the risk and exposure are shown in Fig. 6.2.

6.3.8 Toxicological Studies

The first step in assessing the hazard of an ingredient involves determining its toxic potential through a series of toxicity studies, each targeting specific toxicological endpoints. For finished cosmetic products, the primary toxicological studies to consider generally include acute toxicity, skin absorption, skin irritation, mucous membrane irritation, skin sensitization, sub-chronic toxicity, mutagenicity, phototoxicity, and photomutagenicity (for substances that absorb UV light).

When significant oral intake is anticipated, or when data on skin absorption indicate a substantial amount of ingredients penetrating the skin, it is essential to consider the substance's chemical structure and toxicological profile. This profile includes information on teratogenicity, reproductive toxicity, carcinogenicity,

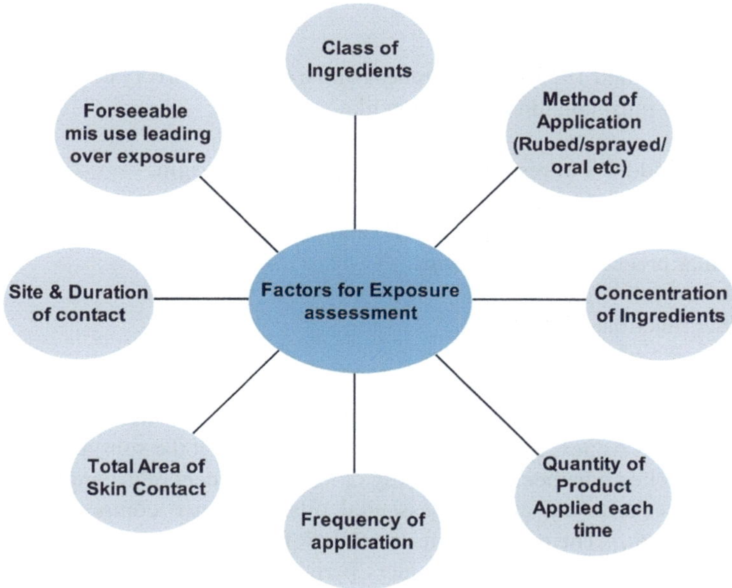

Fig. 6.2 List of factors for exposure assessment

additional genotoxicity, and metabolism. In such cases, studies on metabolism might be necessary to fully demonstrate the safety profile (Pistollato et al. 2021).

6.3.9 Considerations for Cosmetics with Nanomaterials

Based on Article 2.1k of the regulation, nanomaterials are described as an intentionally created insoluble or bio-persistent material with one or more external dimensions or an internal structure on a scale from 1 to 100 nanometres (nm). Nanomaterials are the only topic covered in Article 16 of the rule. For each cosmetic product to be sold in Europe, a "responsible person" must be named (EU 2009). The European Commission is to be notified 6 months in advance of the launch of any cosmetics incorporating nanomaterials, per this representative's request. The following information on the nanomaterials must be included in the information sent to the commission.

(a) Chemical name (IUPAC) and other descriptors for identification of nanomaterials as specified in point 2 of the Preamble to Annexes II–VI.
(b) Nanomaterial specifications like physical and chemical properties, size, etc.
(c) An estimate of the quantity of nanomaterial contained in cosmetic products intended to be placed on the market per year.
(d) The toxicological profile of the nanomaterial.
(e) Safety data.
(f) The reasonably foreseeable exposure conditions.

6.3.10 Labeling

As per Articles 6 and 19, for the awareness of the consumer, every cosmetic product sold in the EU must contain information like.

1. Name and address of responsible person; country of origin.
2. Nominal content (weight or volume).
3. Date of minimum durability (DOMD) and period after opening (PAO).
4. Particular precautions or warnings.
5. Batch number.
6. Product function.

Furthermore, if the concentration of any of the 26 specific substances listed in Annex III of the Regulation is higher than the threshold concentration indicated in column h) of this Annex, it must be indicated, regardless of the substance's function and/or source (i.e., whether added directly or as a component of a complex cosmetic ingredient like botanical extracts, essential oils, fragrance compositions, aroma compositions, etc.). The requirements are 0.01% for rinse-off products and 0.001% for leave-on cosmetics. These parameters pertain to the concentration of the final mixture when applied to the skin in the case of cosmetics that are mixed just before use and comprise many ingredients (EU 2009).

6.4 India

The Drugs and Cosmetics Act of 1940 (D&C Act), rules of 1945, Cosmetics Rules, 2020, and labeling requirements by the Bureau of Indian Standards (BIS) govern cosmetic products in India (Ministry of Health and Family Welfare, Govt. of India 1940, 1945).

The State Licensing Authorities, which are chosen by the individual State Governments, have a system of inspection and licencing in place to govern the manufacture of cosmetics. The Central Drugs Standard Control Organization (CDSCO) regulates the import of cosmetics into India through a system of registration. The primary regulatory body for the cosmetics sector is CDSCO.

6.4.1 Legal Definition of Cosmetics

"Any article intended to be rubbed, poured, sprinkled or sprayed on, or introduced into, or otherwise applied to, the human body or any part thereof for cleansing, beautifying, promoting attractiveness or altering the appearance, and includes any article intended for use as a component of cosmetic" is the definition of cosmetics as given in Section 3 of the Drugs and Cosmetics Act.

A cosmetic product cannot be imported or produced in India if it does not meet the established requirements, standards of quality, or safety. Before a product is imported into the country, it must be registered with the pack size, variant(s), and manufacturing facility if it meets the 1940 Act's definition of a cosmetic. A cosmetic item cannot be imported into India if it is not registered in accordance with the current regulations stipulated in Drugs and Cosmetics Act (Ministry of Health and Family Welfare, Govt. of India 1940).

6.4.2 Assessing the Safety of Cosmetics

Through a set of criteria, the D&C Act, Chapter IV ensures the safety of cosmetics made in India. The manufacturers of cosmetics are required to comply with the Bureau of Indian Standards regulations, and Schedule S has a list of 36 categories of cosmetics and toiletries specifications as shown in Table 6.5. Imported cosmetics must adhere to the specifications outlined in Schedules S and Q as well as any additional quality and safety requirements that may apply to them. The cosmetic must adhere to the requirements under the laws and norms that apply to it in the country of origin if it is not included in Schedule S.

Table 6.5 List of cosmetics in finished form notified by BIS

BIS list of cosmetics in finished form							
1	Skin powders	10	Oxidation hair dyes, liquid	19	Toilet soap	28	Henna powder IS: 11142
2	Skin powder-infants	11	Cologne	20	Liquid toilet soap	29	Bathing bars IS: 13498: 1997
3	Tooth powder	12	Nail polish	21	Baby toilet soap	30	Sindoor IS: 14649: 1999
4	Toothpaste	13	After shave lotion	22	Shaving soap	31	Liquid foundation makeup IS: 14318
5	Skin creams	14	Pomades and Brilliantines	23	Transparent toilet soap	32	Cold wax hair remover IS: 15152
6	Hair oils	15	Depilatories chemical	24	Lip salve IS:10284	33	Face pack IS: 15153
7	Shampoo, soap-based	16	Shaving creams	25	Powder hair dye IS: 10350	34	Kajal IS: 15154
8	Shampoo synthetic-detergent based	17	Cosmetic pencils	26	Bindi (liquid) IS: 10998	35	Oxidation hair dyes IS: 15205
9	Hair creams	18	Lipstick	27	Kum Kum powder IS: 10999	36	Cream bleach IS: 15608

6.4.3 Cosmetics Raw Materials and Adjuncts

The BIS standard 4707, part 2 lists cosmetics raw materials and adjuncts, other than dyes, colors, and pigments into four categories.

(a) **Annex A**: List of substances that must not form part of the cosmetic product composition. They are also known as Generally Not Recognized as Safe (GNRAS) substances. Examples include Methotrexate, Aminocaproic Acid and its salts, epinephrine, etc.
(b) **Annex B**: List of substances that cosmetic products must not contain except subject to the restrictions and conditions laid down like borates, chlorates, etc.
(c) **Annex C**: List of preservatives cosmetic products may contain. Preservatives are substances that may be added to cosmetic products for the primary purpose of inhibiting the development of micro-organisms in such products. Examples include sodium iodates, benzoic acid, salicylic acid, etc.
(d) **Annex D**: List of permitted UV filters cosmetic product may contain. UV filters are substances used in cosmetic sunscreen products which are specifically intended to filter certain UV rays to protect the skin from certain harmful effects of these rays. 3-Benzylidene camphor (2%) and titanium dioxide (25%) are examples.

Additionally, in terms of raw material regulation, hexachlorophene is prohibited in imported cosmetics but permitted in locally produced soaps with a labeling warning that reads, "Contains Hexachlorophene—unsuitable for use on infants."

Lead, arsenic, and mercury compounds are not allowed in finished formulation and testing on animals is completely banned (BIS 2009).

6.4.4 Dyes, Colors, and Pigments

Dyes, colors, and pigments that are generally recognized as safe (GRAS) for use in cosmetics are listed in BIS 4707, Part 1. Additionally, colorants used in preparations for toiletries are included on the list. The listing is based on the application field and is often divided into four divisions.

- Coloring agents allowed in all cosmetics products.
- Coloring agents are allowed in all cosmetics products except those intended to be applied in the vicinity of the eyes, in particular eye makeup and eye makeup remover.
- Coloring agents allowed exclusively in cosmetics products intended not to come into contact into contact with mucous membranes.
- Coloring agents allowed exclusively in cosmetics products intended to come into contact only briefly with the skin.

The maximum quantity of allowed synthetic and natural organic colors in cosmetics is as follows:

1. 2 parts per million of arsenic calculated as arsenic trioxide.
2. 20 parts per million of lead calculated as lead.
3. 100 parts per million of heavy metals other than lead calculated as the total of the respective metals.

All manufacturers and importers are required to adhere to this listing as specified by the schedule Q of the Drugs and Cosmetics Act (BIS 2001).

6.4.5 Mandatory Declarations for Cosmetics Manufactured in India Under Cosmetics Rules, 2020

The labeling requirement is specified in Rule 148 of the Drugs and Cosmetics Act and as per the Cosmetics Rules 2020. A cosmetic product must have labels on both the inner and outer pack, and also accompanied by instruction for use. The information to be printed specifically in these inner and outer pack labels are shown in Table 6.6 (Cosmetics Rules 2020).

Because of their potent chemical constituents and the possibility of reactions, hair dyes need to include extra warnings.

English and local language versions of the following statements must be printed on internal and external labels.

Caution: This product contains components that, in some situations, may cause skin irritation. As such, a preliminary test should be performed in accordance with the

Table 6.6 Labelling information of cosmetics in India

		Placement of information	
S. no.	Category of information	Inner label	Outer label
1	Name of cosmetics	✓	✓
2	Manufacturer name & address	✓	✓
3	Complete address of the premises	✓	✓
4	Best before use date or expiry	✓	✓
5	List of ingredients	✓	✓
6	Distinctive batch/lot code	Either the inner or outer label	
7	Manufacturing license number	Either the inner or outer label	
8	Net contents		✓
9	Number of items, if more than one		✓
10	Directions for use	✓	
11	Warnings and precautions	✓	
12	Hazardous ingredient list with quantity	✓	

instructions that come with it. The use of this product for eyebrow or eyelash coloring is not advised because doing so could result in blindness.

It is important to always perform a preliminary test to rule out any special sensitivity since this preparation has the potential to cause severe skin inflammation in certain individuals. In order to do the test, wash a tiny patch of skin behind the ear or on the inside of the forearm with alcohol or soap and water. Once the hair dye is ready, apply a small amount to the affected area and let it dry. Wash the area gently with soap and water after 24 h. It is possible to conclude that there is no hypersensitivity to the dye if there is no visible irritation or inflammation. However, the test must be run before each and every application. Never use this preparation to dye your eyelashes or eyebrows as it may cause severe irritation of the eyes or possibly blindness (Cosmetics Rules 2020).

6.4.6 Animal Testing Declaration

The Cosmetics Rules do not mandate that the declaration on animal testing be made because animal testing is categorically prohibited for cosmetic items, even though the majority of products do include a statement or emblem indicating that the cosmetic product was not tested on animals. If the firm decides to include the statement, it must be careful not to use any certification insignia, such as Cruelty-Free International's "Leaping Bunny" or PETA's "Beauty without Bunnies," until the certification has been received (Cosmetics Rules 2020).

6.4.7 Required Declarations Under the 2011 Legal Metrology (Packaged Commodities) Rules for Cosmetics Manufactured or Imported into India

The Legal Metrology (Packaged Commodities) Rules, 2011 (the "Packaged Commodity Rules") require the declarations listed on the label in addition to the Cosmetics Rules.

The following declarations would also need to be made:

- Product name in general
- Maximum retail price
- Customer service contact information; and, if relevant, the date of import

6.5 Opportunities and Challenges for Global Cosmetic Regulation

The cosmetics market is a dynamic one, with more products being introduced every year and more competition. The need to innovate and introduce new products puts a lot of pressure on companies to stay up with the ever-changing international regulations. Diverse populations, cultures, national laws, and restrictions on ingredients, labels, and packages all increase the complexity of the regulatory system and raise the cost of compliance. Maintaining a thorough awareness of international regulations will help to manage product development timeframes and avoid regulatory bottlenecks (Beg 2020).

Manufacturers and producers of cosmetic products must constantly innovate to meet changing consumer demands. Gaining a competitive edge in the cosmetics sector requires product innovation. Cosmetic firms are continually developing new product innovations or upgrading existing products due to the limited lifespan of beauty products. Depending on the novelty level, incremental product enhancement, product repositioning, and cost reduction, there are several types of product innovation.

The need to constantly develop new products and a heavy reliance on novel ingredients, particularly essences and functional actives, are further characteristics of the beauty sector. Additionally, brands are essential for creating a competitive edge over rivals. Cosmetic companies expand the potential of current brands and provide new ones. The cosmetics industry is always innovating, so fashions in skincare, make-up, and hair care from 10 years ago are now out of date (Beg 2020; Ferreira et al. 2022).

The products that have studied their environments and use the most recent advancements in ingredient design, product development, and packaging are the ones that are most successful on the market. To evaluate the safety and efficacy of these developments, updated regulations and guidelines are also necessary.

Although product safety and cruelty-free cosmetics have long been the leading regulatory concerns for the cosmetics sector, in 2022 we witnessed lawmakers pay more attention to greenwashing and the environmental effects of the products on the market. Particularly the European Union has begun to enact legislation in this area, drafting new circular economy guidelines and possibly outlawing “greenwashing” (Santosh Bali and Acharya 2021).

Over the past 10 years, consumers’ choices in cosmetics have shifted from conventional to green products containing natural ingredients. Consumers are becoming more conscious of the fact that “natural” does not always equate to “clean” or “nature friendly.” For instance, some natural compounds derived from large-scale crops encourage deforestation and reduce biodiversity. Consumers are now gravitating toward goods with “green” labelling that has a track record of effectiveness. These items have active compounds that were bioengineered to maximize natural components. Ingredients that have been bioengineered are molecules derived from plants, microbes, or algae in which the target molecule has been isolated and improved by utilizing biotechnological methods like bio-fermentation. The cosmetics sector can produce sustainable components, which redefines the idea of natural cosmetics (Santosh Bali and Acharya 2021).

Algae and stem cells are two examples of living cells being investigated as ingredients to boost the bioactive effect. The medical community has been greatly impacted by stem-cell-based technologies, and the availability of these stem-cell skin cosmetic goods is also rather significant. The regulations controlling the internal clinical trials conducted become more significant when stem cell technologies are considered in this. And in some of the jurisdictions, it is unregulated.

Last but not least, animal testing is a crucial topic from a legal and ethical standpoint. The cosmetics industry can be safe without using animals for testing by using in-vitro techniques, as demonstrated by Europe and India. Even though most nations favor the use of in vitro alternatives, they do not outlaw animal testing, hence it is still widely used around the world. While there is a growing demand for cruelty-free cosmetics, authorities are also becoming more interested in outlawing them. However, for the time being, the majority of the regulatory community is still in the middle and will not go in any direction (Filaire et al. 2022).

Globally, the rapidly expanding, highly innovative, scientifically driven cosmetics industry makes a substantial contribution to the social and economic prosperity of national and regional economies. Given the globalization of cosmetics, ensuring that they are safe for people and the environment through regulatory compliance with international rules is the first step in developing or redeveloping goods that meet the constantly shifting demands of consumers in worldwide markets. The cosmetics sector must be able to acclimatize to a framework that is always evolving, from producers to distributors. A more harmonization in the regulation is also expected in the coming years.

6.6 Summary

The cosmetic industry has witnessed tremendous growth over the past years and is a science-driven, fast-paced sector globally. Because of the complexity of the sector, each country has implemented their own set of regulations to ensure the safety and efficacy of cosmetic products developed. Our intention through this chapter is to briefly review the regulatory requirements to assess safety of cosmetic products in the United States (US), European Union (EU), and India. Existing laws, its major provisions, safety considerations, oppertunities and challenges and labeling requirements are all reviewed with focus on cosmetics and nano cosmetics.

References

Beg MR (2020) Cosmetic-regulations, research & marketing challenges and global compliance: an overview. https://www.researchgate.net/publication/344426738

BIS (2001) IS 4707: Classification of raw materials and adjuncts: part 2 dyes, colours and pigments, India

BIS (2009) IS 4707-2: Classification of raw materials and adjuncts: list of materials generally not recognized as safe for use in cosmetics, India

Code of Federal Regulations (2024a) 21 CFR Part 700, Subpart B, US. https://www.ecfr.gov/current/title-21/chapter-I/subchapter-G/part-700?toc=1

Code of Federal Regulations (2024b) 21 CFR Parts 701 and 740, US

Code of Federal Regulations (2024c) 21 CFR Section 740.10

Cosmetics Market (2021) Cosmetics market size, share, industry trends and analysis 2021–2027

Cosmetics Rules (2020) Central Drugs Standard Control Organization, Ministry of Health and Family Welfare, Govt. of India
EU No. 1223/2009. Commission Regulation (EU) on cosmetic products No. 1223/2009. Official Journal of the European Union, 30 Nov 2009, Europe
EU Regulation No. 1531/2022, Europe, Sep 2022
Fair Packaging and Labeling Act 15 U.S.C. 1451–1460, 1966
FD&C Act, Federal Food, Drugs and Cosmetics Act, US, 1938a
FD&C Act, Sec. 201(i), 1938b
FD&C Act, Sec. 601(e); 21 U.S. Code 361(e), 2024a
FD&C Act, Section 721(c); 21 U.S. C. 379e(c), 2024b
FD&C Act, Sec. 201(m); 21 U.S.C. 321(m), US, 2024c
Ferreira M, Matos A, Couras A, Marto J, Ribeiro H (2022) Overview of cosmetic regulatory frameworks around the world. Cosmetics 9(4):72. https://www.mdpi.com/2079-9284/9/4/72
Filaire E, Nachat-Kappes R, Laporte C, Harmand MF, Simon M, Poinsot C (2022) Alternative in vitro models used in the main safety tests of cosmetic products and new challenges. Int J Cosmet Sci 44(6):604–613. https://onlinelibrary.wiley.com/doi/abs/10.1111/ics.12803
Katz LM, Lewis KM, Spence SL, Sadrieh N (2022) Regulation of cosmetics in the United States. Dermatol Clin 40(3):307–318. https://doi.org/10.1016/j.det.2022.02.006
McDevitt JP, Criddle CS, Morse M, Hale RC, Bott CB, Rochman CM (2017) Addressing the issue of microplastics in the wake of the Microbead-Free Waters Act—a new standard can facilitate improved policy. Environ Sci Technol 51(12):6611–6617. https://doi.org/10.1021/acs.est.6b05812
Ministry of Health and Family Welfare, Govt. of India (1940) Drugs and cosmetics act, India
Ministry of Health and Family Welfare, Govt. of India (1945) Drugs and cosmetic rules, India
Pauwels M, Rogiers V (2004) Safety evaluation of cosmetics in the EU: reality and challenges for the toxicologist. Toxicol Lett 151(1):7–17. https://www.sciencedirect.com/science/article/pii/S037842740400089X
Paye M, Maibach HI, Barel AO (2005) Handbook of cosmetic science and technology. Informa Healthcare
Pistollato F, Madia F, Corvi R, Munn S, Grignard E, Paini A et al (2021) Current EU regulatory requirements for the assessment of chemicals and cosmetic products: challenges and opportunities for introducing new approach methodologies. Arch Toxicol 95:1867–1897
Rampal A, Kanti SPY (2020) Cosmetics in US and India: overview of regulations and registration process. Int J Drug Regul Aff 8(4):20–24
Rodriguez-Perez J (2022) Handbook of US cosmetic products regulation. BEC Press
Ruhela M, Nagar L, Gupta A, Professor P, Delhi Pharmaceutical, Manish H, Ruhela C, Popli H (2018) Cosmetics: regulatory and market scenario for us and India. Pharma Innov J 7(6):164–169. www.thepharmajournal.com
Santosh Bali D, Acharya S (2021) Green cosmetic: trends, challenges and future scope in India
Steiling W, Almeida JF, Assaf Vandecasteele H, Gilpin S, Kawamoto T, O'Keeffe L et al (2018) Principles for the safety evaluation of cosmetic powders. Toxicol Lett 297:8–18. https://www.sciencedirect.com/science/article/pii/S0378427418317508
Taylor K, Rego Alvarez L (2020) Regulatory drivers in the last 20 years towards the use of in silico techniques as replacements to animal testing for cosmetic-related substances. Comput Toxicol 13:100112. https://www.sciencedirect.com/science/article/pii/S2468111318301154
The Microbead-Free Waters Act, US, 2015
US FDA (2014) Guidance for industry: safety of nanomaterials in cosmetic products, Jun 2014
US FDA (2024) Cosmetic labelling guide
Williams ES, Panko J, Paustenbach DJ (2009) The European Union's REACH regulation: a review of its history and requirements. Crit Rev Toxicol 39(7):553–575. https://doi.org/10.1080/10408440903036056

A Global Perspective on Safety and Efficacy Testing of Hair Care Products

7

Rashmi Mehta and Malini Kotak

Abstract

Hair care products are widely used to maintain the health, appearance, and condition of hair. Despite their daily use, these products are either unregulated or lightly regulated in many countries. The manufacturer is responsible for ensuring the safety and effectiveness of the products before releasing in the market.

Even though most products use approved ingredients, people can still have specific sensitivities to certain ingredients or combinations. With the increasing availability and use of these products, such sensitivities are becoming more common. This makes it essential to adopt thorough methods to ensure product safety and effectiveness.

This chapter will provide a detailed summary of how to assess the safety and effectiveness of ingredients and finished products. It will include new alternative methods to animal testing, such as in vitro (test tube), ex vivo (outside a living organism), and 3D cell models. It will also cover techniques for evaluating product sensory performance and functional benefits through instrumental measurements and user feedback, helping to create a comprehensive product assessment strategy.

R. Mehta (✉)
Personal Care Products Business, ITC Life Sciences and Technology Centre, Bengaluru, India
e-mail: Rashmi.Mehta@itc.in

M. Kotak
Personal Care Unilever, Mumbai, India

A. Dwivedi et al. (eds.), *Hair Care Products*,
https://doi.org/10.1007/978-981-97-6740-3_7

Keywords

Hair care products · Safety assessment · Efficacy assessment · Alternative methods to animal testing · QSAR · NCTC2544 IL-18 assay · 3T3 NRU PT · 3-Dimensional epidermal-equivalent model · ARE-Nrf2 luciferase test · Direct peptide reactivity assay · Activation of dendritic cells · IL-8 Luc assay · h-CLAT · U-SENS · Genomic allergen rapid detection · Local lymph nodes (LLNA) test · Patch test · ROAT · TRUE test · Soap-chamber test · HRIPT · Atomic Force Microscopy · Transmission Electron Microscopy · Scanning Electron Microscopy · Optical Coherence Tomography · Thermal analysis · Mechanical resistance · Sensory · Anti dandruff · Anti hair fall · Anti pruritus

7.1 Introduction

Hair plays a pivotal role in an individual's image and physical identity and thus a healthy, attractive, and strong hair is a key consumer desire (Bolduc and Shapiro 2001). A large variety of hair care products are available in market and are used by consumers to maintain, manage, or change the physical appearance of hair (Gavazzoni-Dias 2015). Hair Care Products (HCP) are designed to be applied topically on hair and scalp for cleansing, managing appearance of hair, enhancing its attractiveness, preserving scalp and hair health, and protecting from daily wear and tear that is caused by environmental and lifestyle factors (Trüeb 2005; Alessandrini and Piraccini 2016).

Generally, HCP do not get absorbed systemically through skin permeation; however, there is a possibility of systemic absorption of these products due to prolonged use or when the product comes in contact with the mucous membranes during application or rinsing. Even then, serious health perils associated with the use of HCP are infrequent.

Although HCP are generally safe, possibility of allergic reactions, irritation symptoms, comedone formation, toxicity to genes, sensitivity to light, etc. cannot be negated without testing. Moreover, the potential long-term consequences of frequent uses of HCP remain uncertain. Hence, it is essential to establish the safety of HCP before releasing them into the market. This chapter provides a concise overview of assessment processes that are generally adopted across the globe to ensure the safety and efficacy of HCP.

Broadly, HCPs can be classified into two main categories basis the impact they have on the hair:

	Hair care product category	Examples
1.	Products with temporary effect on hair	Shampoos, conditioners, sprays, temporary colors
2.	Products with permanent effect on hair	Permanent waves, relaxers, bleaches, permanent colors

A. ***HCPs with temporary impact on hair***
1. A **Shampoo** comprises of
   (a) *Cleansing agents*: These include surfactants or detergents and foaming agents that help remove sebum (natural oils) and skin flakes from the scalp.
   (b) *Conditioning agents*: These ingredients enhance the softness, smoothness, and manageability of hair (Trüeb 2007; Draelos 2013).
   (c) Active ingredients to address specific scalp conditions like dandruff, seborrheic dermatitis, and scalp psoriasis. These ingredients help treat and manage various scalp issues effectively.
   (d) Additives to enhance their performance and stability, improve the texture, stability, and overall effectiveness of the shampoo.

   All the ingredients in a shampoo must be tested, approved and listed on the label. Most shampoos are gentle and only stay on the scalp for a short time, so adverse reactions are rare. Problems occur if the shampoo accidentally gets in the eyes, nose, or mouth during use. Shampoos are tested for skin irritation by applying them to the skin at different strengths (1–8%) for a certain time, either once for 24 h or multiple times over 3 weeks (Dias 2015).
2. Many shampoos already have **conditioners** in them, but some people like to use extra conditioner for better results to reduce frizz, add shine, and make hair easier to manage, protect hair from damage caused by chemical treatments or styling and help retain hair color and smoothness. The Conditioner's positive charges neutralizes the hair shaft's negative charges, settle the lifted cuticles to align them and reduces the hydrophillicity of the hair fibre, thereby rendering a more aligned, smoother, shinier and well groomed appearance to the hair (De Groot et al. 1994).

   Conditioners largely comprise of:
   (a) Cationic surfactants, polymers that cover the hair shaft defects to create a smooth surface, eliminating the static electricity and enhancing the shine (Trüeb 2007).
   (b) Emollients/oily compounds like natural oils or synthetic oils, esters and waxes, fatty alcohols, waxes, gums, salts and emulsifiers that enable uniform mixing of immiscible liquids (Souza and Rathi 2015).

   There are different ways to use hair conditioners. Instant conditioners are used right after shampooing and rinsed off in a few minutes. Deep conditioners are retained on hair for 20–30 min for better results. Leave-in conditioners are not rinsed out and provide long-lasting effects. Hair glazes or thickening conditioners coat the hair, making each strand thicker.

B. ***HCPs with permanent effect on the hair shaft***
3. **Hair coloring products** come in three types: temporary, semi-permanent, and permanent. They differ in how deeply they penetrate the hair.

   Most hair dyes contain (a) Hydrogen Peroxide to bleach and activate the dye, (b) Ammonia or Ethanolamine to make the dye work better, (c) Chelating agents to retain the dye stability, (d) Solvents to improve the viscosity and solubilize the dye to aid proper mixing and (e) Surfactants to clean the hair (Dias 2015; Angra et al. 2015).

Temporary colors are available in the form of shampoos, lotions, foams, or mousses. They add highlights, remove yellow tones from white hair, or cover grey hair. Owing to high molecular weight, they deposit on the hair surface and get washed out easily (Souza and Rathi 2015). Semi-permanent colors comprise of small dye molecules that disperse into the hair's outer layers of cuticles without binding to the hair protein. They add color without lightening the hair and wash out over time (Neuser and Schlatter 2010).

Permanent colors involve a process that opens the hair shaft, allowing dye molecules to enter and bond with the hair. This type can cause allergies, hence a skin test 48 h before use is pertinent. Long-term use can lead to skin issues like dermatitis or hair loss (Søsted et al. 2004).

4. **Bleaching agents** like hydrogen peroxide, ammonia, and persulfate salts permanently lighten hair. They work by lifting the hair's outer layer the cuticles, entering the cortex, and breaking down melanin, which lightens the hair color. Frequent bleaching can make hair more porous and damaged (Guerra-Tapia and Gonzalez-Guerra 2014).

   Common allergens in hair dyes include PPD (permitted up to 6%) Toluene-2,5-diamine (permitted up to 10%) *p*-Aminophenol and *m*-Aminophenol (permitted up to 2%) (Søsted et al. 2004). Persulfates can cause allergic reactions like rhinitis, asthma, hives, contact urticaria and even severe reactions like anaphylaxis (Hougaard et al. 2012).
5. **Hair styling agents** work by varied methods such as
   (a) ***Permanent waves*** or perms, make hair curly using chemicals like ammonium thioglycolate. These chemical breaks the strong bonds in the hair. Then, a neutralizer with oxidizing agents like hydrogen peroxide reforms these bonds in a different way, making the hair stay curly. Bleaching and perming renders the hair weak and more susceptible to breakage. Trichorrhexis nodosa is a condition where hair becomes weak and breaks easily.
   (b) ***Relaxers*** change curly hair to straight hair by breaking the strong bonds in the hair and then reforming them in a straight shape.
   (c) ***Hair straightening***: At high pH levels (9.0–14.0), straightening products react with keratin in the hair, breaking and rearranging the strong bonds, making the hair soft and stretched (Dias 2015). Common ingredients in these products include sodium hydroxide (lye relaxer), lithium hydroxide, calcium hydroxide, and a mix of guanidine carbonate and calcium hydroxide (no-lye relaxers). Ammonium thioglycolate also weakens the bonds in hair. Straightening hair can cause problems like breakage, frizz, thinning, weakening, greying, split ends, and scalp burns (Shetty et al. 2013). Keratin treatments with formaldehyde or glutaraldehyde, which are harmful chemicals, are limited to 0.2% in cosmetics because they can cause severe skin reactions in some people (Van Lerberghe and Baeck 2014).
6. **Hair extensions**, whether synthetic or natural, can be attached to the original hair following strand-by-strand, weft, or clip-in methods. Common problems with hair extensions include hair loss due to pulling, tenderness, pain,

tangling, and itching (Van Lerberghe and Baeck 2014; Salibi et al. 2014). Some people may also have local allergic reactions to the glues or adhesives used, causing skin irritation or severe allergic reactions (Cogen and Beezhold 2002).

7. **Camouflage and tattooing**, are common approaches that involve the use of hair pieces made of natural or synthetic hair such as wigs, demi wigs, toupees etc to cover up hair loss from conditions like alopecia areata, (Draelos 2011). Sometimes, the glue or latex used to attach them can cause skin allergies. For scalp hair loss in conditions like epidermolysis bullosa or androgenetic alopecia, pigmented hair fibers that stick to hair with static electricity can be used (Harris et al. 2016; Cossman et al. 2013). Scalp micropigmentation (SMP) is a tattooing method that is used to cover bald spots (Rassman et al. 2015). However, some people might experience itching or irritation.

7.2 Safety Assessment of Hair Care Products

The USFDA defines cosmetics as products used on the body to clean, beautify/promote attractiveness, or change appearance without affecting the body's structure or functions. The FDA recommends that manufacturers ensure their products are safe, pure, and of good quality. Manufacturers are responsible for the safety of their products before they are sold (FDA 2017a, b; Bhuvaneswari and Chandan 2018).

In the European Union, regulations are stricter. Manufacturers must have someone in the EU responsible for ensuring the product's safety according to the law before it is sold (EU Regulations 2017).

In India, the Drugs and Cosmetics Act and the Central Drug Standard Control Organization (CDSCO) regulate personal care products. The Bureau of Indian Standards (BIS) sets standards for ingredients and methods to ensure the safety and quality of these products. Other countries have similar regulations to ensure the safety of cosmetic products in their markets (Central Drugs Standard Control Organization, Bureau of Indian Standards (BIS) 2017; Indian Cosmetic Legislation 2017; ICCR 2017).

7.2.1 ICCR and Basic Steps in Safety Assessment Process

According to the International Cooperation on Cosmetics Regulation (ICCR), personal care product ingredients must be tested for safety using a detailed process, including both basic and advanced testing methods, following local and legal requirements (ICCR 2017).

The safety assessment process includes:

- Studying the physical and chemical properties, purity, impurities, and chemical structure of ingredients.
- Reviewing the toxicological potential of all ingredients, including related studies and information on related ingredients.

- Assessing ingredient stability, microbiological status, and overall characteristics.
- Heightened attention to specific ingredients like perfumes, UV filters, and hair dyes.
- Understanding on product application and use, including the amount, frequency, duration, and target users.
- Calculating safety and exposure margins based on product usage data.
- Decisions on whether the product is safe, needs conditions for safe use, or should be rejected.
- Monitoring the product after it is on the market to ensure it remains safe.

The Cosmetic Product Safety Report (CPSR) documents the product's safety and is a key part of the Product Information File (PIF).

7.2.2 Safety Assessment of Cosmetic Ingredients

Before choosing ingredients for a product, companies test them to make sure they are safe and work well with other ingredients in the formula (FDA 2013; Boyer et al. 2017; Rogiers 2017; Nigam 2009).

Tests used in the safety assessment of ingredients are listed below (Bhuvaneswari and Chandan 2018; Rogiers 2017; Nigam 2009; Al-Dayel et al. 2011):

Physico chemical characterization	Toxicological studies
Chemical identity	Toxicokinetics
Physical form	Acute toxicity
Molecular weight	Local toxicity (corrosion, irritation, sensitization)
Purity and isomer composition	Systemic toxicity
Solubility	Repeated dose toxicity
Partition coefficient	Reproductive toxicity
Characterization of impurities and contaminants	Mutagenicity/genotoxicity
Homogeneity and stability	Carcinogenicity
Functions and uses	Photo-induced toxicity

7.2.3 Safety Assessment of HCP

The North American Contact Dermatitis Group (NACDG) reports that about 9% of all allergies are related to hair care products (Alajaji 2024). The most common problems are contact allergic dermatitis and photoallergic dermatitis.

The safety of a cosmetic product depends on the area of skin it comes in contact with, how often it is used and the type of application.

- Skin and hair cleansers cover a large area but are used as diluted and washed off quickly.
- Permanent hair dyes react with hydrogen peroxide, and the final by-product of this reaction can cause allergies.

To assess safety, the amount of a substance that might be ingested, inhaled, or absorbed through the skin or mucous membrane is calculated and converted into a daily exposure amount per kg of body weight, called the Systemic Exposure Dose (SED) (EU Regulations 2017). Various tests, including in vivo (live models) and in vitro (lab models), are used to determine the safety of cosmetic products and ingredients.

The process of skin sensitization (allergic reaction) involves four main key events:

1. KE1: The initial event where potential allergens bind to skin proteins.
2. KE2: Activation of skin cells called keratinocytes.
3. KE3: Activation of immune cells called dendritic cells.
4. KE4: Proliferation (increase) of specific Immune cells called T-cells.

In compliance with the 7th Amendment of the European Cosmetics Directive, which forbids animal testing for cosmetic products and ingredients, new nonanimal testing methods have been developed. These methods include computer models (in silico), chemical tests (in chemico), and lab tests (in vitro) to study each step of the skin sensitization process (Barthe et al. 2021).

New methods to replace animal testing in toxicology are grouped into five systems:

1. **In silico methods**: These use computer simulations and mathematical models like QSAR (Quantitative Structure–Activity Relationship) to predict the behavior of chemicals based on their structure (European Food Safety Authority 2014; Kandárová and Letašiová 2011; Kue and Kumar 2023).
2. **In chemico methods**: These are lab-based chemical reactions to test substances, especially for skin sensitization. Some examples include:
   (a) Direct Peptide Reactivity Assay (DPRA): Tests how chemicals bind to proteins, a vital step to elicit allergic reactions (Kandárová and Letašiová 2011; OECD 2022a; Lalko et al. 2012).
   (b) Amino Acid Derivative Reactivity Assay (ADRA): Uses synthetic reagents to test chemical reactivity (Fujita et al. 2014; Yamamoto et al. 2015; Akimoto et al. 2020).
   (c) Kinetic Direct Peptide Reactivity Assay (kDPRA): Studies the reactive potential of a substance to nucleophillic peptides (OECD 2022a; Wareing et al. 2020).
   (d) SH-test: Uses a special compound to detect changes in cell-surface proteins caused by chemicals (Gądarowska et al. 2022; Suzuki et al. 2009; Imai et al. 2021).
3. **In vitro methods**: These include cell culture based assays to test for skin irritation, corrosion, sensitization, and other toxic effects (Caloni et al. 2022). The second key event in skin sensitization involves keratinocytes (skin cells) and inflammatory responses and gene expressions related to specific cell signaling pathways (Gusev and Zhuravleva 2022).

The Keratinocyte activation-based assays include the following:

(a) **ARE-Nrf2 luciferase test**: This test checks if a substance can activate a protective pathway in skin cells. It uses a modified cell line that lights up (luminescence) when the pathway is activated. If the substance causes a significant light signal, it is classified as a skin sensitizer (OECD 2022b). There are two main methods for this test: KeratinoSens™ and LuSens.
   - KeratinoSens™ test: Uses human skin cells with a luciferase gene. When exposed to a sensitizing substance, the cells light up, indicating activation of the protective pathway (Nishijo et al. 2019).
   - LuSens test: Similar to KeratinoSens™, albeit, deployment of a genetically altered human skin cells. It measures the light signal after 48 h of exposure to the test substance (Ramirez et al. 2014).

   These tests are more sensitive than older methods and can detect even small amounts of sensitizers.

(b) **The epidermal sensitization assay (EpiSensA)**, uses reconstructed human skin to measure how certain genes (ATF3, IL-8, DNAJB4, GCLM) respond to a sensitizing substance. These genes are linked to inflammation and protective responses in skin cells. EpiSensA is more sensitive and accurate than other tests like DPRA, KeratinoSens, and h-CLAT (Saito et al. 2017).

(c) **In vitro skin sensitization assays based on Key Event 3 (KE3) dendritic cells activation following exposure to skin sensitizers**

   These tests focus on how dendritic cells (DCs) react to skin sensitizers. DCs process and present antigens to lymphocytes with the help of a major histocompatibility complex (MHC). When exposed to sensitizers, DCs move from the skin to lymph nodes and become efficient at presenting antigens. They increase the expression of cell markers like CD40, CD54, CD80, CD83, and CD86 and produce inflammatory cytokines like TNFα, IL-6, IL-8, and IL-1β (Lukas et al. 1996; Gober and Gaspari 2008; Stępnik and Arkusz 2003).

   The OECD guideline No. 442E includes several methods to identify skin sensitizers:
   - **Human Cell Line Activation Test (h-CLAT)**: This test involves exposure of cells to substances for 24 h, causing dendritic cells and monocytes to express markers CD86 and CD54. Flow cytometry measures these markers to determine sensitization (OECD 2022c).
   - **U937 Cell Line Activation Test (U-SENS™)**: Uses human myeloid cells exposed to chemicals and analyzed for CD86 expression and gene expressions of IL-1β and IL-8 (Stępnik and Arkusz 2003; Python et al. 2007; Alépée et al. 2017).
   - **Interleukin-8 Reporter Gene Assay (IL-8 Luc Assay)**: Measures changes in IL-8 expression using a luciferase gene to identify sensitizers (OECD 2022c; Takahashi et al. 2011).
   - **Evaluation of Pro-inflammatory Cytokines**: Tests for IL-18 production in skin cells exposed to sensitizing chemicals (Corsini et al. 2013).
   - **Genomic Allergen Rapid Detection (GARD™) Test**: Uses a cell line that mimics dendritic cells to study immune response activation (Johansson et al. 2019).

- **3T3 Neutral Red Uptake Phototoxicity Test (3T3 NRU PT)**: Measures cytotoxicity of substances on mouse fibroblast cells using Neutral Red dye (WebMD 2017; Hans et al. 2008).
- **In vitro tests for genotoxicity**:
  - Ames test: Tests mutagenic potential using bacteria that need external amino acids to survive (OECD 2020).
  - Mammalian cell chromosome aberration test: Measures chromosome aberrations in mammalian cells (OECD 2016).
  - Micronucleus test: Identifies substances causing genotoxic damage by detecting micronuclei in cells (OECD 2023).
- **Epidermal equivalent model**: Uses a 3D skin model to assess allergenicity of substances.
  - SenCeeTox method: Uses a reconstituted human epidermis model for potency assessment (Spiekstra et al. 2009; Galbiati et al. 2018).
  - SENS-IS assay: Analyzes genes involved in redox protection and inflammation to predict sensitizer potency (Cottrez et al. 2015; Uruno and Motohashi 2011).

4. **Ex vivo methods** use isolated animal tissues and organs to test the corrosive potential of Cosmetic ingredients or products
   (a) Ocular irritation/corrosion: Product is tested on the eyes of slaughtered fowl (OECD TG 438) or cattle (OECD TG 437) (OECD 2009, 2018).
   (b) Skin corrosion: Product is tested on rat skin using the transcutaneous electrical resistance (TER) test (OECD TG 430) (OECD 2015).
5. **In vivo assessment methods**
   (a) Murine Local Lymph Node Assay (LLNA): This test measures the activation and growth of T lymphocytes in lymph nodes after exposure to sensitizers. It helps identify allergens and their strength. The test is recognized by the OECD (guideline 429) (Geobrick et al. 2007; Kimber and Dearman 2010).
   (b) Patch Test: This test checks if a product causes allergies or irritation. A small amount of the product is applied to the skin, and the area is observed for reactions (IS 2001, 2018).
   (c) Photo-Patch Test: This test is for products with photosensitive ingredients. The product is applied to the skin, exposed to UV light, and checked for dermatitis (IS 2018).
   (d) Repeated Open Application Test (ROAT): This test involves applying the product to the skin daily for 15 days to see if it causes irritation, simulating real-life use (Khambra et al. 2020; Ryder 2014; Hannuksela and Salo 1986).
   (e) Thin-layer Rapid Use Epicutaneous Test (TRUE): This ready-to-use patch test applies allergens to the skin to check for allergic reactions (Krob et al. 2004; Jerajani and Melkote 2007).
   (f) Dimethylgloxime Test: This test detects nickel in products by causing a red precipitate to form if nickel is present (Shore et al. 1977).
   (g) Tests of Irritancy and Sensitivity:

- Soap-Chamber Test: Enables detects low levels of skin toxicity by exposing the skin to the product multiple times (Frosch and Kligman 1979; Barel et al. 2001).
- Human Repeat Insult Patch Test (HRIPT): This is used to detect product's potential to cause skin irritation, sensitization or elicit allergic reactions following repeated exposure to the product (IS 2018).

(h) Tests of Comedogenicity: Aids assess the tendency of HCP to clog pores and causes comedones (blocked pores) (Calvarese 2014).

(i) Tests for Photosensitivity:
- Phototoxic Reactions: Tests if sunlight activates the ingredient, causing skin damage.
- Photo Allergy Reactions: Tests if UV exposure changes the ingredient, causing allergic reactions (WebMD 2017).

7.2.4 Efficacy Assessment of HCP

7.2.4.1 In Vitro Methods to Evaluate the Efficacy of Hair Care Products on Hair Fibres

Hair has two main parts: the root, which is inside the follicle, and the shaft, which sticks out of the scalp (Lai-Cheong and McGrath 2009). The shaft has three layers, Medulla: the innermost part, is composed of soft thin core of transparent cells and air spaces, Cortex: the middle, thicker layer comprises of proteins called keratin filaments that impart bulk, strength and elasticity and melanin, responsible for hair color, Cuticle: The outermost layer made of overlapping keratinized cells that protects the hair from damage. The way the cuticle scales are arranged affects how smooth, shiny, and easy to comb the hair is (Bolduc and Shapiro 2001; Gavazzoni-Dias 2015; Tomes et al. 2007).

Consumers actively seek information regarding evidence on the acclaimed benefits of the HCP.

Sophisticated equipments are involved in the in vitro assessment methodologies used to evaluate the effects of hair care products on hair fiber, such as:

Scanning Electron Microscopy (SEM) (Tomes et al. 2007)	Protein loss (Faria et al. 2013; Sá-Dias et al. 2008)
Transmission Electron Microscopy (TEM) (Ahn and Lee 2002)	Electrophoresis (Ruetsch et al. 2003)
Atomic Force Microscopy (AFM) (Dupres et al. 2004)	Color and brightness (Dario et al. 2013)
Optical Coherence Tomography (OCT) (Velasco et al. 2009)	Thermal analysis (Gama et al. 2011)
Infrared Spectroscopy (IRS) (Pande and Yang 2000)	Mechanical resistance to combing and elasticity (Sá-Dias et al. 2008; Ruetsch et al. 2003; Dario et al. 2013; Gama et al. 2011; Velasco et al. 2015)
Raman Spectroscopy (RS) (Kuzuhara 2006)	

Accuracy of the equipment is most important to ensure reproducibility, repeatability, and ease of use for analytical methods.

1. ***Image analysis***
   (a) ***Scanning Electron Microscopy (SEM)*** creates highly detailed and magnified images (up to 300,000 times) by scanning an object's surface with a focused beam of electrons in a vacuum. The electrons interact with the object's atoms, producing signals that are converted into 3D images. These images provide the surface details and composition of the object (Zhou et al. 2007; Duarte et al. 2003). SEM is useful for examining the surface of the hair cuticle in detail. It can be used to study hair properties like overall appearance, particle deposition, and cuticle surface and affinity to ingredients of HCPs (Ali et al. 2015; Monteiro et al. 2003; Camacho-Bragado et al. 2015; Richena and Rezende 2015).
   (b) ***Transmission Electron Microscopy (TEM)*** is a high-resolution technique that produces detailed images, down to about 0.19 nm. It works by passing electrons through very thin samples and detecting the signals produced. This allows scientists to study the tiny structures and chemical features of materials at nearly atomic levels (TEM n.d.). TEM can be used to examine the fine details of changes in hair fibers after cosmetic treatments (Ahn and Lee 2002; Pande and Yang 2000; Kuzuhara 2006; Thibaut et al. 2007; Kim 2011).
   (c) ***Atomic Force Microscopy (AFM)*** is a high-resolution technique that uses a sensitive probe to scan the surface of a sample (Rao and Costa 2014). And create detailed images down to fractions of an Angstrom. AFM measures properties like height, magnetic force, surface potential, friction, and intermolecular forces. AFM uses a fine Sillicon tip, attached to a cantilever to move across the sample's surface. The interaction between the tip and the surface is measured by a laser beam reflecting off the cantilever. This creates a high-resolution, 3D image of the surface (Monteiro et al. 2003; Thibaut et al. 2007). AFM can be used to study hair properties, such as particle deposition, ingredient affinity, and changes due to grooming products and techniques (Longo et al. 2006; Edstrom et al. 1990; Swift and Smith 2000; Gurden et al. 2004; Smith et al. 2013; Seshad and Bhushan 2008).
   (d) ***Optical Coherence Tomography (OCT)*** is a noninvasive imaging technique that deploys low-coherence interferometry. It uses Near Infrared light to capture high-resolution (micrometer resolution) 2D and 3D cross-sectional images of a specimen, providing a clear view of its internal structures (Velasco et al. 2009; Freitas et al. 2010; Fujimoto et al. 2000). OCT is useful to study the structural and morphological changes in hair caused by (HCP) and grooming techniques (Souza and Rathi 2015; Monteiro et al. 2003; Velasco et al. 2009, 2016).
2. ***Spectroscopy analysis***
   (a) ***Infrared Spectroscopy (IRS)***: This technique studies how molecules absorb specific frequencies of infrared light, which helps identify compounds and

understand their composition (Rodrigues and Galzerani 2012; Miyamae et al. 2007).

(b) ***Near-Infra Red Spectroscopy (NIRS)***: Used to study the effects of hair care products on hair's physical and structural properties (Pande and Yang 2000; Miyamae et al. 2007).

(c) ***Fourier Transformed Infra-Red Spectroscopy (FT-IR)***: Identifies chemical bonds in molecules by creating an infrared absorption spectrum, providing detailed chemical information about hair (Berthomieu and Hienerwadel 2009).

(d) ***Attenuated Total Reflection Fourier Transform Infrared (ATR FT-IR) Spectroscopy***: A surface-sensitive technique that captures the surface spectrum of hair (Pienpinijtham et al. 2018).

(e) ***Raman Spectroscopy (RS)***: Measures vibrational energy modes of a sample based on scattered light, providing chemical and structural information. It helps study changes in hair strength and structure due to aging or chemical or grooming treatments (Raman and Krishnan 1928; Smith and Dent 2005; Kuzuhara 2007, 2016; Kuzuhara et al. 2007; Rodrigues and Galzerani 2012).

(f) ***Photoluminescence Spectroscopy***: Measures light emitted from a material when excited by photons. It assesses hair health and changes in hair matrix by measuring tryptophan and its by-products, (Dyer et al. 2006; Longo et al. 2013; Signori 2004; Dario et al. 2015; Fernández et al. 2012; Jachowicz and McMullen 2011; Evans 2013).

(g) ***Diffuse Reflectance Spectrophotometry (DRS)***: Studies the spectral characteristics of objects / materials like hair by measuring the light reflected from the surface. It is used to measure hair luster and color (Scanavez et al. 2000).

3. ***Protein loss***: Changes in hair protein composition due to physical and chemical treatments are measured by analyzing the amino acids in hair using methods like the Lowry, Bradford and Bicinchoninic acid (BCA) assays (França-Stefoni et al. 2015; Nogueira and Joekes 2004; Sandhu and Robbins 1993; Sandhu et al. 1995; Silva et al. 2004; Zhang et al. 2023).
4. ***Electrophoresis*** is a technique that separates particles like nucleic acids, proteins, and enzymes based on their size, weight, and charge using an electric field (Westermeier 2005). It helps study how hair care product (HCP) treatments, light exposure, and their combinations affect hair protein composition. It also helps quantify proteins, peptides, and amino acids in hair (Ruetsch et al. 2003; Kim 2011).
5. ***Thermal analysis***: The method enables study temperature induced changes in the material's properties and reaction by products. It includes several techniques Thermogravimetry (TG), Differential Thermogravimetry (DTG), Differential Thermal Analysis (DTA), Differential Scanning Calorimetry (DSC), Thermomechanical Analysis (TMA) (Silva et al. 2007; Wendlandt 1986). (a) Differential Scanning Calorimetry (DSC) measures the thermal energy differences between a test material and a reference. When a material undergoes a physical or chemical change, its heat capacity changes. This change is detected and recorded, creating a thermogram that shows heat flow versus temperature (Silva et al. 2007; Wendlandt 1986; Stevens 1999).

6. ***(b) Thermogravimetric analysis (TGA)*** is used to study the thermal stability of materials, like human hair, when heated in presence of air or an inert gas. With the increase in temperature, the material looses mass due to moisture evaporation, decomposition, degrdation or other reactions leading to loss of mass. Mass loss over a range of temperature is used to characterize the material (Gama et al. 2011; Lima et al. 2016; Éhén et al. 2004; Wortmann et al. 2002, 2006; Belleti et al. 2003).

   Combining TGA with another method called Differential Scanning Calorimetry (DSC) gives a better understanding of the material. TGA measures the mass loss, while DSC measures the heat changes associated with this mass loss.
7. ***Mechanical resistance/strength***: Hair strength, is measured using tools like the Instron Tensile Tester, Dia-Stron MTT, Texturometer, or a Tensile tester. These instruments measure the force required to stretch and break the hair strand and also the elongation of hair before it breaks. This information is converted to assess the tensile strength property of the hair.

   Treatments like straightening, bleaching, and back combing can weaken hair, making it easier to break (Dario et al. 2013; Velasco et al. 2015; Ruetsch and Weigmann 1996; Robbins and Crawford 1991).
8. ***Combing resistance analyses***: Combability is ease of combing hair. To measure combability, tools like a Texturometer or Dynamometer are used which measure the resistance force exerted by hair when a non-metallic comb is moved through from the root to tip ends at a standard speed (Garcia and Diaz 1976; Evans 2011). Dry or damaged hair tangles easily because of raised cuticle scales, static energy, or products that make hair stick together. For testing, a non-metallic comb is moved through a hair sample from root to tip at a standard speed. Treatments like straightening, smoothing, and conditioning can make hair easier to comb by reducing resistance.

In vitro methodologies to assess efficacy of hair care products

Claims	Methodologies
Hair damage reduction/prevention	Protein Loss, Electrophoresis, Scanning Electron Microscopy, Transmission Electron Microscopy, Atomic Force Microscopy, Optical Coherence Tomography, RAMAN Spectroscopy
Hair shine	Diffuse reflectance spectrophotometry
Combability/ease of comb	Resistance to combing force
Hair strength	Resistance to breakage
Photoprotection	Photoluminescence spectroscopy
Color longevity	Diffuse reflectance spectrophotometry
Heat protection/ resistance	Differential scanning colorimetry, thermogravimetry
Moisturized hair	Near infrared spectroscopy, differential scanning colorimetry, thermogravimetry

Conclusion

The methodologies described above are commonly used to study a single attribute of Hair, the result from each method can often be used complementarily. An appropriate instrument for testing hair is selected based on the parameter being studied (sensorials/resistance to rupture/elasticity/integrity of hair strain/shine/color, etc.), availability, cost, maintenance, accuracy, reproducibility, and ease of operation.

7.2.4.2 In Vivo Methods to Evaluate the Efficacy of Hair Care Products

Sensory Assessments

Sensory analysis consists of methods to evaluate the characteristics of a product or hair before and after use of the test product primarily using the four senses, namely,

(a) Vision/Sight: Product color, appearance, Consistency, Foam quantity, Foam quality, Foam density, Foamability, Rinsability, etc.; Hair Appearance, Thickness, Frizz, Volume, Alignment, Color, Shine, Static, Style retention, etc.
(b) Olfactory/Smell: fragrance of the product, fragrance of product on hair.
(c) Auditory/Sound: sound of product, hair roughness, etc.
(d) Touch/Tactile: Feel of the product (Consistency, Stickiness, Oiliness, Residue, Wetness), Feel of hair (Dryness, Softness, Smoothness, Friction, etc.).

Finished products are often tested by either trained experts or regular naive consumers to study the products for how they feel, smell, look, and perform. Trained professional sensory experts follow specific test procedures and use standard terms to describe their experiences. This ensures consistency in their assessments. Untrained / naive consumers use the product either under a controlled setting or at home as they normally would. Their feedback helps understand how the product performs in real-life situations.

Sensory Parameters

Product application feel characteristic		Wet hair characteristic	Dry hair characteristic	
Ease of spread Bubble size Amount of foam Speed of foam	Foam quality Wetness of foam Cushion of foam Ease of rinse	Slipperiness Ease of detangling Force to comb	Slipperiness Ease of detangling Force to comb Residue Static Volume	Gloss Softness Smoothness Silkiness Moisturization Likability

Sensory analysis is a crucial tool for identifying the key features of a product that make people happy and encourage them to use and buy it again. Sensory analysis helps find the features that people like the most. Trained experts provide detailed scores, while naive consumers give their overall preferences and likability scores. By comparing the scores of Sensory experts and Consumers, researchers can create maps to identify consumers' preferences and product likability drivers and thus aid the design and development of Consumer preferred products.

Efficacy Assessments

The benefits or effectiveness of HCPs, are assessed by deploying customized tests designed based on the product's ingredients, expected usage duration and frequency and related factors. Here are some common benefits tested in hair care products:

- Anti-dandruff efficacy: Measures how well the product reduces or eliminates dandruff.
- Anti-hair fall efficacy: Assesses the product's ability to control /reduce and / or prevent hair loss.
- Anti-pruritus efficacy: Evaluates how effectively the product reduces itching of the scalp.

These tests help assess the product's efficacy to deliver the promised benefits to consumers.

Anti Dandruff Efficacy Assessment

Dandruff is a common and prevalent scalp condition that affects about 50% of people worldwide. It appears as flakes on the scalp and is influenced by factors like seborrhea (excess oil) and scalp microbes (Chaisripipat et al. 2015). Dandruff management products, such as shampoos, often contain antifungal agents to reduce flaking and other symptoms.

The anti-dandruff efficacy of products is assessed through measurement of symptoms like scalp flaking and itching in people with mild to moderate dandruff, using both objective measurements and self-evaluations over a period of 21 days to 6 months (CTRI 2018a, 2021, 2022a; Ranganathan and Mukhopadhyay 2010; Barak-Shinar and Green 2018). The study usually includes:

- Washout phase: 2–3 weeks where participants stop using any anti-dandruff products.
- Treatment phase: 4 weeks of using the test product.
- Regression phase: 4 weeks to observe any changes after stopping the product.

Dandruff Severity Assessment: The most common ways to evaluate the severity of dandruff include: (a) Visual scoring systems like Total severity assessment and Adherent Flakes severity score, Scalp Photographic Index (SPI) (Barak-Shinar and Green 2018; Zhu et al. 2017; Kim et al. 2023) and (b) Bio instrumental techniques like Squamometry, Photography, Image analysis, for Dandruff quantification (Kim et al. 2023; Dhurat and Saraogi 2009; Kolasinski 2009). These methods help measure and understand the extent of dandruff.

Dandruff severity visual assessment scale		
Erythema severity 0 = None 1 = Mild 2 = Moderate 3 = Marked 4 = Severe	**Loose flakes severity** 0 = None 1 = Mild 2 = Moderate 3 = Severe 4 = Very Severe	**Adherent scalp flaking score**: 0 = No Flaking 1–2 = Small powdery grayish coarse scales 3–4 = Small to moderate flakes 5–6 = Large and thin flakes loosely attached to the scalp 7–8 = Adherent flakes 9–10 = Thicker white-yellow scales closely attached to the scalp
Total Severity Score = Erythema Severity Score + Loose Flakes Severity + Adherent Scalp Flaking Score		

Barak-Shinar and Green (2018), Zhang et al. (2023) and Zhu et al. (2017)

Anti Hair Fall/Hair Thinning Assessment

Hair fall and hair thinning can be assessed using various methods. The general process involves:

an Assessment duration ranging from 45 days to 6 months and includes a washout period, treatment period and assessment period.

Assessment Method categories:

- Noninvasive methods: These methods don't harm the scalp and include questionnaires, counting fallen hairs, standardized wash tests, comb tests, pull tests, global photographs, dermoscopy, gravimetric measurements, contrasting felt examination, hair feathering tests, phototrichograms, and TrichoScan (Dhurat and Saraogi 2009).
- Semi-invasive methods: These methods include trichograms and unit area trichograms (Dhurat and Saraogi 2009).
- Invasive methods: These methods involve more significant procedures like scalp biopsies (Dhurat and Saraogi 2009).

TrichoScan® (Tricholog GmbH & Datinf GmbH) is a software that helps in objectively measuring hair parameters like density, diameter, growth rate, and the ratio of different hair types using digital images. It distinguishes between growing (anagen) and non-growing (telogen) hair based on their length (Kolasinski 2009).

Dermatologists tests: Dermatologists use comb tests and pull tests to evaluate hair thinning and hair fall (Majeed et al. 2020; CTRI 2018b, 2019, 2022b, c; Hamilton 1951).

Baldness pattern Classification system: Various systems are used to grade hair loss patterns. The Hamilton-Norwood system is popular for male baldness, (Norwood 1975; Ludwig 1977) and the Ludwig system is widely used for female baldness pattern (Gupta and Mysore 2016; Lee et al. 2007). The BASP classification is a newer, more detailed system that considers the hairline's shape and hair quantity on different scalp areas (Kilic et al. 2018; Reich et al. 2017) and is gender neutral.

Anti Pruritus Efficacy

Pruritus, or itching, is common in skin conditions like eczema, psoriasis and dermatitis and it can get worse with dry skin. The anti-itch efficacy of products, is studied on adults with dry, itchy scalps prone to dermatitis. Participants continue using their usual scalp and hair care products and hair grooming practices. As part of baseline evaluation, intensity and frequency of itching are recorded before starting the test product use. Before each assessment, participants get acclimated to the controlled room environment. For safety and tolerance evaluation, the scalp is checked for redness, swelling, dryness, scaling, cracks, bumps, pustules, itching, burning, and other discomforts at the start and at set intervals during the study. Pruritus severity is assessed on various parameters listed below.

Pruritus Assessment

Pruritus intensity

Score	Description
1	None
2	Very slightly perceptible
3	Slightly perceptible
4	Clearly perceptible
5	Strongly perceptible
6	Very strongly perceptible

Pruritus persistence

Score	Description
1	Never
2	Only short term
3	Only hourly
4	Constantly

Pruritus incidence

Score	Description
1	Never
2	Only daytime
3	Only nighttime
4	Day and night
5	Only in cold conditions
6	Only in warm conditions
7	During cold and warm conditions
8	After showering
9	Other cause (to be specified)

Pruritus burden

Score	Description
1	None (no longer exists)
2	No longer disturbing
3	Does not bother very much
4	Bothers strongly
5	Bothers very strongly
6	Unbearable

Anti-pruritic efficacy assessment of the Product involves assessing the time taken for anti-pruritic effect to manifest and longevity of the efficacy.

Antipruritic effect occurrence

Score	Description
1	Immediately post first application
2	Only after several days of product use
3	Not perceptible

Anti pruritic effect longevity

Score	Description
1	More than 24 h
2	Only overnight
3	Only during daytime
4	Only few hours (1–3 h)
5	Short term (<1 h)
6	Not at all

Courtesy: Kilic et al. (2018)

The 12-item Pruritus Severity Scale (PSS) assesses various aspects of pruritus, including intensity, duration, frequency, impact on daily activities, mood, and assessment of scratching.

Score	Description	Response options	Scores
1	How often did you feel pruritus within the last 3 days?	All the time	3
		All morning/afternoon/evening/night: long itch episodes	2
		Occasionally short itch episodes	1
2	Did pruritus hinder your ability to do simple things like watching TV, hearing music, etc.?	Yes	1
		No	0
3	Did you feel irritated or nervous because of your itching?	Yes	1
		No	0

Score	Description	Response options	Scores
4	Did pruritus cause depression?	Yes	1
		No	0
5	Did your pruritus impede your work or learning abilities?	Yes	1
		No	0
6	Did you scratch your skin because of itching?	Yes	1
		No	0
7	Did scratching cause relief?	Yes	1
		No	0
8	Were you able to refrain from scratching?	Yes	0
		No	1
9	Did you wake up during last night because of pruritus?	No	0
		Yes, 1–2 times	1
		Yes 3–4 times	2
		Yes 5 or more times	3
10	Can you assess the severity of your pruritus during the last 3 days?	Very mild	1
		Mild	2
		Moderate	3
		Severe	4
		Very severe	5
11	Please indicate location of pruritus	Single location	1
		Large areas	2
		Generalized	3
12	Are excoriations or other scratch lesions present?	Yes	1
		No	0

Courtesy: Reich et al. (2017)

7.2.5 Conclusion

An overview of the various approaches most commonly deployed to study the safety and efficacy of ingredients used in HCP or the finished product through in silico, in vitro, and in vivo assessments is provided in this chapter. The list of methods covered in the sections above is not comprehensive but is just indicative of various methods followed by researchers globally. The methods are not standard or mandated procedures but can serve as a guideline that can be either directly adopted or modified to suit the research objectives.

References

Al-Dayel O, Hefne J, Al-Ajyan T (2011) Human exposure to heavy metals from cosmetics. Orient J Chem 27(1):1–11

Ahn HJ, Lee W (2002) An ultrastructural study of hair fiber damage and restoration following treatment with permanent hair dye. Int J Dermatol 7:88–92

Akimoto M, Yamamoto Y, Watanabe S, Yamaga H, Yoshida K, Wakabayashi K, Tahara Y, Horie N, Fujimoto K, Kusakari K et al (2020) Oxidation of a cysteine-derived nucleophilic reagent by dimethyl sulfoxide in the amino acid derivative reactivity assay. J Appl Toxicol 40:843–854

Alajaji AN (2024) Hair product allergy: a review of epidemiology and management. Cureus 16(4):e58054

Alépée N, Piroird C, Nardelli L (2017) U-SENS™: a U937 cell line activation test for skin sensitization. In: Eskes C, van Vliet E, Maibach H (eds) Alternatives for dermal toxicity testing. Springer, Cham. https://doi.org/10.1007/978-3-319-50353-0_22

Alessandrini A, Piraccini BM (2016) Essentials of hair care cosmetics. Cosmetics 3(4):34. https://doi.org/10.3390/cosmetics3040034

Ali N, Zohra RR, Qader SA, Mumtaz M (2015) Scanning electron microscopy analysis of hair index on Karachi's population for social and professional appearance enhancement. Int J Cosmet Sci 37:312–320

Angra K, LaSenna CE, Nichols AJ, Tosti A (2015) Hair dye: a trichoscopy pitfall. J Am Acad Dermatol 72:e101–e102

Barak-Shinar D, Green LJ (2018) Scalp seborrheic dermatitis and dandruff therapy using a herbal and zinc pyrithione-based therapy of shampoo and scalp lotion. J Clin Aesthet Dermatol 11(1):26

Barel AO, Lambrecht R, Clarys P, Morrison BM Jr, Paye M (2001) A comparative study of the effects on the skin of a classical bar soap and a syndet cleansing bar in normal use conditions and in the soap chamber test. Skin Res Technol 7(2):98–104

Barthe M, Bavoux C, Finot F, Mouche I, Cuceu-Petrenci C, Forreryd A, Chérouvrier Hansson A, Johansson H, Lemkine GF, Thénot J-P (2021) Safety testing of cosmetic products: overview of established methods and new approach methodologies (NAMs). Cosmetics 8:50

Belleti KMS, Fererman IH, Mendes TRO, Piacekeski AD, Monteiro VF, Carreño NV, Valentini A, Leite ER, Longo E (2003) Evaluation of hair fiber hydration by differential scanning calorimetry, gas chromatography and sensory analysis. J Cosmet Sci 54:527–535

Berthomieu C, Hienerwadel R (2009) Fourier transform infrared (FTIR) spectroscopy. Photosynth Res 101(2–3):157–170

Bhuvaneswari VS, Chandan RS (2018) Review on safety assessment of a cosmetic product. Int J Curr Pharm Res 10(3):1–6

Bolduc C, Shapiro J (2001) Hair care products: waving, straightening, conditioning and coloring. Clin Dermatol 19:431–436

Boyer IJ, Bergfeld WF, Heldreth B, Fiume MM, Gill LJ (2017) The cosmetic ingredient review program—expert safety assessments of cosmetic ingredients in an open forum. Int J Toxicol 36(5 Suppl 2):13–15

Caloni F, De Angelis I, Hartung T (2022) Replacement of animal testing by integrated approaches to testing and assessment (IATA): a call for in vivitrosi. Arch Toxicol 96:1935–1950. https://doi.org/10.1007/s00204-022-03299-x

Calvarese M (2014) Comedogenicity—a complicated conversation. www.skininc.com. Last accessed 10 Dec 2017

Camacho-Bragado GA, Balooch G, Dixon-Parks F, Porter C, Bryant H (2015) Understanding breakage in curly hair. Br J Dermatol 173:10–16

Central Drugs Standard Control Organization, Bureau of Indian Standards (BIS). http://www.cdsco.nic.in/forms/Default.aspx, http://www.bis.gov.in/. Last accessed 10 Dec 2017

Chaisripipat W, Lourith N, Kanlayavattanakul M (2015) Anti-dandruff hair tonic containing lemongrass (Cymbopogon flexuosus) oil. Forsch Komplementmed 22:226–229

Cogen FC, Beezhold DH (2002) Hair glue anaphylaxis: a hidden latex allergy. Ann Allergy Asthma Immunol 88(1):61–63

Corsini E, Galbiati V, Mitjans M, Galli CL, Marinovich M (2013) NCTC 2544 and IL-18 production: a tool for the identification of contact allergens. Toxicol In Vitro 27(3):1127–1134

Cossman JP, Ladizinski B, Lee KC (2013) Pigmented concealing powders for the hair loss patient. J Cosmet Dermatol 12:322–324

Cottrez F, Boitel E, Auriault C, Aeby P, Groux H (2015) Genes specifically modulated in sensitized skins allow the detection of sensitizers in a reconstructed human skin model. Development of the SENS-IS assay. Toxicol In Vitro 29:787–802
CTRI (2018a) Evaluating anti dandruff efficacy of hair leave on products on healthy human subjects based on half head application design. CTRI/2018/05/013774
CTRI (2018b) A single blinded (investigator blinded), randomized study to evaluate and compare the efficacy of the hair leave-on product with a comparator product for scalp hair growth and density. CTRI/2018/03/012308
CTRI (2019) An open-label, single-center, safety and efficacy study of a hair care regimen consisting of hair oil, shampoo, hot oil treatment and oil replacement used in combination in treatment of hair fall control in healthy adult human subjects. CTRI/2019/09/021018
CTRI (2021) Anti-dandruff efficacy of Ayurvedic product. CTRI/2021/12/038437
CTRI (2022a) An open label, single-arm, clinical study to evaluate efficacy and safety of Vatika Dandruff Guard Shampoo in subjects with moderate to severe scalp dandruff. CTRI/2022/03/040886
CTRI (2022b) A proof-of-concept, safety and efficacy study of hair growth promoting product in healthy adult human subjects with hair fall complaints. CTRI/2022/11/047330
CTRI (2022c) An open label, single-arm, clinical study to evaluate efficacy and safety of chewable gummies in healthy adult subjects with hair fall. CTRI/2022/08/044809
Dario MF, Pahl R, Castro JR, Lima FS, Kaneko TM, Pinto CASO, Baby AR, Velasco MVR (2013) Efficacy of Punica granatum L. hydroalcoholic extract on properties of dyed hair exposed to UVA radiation. J Photochem Photobiol B Biol 120:142–147
Dario MF, Baby AR, Velasco MV (2015) Effects of solar radiation on hair and photoprotection. J Photochem Photobiol B 153:240–246
De Groot AC, Weyland JW, Nater JP (1994) Unwanted effects of cosmetics and drugs used in dermatology. Elsevier, Amsterdam, pp 473–476
Dhurat R, Saraogi P (2009) Hair evaluation methods: merits and demerits. Int J Trichology 1(2):108–119
Dias MFRG (2015) Hair cosmetics: an overview. Int J Trichology 7:2–15
Draelos ZD (2011) Camouflage technique for alopecia areata: what is a patient to do? Dermatol Ther 24:305–310
Draelos ZD (2013) Shampoos, conditioners, and camouflage techniques. Dermatol Clin 31:173–178
Duarte LC, Juchen PL, Pulz GM, Brum TM, Chodur N, Liccardo A, Fischer AC, Acauan RB (2003) Aplicações de microscopia eletrônica de varredura (MEV) e sistema de energia dispersiva (EDS) no Estudo de Gemas: Exemplos brasileiros. Pesquisas Geociências 30:3–15 (in Portuguese)
Dupres V, Camesano T, Langevin D, Checco A, Guenoun P (2004) Atomic force microscopy imaging of hair: correlations between surface potential and wetting at the nanometer scale. J Colloid Interface Sci 269:329–335
Dyer JM, Bringans SD, Bryson WG (2006) Determination of photo-oxidation products within photo-yellowed bleached wool proteins. Photochem Photobiol 82:551–557
Edstrom RD, Meinke MH, Yang X, Yang R, Evans DF (1990) Scanning tunneling microscopy of the enzymes of muscle glycogenolisis. Ultramicroscopy 33:99–106
Éhén Z, Novák C, Sztatisz J, Bene O (2004) Thermal characterization of hair using TG-MS combined thermoanalytical technique. J Therm Anal Calorim 78:427–440
EU Regulations for premarket release of cosmetic products. http://eur-lex.europa.eu/legal-content. Last accessed 10 Dec 2017
European Food Safety Authority (2014) Modern methodologies and tools for human hazard assessment of chemicals. EFSA J 12(4):3638
Evans T (2011) Evaluating hair conditioning with instrumental combing. Cosmet Toilet 126:558–563

Evans T (2013) Measuring hair strength, part I: stress-strain curves, 8 Aug 2013. https://www.cosmeticsandtoiletries.com/testing/efficacy/blog/21837724/measuring-hair-strength-part-i-stressstrain-curves
Faria PM, Camargo LN, Carvalho RSH, Paludetti LA, Velasco MVR, Gama RM (2013) Protective effect of argan oil (Argania spinosa kernel oil) and cupuassu butter (Theobroma grandiflorum seed butter) post treatment with hair dye. J Cosmet Dermatol Sci Appl 3:40–44
FDA (2013) Guidance for industry. Cosmetic good manufacturing practices
FDA (2017a) FDA regulations on product testing before release into market. https://www.fda.gov/cosmetics/scienceresearch/producttesting. Last accessed 10 Dec 2017
FDA (2017b) Regulations on premarket release of cosmetic products in US. https://www.fda.gov/Cosmetics/GuidanceRegulation/lawsRegulations. Last accessed 10 Dec 2017
Fernández E, Martínez-Teipel B, Armengol R, Barba C, Coderch L (2012) Efficacy of antioxidants in human hair. J Photochem Photobiol B 117:146–156
França-Stefoni SA, Dario MF, Sá-Dias TC, Bedin V, Almeida AJ, Baby AR, Velasco MV (2015) Protein loss in human hair from combination straightening and coloring treatments. J Cosmet Dermatol 14:204–208
Freitas AZ, Amaral MM, Raele MP (2010) Optical coherence tomography: development and applications. In: Duarte FJ (ed) Laser pulse phenomena and applications. InTech, Rijeka, pp 409–432
Frosch PJ, Kligman AM (1979) The soap chamber test. A new method for assessing the irritancy of soaps. J Am Acad Dermatol 1(1):35–41
Fujimoto JG, Pitris C, Boppart SA, Brezinski ME (2000) Optical coherence tomography: an emerging technology for biomedical imaging and optical biopsy. Neoplasia 2(1–2):9–25
Fujita M, Yamamoto Y, Tahara H, Kasahara T, Jimbo Y, Hioki T (2014) Development of a prediction method for skin sensitization using novel cysteine and lysine derivatives. J Pharmacol Toxicol Methods 70:94–105
Gądarowska D, Kalka J, Wójcik AD, Mrzyk I (2022) Alternative methods for skin-sensitization assessment. Toxics 10(12):740
Galbiati V, Gibbs S, Roggen E, Corsini E (2018) Development of an in vitro method to estimate the sensitization induction level of contact allergens. Curr Protoc Toxicol 75:20.15.1–20.15.20
Gama RM, Balogh TS, França S, Sá-Dias TC, Bedin V, Baby AR, Velasco MVR (2011) Thermal analysis of hair treated with oxidative hair dye under influence of conditioners agents. J Therm Anal Calorim 106:399–405
Garcia ML, Diaz J (1976) Combability measurements on hair. J Soc Cosmet Chem 27:379–398
Gavazzoni-Dias MR (2015) Hair cosmetics: an overview. Int J Trichology 7:2–15
Geobrick GF, Ryan CA, Dearman RJ, Kimber I (2007) Local lymph node assay (LLNA) for detection of sensitization capacity of chemicals. Methods 41:54–60
Gober MD, Gaspari AA (2008) Allergic contact dermatitis. Dermatol Immun 10:1–26
Guerra-Tapia A, Gonzalez-Guerra E (2014) Hair cosmetics: dyes. Actas Dermosifiliogr 105:833–839
Gupta M, Mysore V (2016) Classifications of patterned hair loss: a review. J Cutan Aesthet Surg 9:3–12
Gurden SP, Monteiro VF, Longo E, Ferreira MM (2004) Quantitative analysis and classification of AFM images of human hair. J Microsc 215:13–23
Gusev E, Zhuravleva Y (2022) Inflammation: a new look at an old problem. Int J Mol Sci 23:4596
Hamilton JB (1951) Patterned loss of hair in man: types and incidence. Ann N Y Acad Sci 53:708–728
Hannuksela M, Salo H (1986) The repeated open application test (ROAT). Contact Derm 14:221–227
Hans RK, Agrawal N, Verma K, Misra RB, Ray RS, Farooq M (2008) Assessment of the phototoxic potential of cosmetic products. Food Chem Toxicol 46:1653–1658
Harris AG, Kim M, Murrell DF (2016) Pigmented hair-thickening fibers: a camouflage technique for alopecia in patients with epidermolysis bullosa. Skin Appendage Disord 1:153–155

Hougaard MG, Menné T, Søsted H (2012) Occupational eczema and asthma in a hairdresser caused by hair-bleaching products. Dermatitis 23:284–287

ICCR (International Cooperation on Cosmetic Regulation). http://www.iccr-cosmetics.org. Last accessed 10 Dec 2017

Imai N, Takeyoshi M, Aizawa S, Tsurumaki M, Kurosawa M, Toyoda A, Sugiyama M, Kasahara K, Ogata S, Omori T, Hirota M (2021) Improved performance of the SH test as an in vitro skin sensitization test with a new predictive model and decision tree. J Appl Toxicol 42(6):1029–1043. https://doi.org/10.1002/jat.4275

Indian Cosmetic Legislation, Chapter 6. https://cosmetic.chemlinked.com/chempedia/india-cosmetic-legislation. Last accessed 10 Dec 2017

IS 13424-2001: Safety evaluation of bathing bars and toilet soaps—methods of test (First Revision)

IS 4011:2018: Methods of test for safety evaluation of cosmetics (Third Revision)

Jachowicz J, McMullen R (2011) Tryptophan fluorescence in hair-examination of contributing factors. J Cosmet Sci 62:291–304

Jerajani HR, Melkote S (2007) Thin-layer rapid-use epicutaneous test (TRUE test). Indian J Dermatol Venereol Leprol 73:292–295

Johansson H, Gradin R, Johansson A, Adriaens E, Edwards A, Zuckerstätter V, Jerre A, Burleson F, Gehrke H, Roggen EL (2019) Validation of the GARD™ skin assay for assessment of chemical skin sensitizers: ring trial results of predictive performance and reproducibility. Toxicol Sci 170(2):374–381

Kandárová H, Letašiová S (2011) Alternative methods in toxicology: pre-validated and validated methods. Interdiscip Toxicol 4:107–113

Khambra A, Ramesh V, Kathuria S (2020) Correlation of patch test with repeated open application test in patients with suspected hair-dye dermatitis. Indian J Dermatol Venereol Leprol 86:733–736

Kilic A, Harder A, Reich H, Knie U, Masur C, Abels C (2018) Efficacy of hydrophilic or lipophilic emulsions containing Echinacea purpurea extract in treatment of different types of pruritus. Clin Cosmet Investig Dermatol 11:591

Kim M (2011) Effect of procyandin oligomers on oxidative hair damage. Skin Res Technol 17:108–118

Kim BR, Won SH, Kim JW, Kim M, Jeong J-I, Shin J-W, Huh C-H, Na J-I (2023) Development of a new classification and scoring system for scalp conditions: Scalp Photographic Index (SPI). J Dermatol Treat 34(1):2181655

Kimber I, Dearman RJ (2010) The local lymph node assay and skin sensitization testing. Methods Mol Biol 598:221–231

Kolasinski J (2009) TrichoScan®: a useful method for measuring hair growth parameters and enhancing patient selection for hair restoration surgery. Hair Transplant Forum International

Krob HA, Fleischer AB, D'Agostino R, Haverstock CL, Feldman S (2004) Prevalence and relevance of contact dermatitis allergens: a meta-analysis of 15 y of published TRUE test data. J Am Acad Dermatol 51:349–353

Kue CS, Kumar S (2023) Non mammalian models in toxicology screening. In: Reference module in biomedical sciences

Kuzuhara A (2006) Analysis of structural changes in bleached keratin fibers (black and white human hair) using Raman spectroscopy. Biopolymers 81:506–514

Kuzuhara A (2007) Analysis of structural changes in permanent waved human hair using Raman spectroscopy. Biopolymers 85:274–283

Kuzuhara A (2016) Internal structural changes in keratin fibres resulting from combined hair waving and stress relaxation treatments: a Raman spectroscopic investigation. Int J Cosmet Sci 38:201–209

Kuzuhara A, Fujiwara N, Hori T (2007) Analysis of internal structure changes in black human hair keratin fibers with aging using Raman spectroscopy. Biopolymers 87:134–140

Lai-Cheong JE, McGrath JA (2009) Structure and function of skin, hair and nails. Medicine 37:223–226

Lalko JF, Kimber I, Gerberick GF, Foertsch LM, Api AM, Dearman RJ (2012) The direct peptide reactivity assay: selectivity of chemical respiratory allergens. Toxicol Sci 129(2):421–431

Lee WS, Ro BI, Hong SP, Bak H, Sim WY, Kim DW et al (2007) A new classification of pattern hair loss that is universal for men and women: basic and specific (BASP) classification. J Am Acad Dermatol 57:37–46

Lima CRR, Almeida MM, Velasco MVR, Matos JR (2016) Thermoanalytical characterization study of hair from different ethnicities. J Therm Anal Calorim 123:2321–2328

Longo VM, Monteiro VF, Pinheiro AS, Terci D, Vasconcelos JS, Paskocimas CA, Leite ER, Longo E, Varela JA (2006) Charge density alterations in human hair fibers: an investigation using electrostatic force microscopy. Int J Cosmet Sci 28:95–101

Longo VM, Silva-Pinheiro A, Sambrano JR, Agnelli JAM, Longo E, Varela JA (2013) Towards an insight on photodamage in hair fibre by UV-light: an experimental and theoretical study. Int J Cosmet Sci 35:539–545

Ludwig E (1977) Classification of the types of androgenetic alopecia (common baldness) occurring in the female sex. Br J Dermatol 97:247–254

Lukas M, Stössel H, Hefel L, Imamura S, Fritsch P, Sepp NT, Schuler G, Romani N (1996) Human cutaneous dendritic cells migrate through dermal lymphatic vessels in a skin organ culture model. J Invest Dermatol 106:1293–1299

Majeed M, Majeed S, Nagabhushanam K, Mundkur L, Neupane P, Shah K (2020) Clinical study to evaluate the efficacy and safety of a hair serum product in healthy adult male and female volunteers with hair fall. Clin Cosmet Investig Dermatol 13:691–700

Miyamae Y, Yamakawa Y, Ozaki Y (2007) Evaluation of physical properties of human hair by diffuse reflectance near-infrared spectroscopy. Appl Spectrosc 61:212–217

Monteiro VF, Natal AMD, Soledade LEB, Longo E (2003) Morphological analysis of polymers on hair fibers by SEM and AFM. Mater Res 6:501–506

Neuser F, Schlatter H (2010) Hair dyes. In: Draelos ZD (ed) Cosmetic dermatology: products and procedures. Wiley-Blackwell, Oxford, pp 256–310

Nigam PK (2009) Adverse reactions to cosmetics and methods of testing. Indian J Dermatol Venereol Leprol 75:19–10

Nishijo T, Miyazawa M, Saito K, Otsubo Y, Mizumachi H, Sakaguchi H (2019) Sensitivity of KeratinoSens™ and h-CLAT for detecting minute amounts of sensitizers to evaluate botanical extract. J Toxicol Sci 44:13–21

Nogueira ACS, Joekes I (2004) Hair color changes and protein damage caused by ultraviolet radiation. J Photochem Photobiol B 74:109–117

Norwood OT (1975) Male pattern baldness: classification and incidence. South Med J 68:1359–1365

OECD (2009) OECD Test No. 437: Bovine corneal opacity and permeability test method for identifying ocular corrosives and severe irritants

OECD (2015) OECD Test No. 430: In vitro skin corrosion: transcutaneous electrical resistance test method (TER)

OECD (2016) OECD Guidelines Test No. 473: In vitro mammalian chromosomal aberration test

OECD (2018) OECD Test No. 438: Isolated chicken eye test method for identifying (i) Chemicals inducing serious eye damage and (ii) Chemicals not requiring classification for eye irritation or serious eye damage

OECD (1997-corrected 26 June 2020). Test No. 471. Bacterial reverse mutation test, OECD guideline for the testing of chemicals. Accessed Jun 2021

OECD (2022a) OECD Test No. 442C: In chemico skin sensitisation

OECD (2022b) OECD Test No. 442D: In vitro skin sensitisation. Accessed 30 Jun 2022

OECD (2022c) OECD Test No. 442E: In vitro skin sensitisation

OECD (2023) OECD Test No. 487: In vitro mammalian cell micronucleus test, Jul 2023

Pande CM, Yang B (2000) Near-infrared spectroscopy: applications in hair research. Int J Cosmet Sci 51:183–192

Pienpinijtham P, Thammacharoen C, Naranitad S, Ekgasit S (2018) Analysis of cosmetic residues on a single human hair by ATR FT-IR microspectroscopy. Spectrochim Acta A Mol Biomol Spectrosc 197(15):230–236

Python F, Goebel C, Aeby P (2007) Assessment of the U937 cell line for the detection of contact allergens. Toxicol Appl Pharmacol 220:113–124

Raman CV, Krishnan KS (1928) A new type of secondary radiation. Nature 121:501–502

Ramirez T, Mehling A, Kolle SN, Wruck CJ, Teubner W, Eltze T, Aumann A, Urbisch D, van Ravenzwaay B, Landsiedel R (2014) LuSens: a keratinocyte based ARE reporter gene assay for use in integrated testing strategies for skin sensitization hazard identification. Toxicol In Vitro 28(8):1482–1497

Ranganathan S, Mukhopadhyay T (2010) Dandruff: the most commercially exploited skin disease. Indian J Dermatol 55(2):130–131

Rao S, Costa KD (2014) Chapter 2: Atomic Force Microscopy (AFM) in biomedical research. In: Biomedical imaging: applications and advances. Woodhead Publishing, pp 41–64

Rassman WR, Pak JP, Kim J, Estrin NF (2015) Scalp micropigmentation: a concealer for hair and scalp deformities. J Clin Aesthet Dermatol 8:35–42

Reich A, Bozek A, Janiszewska K, Szepietowski JC (2017) 12-Item Pruritus Severity Scale: development and validation of new itch severity questionnaire. Biomed Res Int 2017:3896423

Richena M, Rezende CA (2015) Effect of photodamage on the outermost cuticle layer of human hair. J Photochem Photobiol B 153:296–304

Robbins CR, Crawford RJ (1991) Cuticle damage and the tensile properties of human hair. J Soc Cosmet Chem 42:49–58

Rodrigues AG, Galzerani JC (2012) Espectroscopias de infravermelho, Raman e de fotoluminescência: Potencialidades e complementaridades. Rev Bras Ensino Fis 34:1–9 (in Portuguese)

Rogiers V (2017) Safety assessment of cosmetics and their ingredients in the EU. Department of Toxicology, Vrije Universiteit, Brussel

Ruetsch SB, Weigmann HD (1996) Mechanism of tensile stress release in the keratin fiber cuticle: I. J Soc Cosmet Chem 47:13–26

Ruetsch SB, Yang B, Kamath YK (2003) Chemical and photo-oxidative hair damage studied by dye diffusion and electrophoresis. J Cosmet Sci 54:379–394

Ryder E (2014) Repetitive open application test (ROAT). https://www.dermnetnz.org/topics/open-application-test/. Last accessed 10 Dec 2017

Sá-Dias TC, Baby AR, Kaneko TM, Velasco MVR (2008) Protective effect of conditioning agents on Afro-ethnic hair chemically treated with thioglycolate-based straightening emulsion. J Cosmet Dermatol 7:120–126

Saito K, Takenouchi O, Nukada Y, Miyazawa M, Sakaguchi H (2017) An in vitro skin sensitization assay termed EpiSensA for broad sets of chemicals including lipophilic chemicals and pre/prohaptens. Toxicol In Vitro 40:11–25

Salibi A, Souéid A, Dancey A (2014) Hair braiding (plaiting) and hair extensions: an underestimated danger! J Plast Reconstr Aesthet Surg 67:e206–e207

Sandhu SS, Robbins CR (1993) A simple and sensitive technique, based on protein loss measurements, to assess surface damage to human hair. J Soc Cosmet Chem 44:163–175

Sandhu SS, Ramachandran R, Robbins CR (1995) A simple and sensitive method using protein loss measurements to evaluate damage to human hair during combing. J Soc Cosmet Chem 46:39–52

Scanavez C, Zoega M, Barbosa A, Joekes I (2000) Measurement of hair luster by diffuse reflectance spectrophotometry. J Cosmet Sci 55:289–302

Seshad IP, Bhushan B (2008) Effect of ethnicity and treatments on in situ tensile response and morphological changes of human hair characterized by atomic force microscopy. Acta Mater 56:3585–3597

Shetty VH, Shetty NJ, Nair DG (2013) Chemical hair relaxers have adverse effects a myth or reality. Int J Trichology 5:26–28

Shore RN, BinnickShore SRN, Binnick S (1977) Dimethylglyoxime stick test for easier detection of nickel. Arch Dermatol 113(12):1734

Signori V (2004) Review of the current understanding of the effect of ultraviolet and visible radiation on hair structure and options for photoprotection. J Cosmet Sci 55:95–113

Silva AL, Nunes AS, Gesztesi JL (2004) Protein loss quantification of abraded virgin and abraded bleached hair according to Bradford assay. J Cosmet Sci 55(Suppl. 2):S175–S179
Silva EC, Velasco-de-Paola MVR, Matos JR (2007) Análise térmica aplicada à cosmetologia. Rev Bras Cienc Farm 43:347–357 (in Portuguese)
Smith E, Dent G (2005) Modern Raman spectroscopy: a practical approach, 1st edn. Wiley
Smith J, Tsibouklis J, Nevell T, Breakspear S (2013) AFM friction and adhesion mapping of the substructures of human hair cuticles. Appl Surf Sci 285:638–644
Søsted H, Rastogi SC, Andersen KE, Johansen JD, Menné T (2004) Hair dye contact allergy: quantitative exposure assessment of selected products and clinical cases. Contact Dermatitis 50:344–348
Souza P, Rathi SK (2015) Shampoo and conditioners: what a dermatologist should know? Indian J Dermatol 60:248–254
Spiekstra SW, dos Santos GG, Scheper RJ, Gibbs S (2009) Potential method to determine irritant potency in vitro—comparison of two reconstructed epidermal culture models with different barrier competency. Toxicol In Vitro 23(2):349–355
Stępnik M, Arkusz J (2003) Molecular events associated with dendritic cells activation by contact sensitizers. Int J Occup Med Environ Health 16:191–199
Stevens MP (1999) Polymer chemistry, 3rd edn. Oxford University Press, New York
Suzuki M, Hirota M, Hagino S, Itagaki H, Aiba S (2009) Evaluation of changes of cell-surface thiols as a new biomarker for in vitro sensitization test. Toxicol In Vitro 23:687–696
Swift JA, Smith JR (2000) Atomic force microscopy of human hair. Scanning 22:310–318
Takahashi T, Kimura Y, Saito R, Nakajima Y, Ohmiya Y, Yamasaki K, Aiba S (2011) An in vitro test to screen skin sensitizers using a stable THP-1–derived IL-8 reporter cell line, THP-G8. Toxicol Sci 124:359–369
TEM (n.d.). https://www.nottingham.ac.uk/nmrc-commercial/nmcs-facilities/tem.aspx
Thibaut S, Barbarat P, Leroy F, Bernard BA (2007) Human hair keratin network and curvature. Int J Dermatol 46:7–10
Tomes C, Jones JT, Carr CM, Jones D (2007) Three-dimensional imaging and analysis of the surface of hair fibers using scanning electron microscopy. Int J Cosmet Sci 29:293–229
Trüeb RM (2005) Dermocosmetic aspects of hair and scalp. J Investig Dermatol Symp Proc 10:289–292
Trüeb RM (2007) Shampoos: ingredients, efficacy and adverse effects. J Dtsch Dermatol Ges 5:356–365
Uruno A, Motohashi H (2011) The Keap1–Nrf2 system as an in vivo sensor for electrophiles. Nitric Oxide 25:153–160
Van Lerberghe L, Baeck M (2014) A case of acute contact dermatitis induced by formaldehyde in hair-straightening products. Contact Dermatitis 70:384–386
Velasco MVR, Baby AR, Sarruf FD, Kaneko TM, Samad RE, Vieira Junior ND, Freitas AZ (2009) Prospective ultramorphological characterization of human hair by optical coherence tomography. Skin Res Technol 15:440–443
Velasco MVR, Balogh TS, Kagiyama JW, Dario MF, Gama RM, Bedin V, Baby AR (2015) Influence of Brazilian vegetable oils on mechanical resistance of hair fiber. Biomed Biopharm Res 12:99–106
Velasco MVR, Abreu SRP, Freitas AZ, Bedin V, Baby AR, Gama RM (2016) Optical coherence tomography to evaluate the effects of oxidative hair dye on the fiber. Skin Res Technol 22:430–436
Wareing B, Kolle SN, Birk B, Alépée N, Haupt T, Kathawala R, Kern PS, Nardelli L, Raabe H, Rucki M, Ryan CA, Verkaart S, Westerink WMA, Landsiedel R, Natsch A (2020) The kinetic direct peptide reactivity assay (kDPRA): intra- and inter-laboratory reproducibility in a seven-laboratory ring trial. ALTEX 37(4):639–651
WebMD (2017) Phototoxicity and photo allergic reactions. www.webmd.com
Wendlandt WW (1986) Thermal analysis, 3rd edn. Wiley, New York, p 1
Westermeier R (2005) Electrophoresis in practice. Wiley-VCH, Weinheim, p 427

Wortmann FJ, Springob C, Sendelbach G (2002) Investigations of cosmetically treated human hair by differential scanning calorimetry in water. J Cosmet Sci 53:219–228

Wortmann FJ, Stapels M, Elliott R, Chandra L (2006) The effect of water on the glass transition of human hair. Biopolymers 81:371–375

Yamamoto Y, Tahara H, Usami R, Kasahara T, Jimbo Y, Hioki T, Fujita M (2015) A novel in chemico method to detect skin sensitizers in highly diluted reaction conditions. J Appl Toxicol 35(11):1348–1360

Zhang F, Li Y, Ren W, Li S (2023) Establishment of clinical evaluation criteria for scalp seborrheic dermatitis. J Cosmet Dermatol 22:3042–3046

Zhou W, Apkarian R, Wang ZL, Joy D (2007) Fundamentals of scanning electron microscopy (SEM). In: Zhou W, Wang ZL (eds) Scanning microscopy for nanotechnology: techniques and applications. Springer, New York, pp 1–39

Zhu XX, Jiang WJ, Zhu W (2017) Research advances of pathogenesis of seborrheic dermatitis. J Pract Dermatol 10(1):41–43

Microbiome and Hair Care Products: Opportunities and Challenges

8

Shalmali Kamat

Abstract

Hair care is an important part of an individual's personal grooming regime. Hair grooming products have experienced a boom in the past few years, featuring a wide range of natural and chemical options. These products can be a major influence on the natural microbial inhabitants, they can cause imbalance in the scalp microbiome and in turn affect the scalp health. This chapter provides an overview of the role of natural microflora in maintaining a healthy human scalp, importance of their composition, and the consequences of dysbiosis. It also highlights the lack of extensive research in this field making it a potential area for further investigation. This chapter can serve as a preliminary source of information for pharmaceutical companies and cosmetic industries interested in developing hair care products taking natural microflora and its importance into consideration.

Keywords

Hair care · Hair microbiome · Skin microbiome · Dandruff · Scalp dysbiosis · Hair treatment

Hair forms a significant part of a person's appearance and personality. Hair quality and thickness can influence an individual's personality and can affect their emotional state. Haircare products flood the markets promising a variety of benefits for their users. Do these products work? Do they cause more harm than benefit? Which

---

S. Kamat (✉)
Department of Microbiology, St Joseph's University, Bengaluru, Karnataka, India
e-mail: shalmalikamat@sju.edu.in

A. Dwivedi et al. (eds.), *Hair Care Products*,
https://doi.org/10.1007/978-981-97-6740-3_8

is the most suitable product? Before these questions can be answered, it is important to understand what our scalp is made of and how these products might influence it.

Human scalp constitutes multiple layers; the thick outermost layer made of epidermis and dermis holds the root of the hair and the hair shaft extends outward forming the part that is visible on the head. Surrounding the hair root is the hair follicle, a flask-shaped structure embedded into the skin. The follicle is made of four main regions, starting from the upper surface- the infundibulum, isthmus, supra bulbar region, and the bulb (Polak-Witka et al. 2020). It is attached to a sebaceous gland which produces sebum, an oily substance that lubricates the scalp and prevents the scalp from drying (Martel et al. 2023; Blume-Peytavi and Vogt 2011). Apart from the above-mentioned properties, the ATP content of the hair follicle is an important factor in determining the health of the hair follicle. Adenosine promotes the anagen(growth) phase of hair follicles and is also essential for wound-healing process (Oura et al. 2008; Mikhal'chik et al. 2014). However, this is not all that makes up the scalp. Our scalp houses several types of microorganisms including bacteria, fungi, and viruses and they constantly impact the health of our scalp. These microorganisms inhabit different regions of the scalp including the outermost layer "Stratum corneum" and the hair follicle. Understanding the microbiome, how it interacts with the scalp and hair follicles and subsequently promoting growth and maintenance of a healthy scalp microflora may potentially improve the quality of our hair (Saxena et al. 2018; Turner et al. 2012; Grimshaw et al. 2019).

8.1 Microflora of Human Scalp

Humans have evolved alongside microorganisms and have developed symbiotic relationships with them. Human beings harbor a large repertoire of microbes on and inside their bodies. Different regions of the body are colonized by different kinds of microorganisms depending on host physiology-influenced factors like the pH, moisture, oxygen content, and nutrient content of the local environment. Human beings along with their microbial communities are termed as "meta-organism" and humans cannot lead a normal life without their microbiomes (Polak-Witka et al. 2020; Bosch and McFall-Ngai 2011).

Much research has been conducted on microbiomes of the gut and skin. However, microbiome of the scalp needs to be explored more extensively; some of the emerging evidence suggests an increasing role of microorganisms in hair-related disorders and diseases such as alopecia areata, androgenic alopecia, dandruff, etc. (Grimshaw et al. 2019; Ho et al. 2019; Pinto et al. 2019). Human scalp is a different microenvironment from the rest of the skin surface. In comparison to dry and exposed skin surface, the scalp is moister and relatively protected from UV/sunlight exposure. Sebaceous glands associated with the hair follicles are the source of the constant lubrication provided to the scalp. Moreover, the follicles themselves are complex stratified structures that offer specific microenvironments for microbial survival. Hair follicles offer an ideal environment for growth of microbes. Maximum colonization of microorganisms is seen near the infundibulum. The bulb of the follicle also offers an ideal environment for microbial existence by being an immune-privileged site, yet microbial presence in the blub is not very evident (Lousada et al. 2021). Immune function in this region is restricted by the follicular epithelium, furthermore, antigen presentation is suppressed by the keratinocytes of this region making it a desirable environment for microorganisms (Polak-Witka et al. 2020; Paus et al. 2003). Androgenetic alopecia patients showed the presence of microorganisms in the middle and lower regions of hair follicles (Ho et al. 2019).

Studies conducted worldwide indicate that the human scalp does not harbor a wide diversity of microorganisms. Three main genera of microorganisms have been commonly found colonizing the scalp in healthy individuals. *Staphylococcus* spp. and *Cutebacterium* spp. were found to be the most abundant bacterial species and *Malassezia* spp. was the most common fungus. *C. acnes* was the major bacterial colonizer followed by *S. epidermidis*, whereas *M. restricta* and *M. globosa* were the most abundant fungal (yeast) species (Saxena et al. 2018). Apart from these some other bacterial species like Corynebacterium, Brevibacterium, Deinococcus, and fungal species like Aspergillus and Cladosporium were identified in lesser abundance from scalps of healthy individuals (Grimshaw et al. 2019; McGinley et al. 1975). A study from the Indian subcontinent also revealed the presence of similar bacterial and fungal species inhabiting healthy scalps as well as those affected by dandruff. However, there was a marked difference in the percentage of colonization by these organisms, as demonstrated by two independent studies lead by Grimshaw et al. and Saxena et al. Both studies noted a decrease in the Cutebacterium:Staphylococcus ratio in dandruff-affected scalp. A standalone study has reported for the first time, presence of *Leuconostoc holzapfelii* as a natural inhabitant of human scalp (Yoon et al. 2022). With regard to *Malassezia*, increase in an uncharacterized *Malassezia* spp. over the natural commensal *Malassezia* was observed in dandruff affected scalp. In addition, healthy scalps showed a lower ratio of *M. restricta* to *M. globosa* than dandruff-affected scalps wherein the ratio had increased (Saxena et al. 2018). Bacteria and fungi are not the only inhabitants of the scalp and hair follicles, mites such as *Demodex folliculorum* and *Dermatophagoides* were found in the infundibular region and sebaceous glands, respectively. Moreover, viruses like human papilloma virus are natural inhabitants of healthy scalp. Unlike the popular belief that dandruff is caused by fungi, emerging evidence suggests that it is not caused solely by either fungi or bacteria but is a result of an imbalance between the resident natural microflora (Polak-Witka et al. 2020). A study was also conducted on hair shaft colonization by bacteria which showed that growth of gram-positive bacteria *S. aureus* and *S. epidermidis* was inhibited whereas gram-negative bacteria *Pseudomonas aeruginosa* and *E. coli* showed growth as well as biofilm formation on the hair shafts (Kerk et al. 2018).

8.2 Interactions of Microbiome and Host

There is constant interplay between the host surfaces and the microbiome. In healthy individuals this interaction is controlled and helps maintain homeostasis. Both the host tissues and its commensal microbes are essential to maintain the balance. However, this relationship is susceptible to environmental influences and host-related intrinsic factors, if the balance is disturbed it causes dysbiosis and results into diseases and disorders like dandruff, seborrheic dermatitis, and tinea capitis to name a few.

The skin microbiome has not been studied in as much detail as the gut microbiome, and studies on the scalp microbiome are even lesser than the former (Polak-Witka et al. 2020). However, we can draw significant insights from the skin-microbiota interactions about the influences each one has on the other.

8.2.1 How Does Skin Manage Its Microbiome?

The skin and its microbiome have important functions to perform viz. it acts as the first line of defence against invading pathogens, communicates with the immune system, and maintains its activity in the vicinity and metabolism of natural compounds to yield beneficial by-products (Dréno et al. 2016; Byrd et al. 2018). Skin microbiome can mostly survive harsher environments on the skin like the acidic pH and cooler temperatures at the surface than inside the body. They are also capable of surviving on nutrition available in sweat, sebum, and other oily skin exudates. Bacteria like *S. aureus* have evolved to become halotolerant as a result of adaption to survive high salt content from the sweat. These biochemical conditions of the skin usually deter most invading organisms from colonizing the skin and keep it from infections from pathogens. Furthermore, the skin produces Anti-microbial peptides (AMPs) like cathelicidin and defensins which further inhibit microbial growth. AMP genes are being characterized extensively and recent studies suggest that they have a high level of specificity toward their target bacteria suggesting they can selectively eliminate them without harming the natural microbiota (Lousada et al. 2021; Byrd et al. 2018; Lazzaro et al. 1979; Sanford and Gallo 2013).

The commensals occupy space on the skin surface and consume nutrients, thereby competing against external microbes looking for colonization. They also produce antimicrobials that kill other bacteria which may cause pathogenesis. For example, *S. epidermidis* produces serine proteases and phenol-soluble modulins (PSMs) which prevent *S. aureus* from forming biofilms and kills skin pathogens like *Streptococcus pyogenes* and *S. aureus,* respectively (Sanford and Gallo 2013). Cogen et al. (2010) showed that *S. epidermidis* PSMs augment host defences by acting synergistically with host-derived antimicrobial peptides.

Skin commensals enhance signaling of IL-1 pathway, which is implicated in various skin-related diseases. In absence of the commensals like *S. epidermidis*, this pathway is seen to be downregulated (Naik et al. 2012). As observed in mice; the microbes also regulate local inflammatory response by enhancing proliferation of

T-regulatory lymphocyte function in the epidermis helping the colonization of the epidermis with commensal microorganisms. Thus, it is clear that the normal microbiota keeps the host immunity active and also regulates its activity.

Another group of scientists studying hair microbiome and its effect on the scalp conducted studies which showed that skin microbiota promotes regeneration of hair follicles and skin tissue by glutamine metabolism in keratinocytes. This shows that harsh washing products used to get rid of the scalp bacteria completely, might have detrimental effects on the health of the scalp (Wang et al. 2023).

While scalp microbiome in healthy scalps is mostly beneficial, in the event of dysbiosis the same beneficial microorganisms can over-colonize and turn detrimental, resulting into diseased scalp conditions like dandruff, alopecia areata, bacterial folliculitis etc. (Saxena et al. 2018; Grimshaw et al. 2019).

8.3 Hair Care Products and Their Effect on Scalp Microbiome

Several haircare products are available in the market promising variety of benefits. However, in depth research on how they affect the scalp and its microbiome has not been conducted.

Hair care products are formulated with different objectives depending on their purpose. Some products like shampoos are used for cleaning purposes, others like conditioners and waxes are used for softening and moisturizing or providing textures, and some others are used for coloring and hair oils and serums are used with the objective of nourishment of the hair and scalp. Some have a temporary effect on the hair while others have permanent effects like bleaching and straightening or curling. While some products are beneficial to hair and scalp indiscreet use of some might harm causing permanent damage or disease (Alessandrini and Piraccini 2016; Gavazzoni Dias 2015). Shampoos in particular are the most commonly used hair products. They come in a variety of formulations. Commercial shampoo pH ranges from anywhere between 3 and 9 and there is no standardized range of pH determined for shampoos. This can be one important factor that can affect the microbiome of the scalp. Unfortunately, sufficient data is not available on natural hair pH and desirable pH of shampoos (Dias et al. 2014).

It is important to understand that hair products can have an effect on the resident microbiota of the scalp and their overuse leading to imbalance in their numbers can cause dysbiosis which in turn deteriorates the hair quality (Holland and Bojar 2002). On the other hand, products like anti-dandruff shampoos and creams can keep unwanted microorganisms in check and help maintain cleaner scalps (Turner et al. 2013). Appropriate usage of hair care products is advisable for them to work beneficially.

Most commercially available hair care products are formulated to have antimicrobial properties. However, now we have realized that getting completely rid of the natural microflora inhabiting the body would cause much more harm than gain regardless of where they colonize in or on our bodies (Wang et al. 2023).

Hair experts and researchers have now realized the importance of developing hair care products that are microbiome-friendly. Scientists are testing microbially enhanced products to examine their effect on human subjects for hair loss control and hair gain. Shampoos and lotions containing microbially derived constituents like fermented papaya and mangosteen were tested on patients suffering from alopecia for 3 months. The hair and scalp quality were assessed for several parameters including microbiota and ATP content of hair follicle. Significant hair gain and increase in ATP content of the hair follicle were observed and the microbial profile of the scalp also showed improvement (Mayer et al. 2023).

Coconut oil is traditionally used as hair nourishment in many countries. Saxena et al. (2021) have studied the effect of coconut oil application on scalp microbiota on 140 Indian women with healthy as well as dandruff-affected scalps. The results were rather interesting, the ratio of *M. restricta* to *M. globosa* reduced significantly in both dandruff and healthy scalps. Coconut oil application showed significant increase in *C. acnes* to *S. epidermidis* ratio suggesting that coconut oil promotes establishment of healthy microbiome in dandruff affected scalps while enhancing the microbiome effect in healthy scalps. Second important observation was the upregulation of biotin metabolism by the commensal bacteria on treatment with coconut oil (Saxena et al. 2018). Biotin is an important precursor for many enzymes and is known to reduce cellular inflammation, and improve scalp quality (Patel et al. 2017).

Another study showed that a shampoo formulation made from a particular heat-killed *Lacticaseibacillus paracasei* had beneficial effects on scalp microbiome of individuals with scalp conditions like dandruff, high oil secretion, and thinning hair, it restored the balance between the scalp microbial communities but did not show much effect on healthy scalps (Tsai et al. 2023).

These studies show that maintenance of healthy microbiome can have multiple benefits and conducting further research on developing products containing ingredients that enhance beneficial microbiomes is the path forward.

Attempts have to be made to develop hair care products that are sensitive to the microbiome, in order to promote the growth and regulation of a healthy microbiome and simultaneously lower the number of disease-causing microbes. Hair microbiome research is still in its infancy and requires structured attempts in order to understand the kind of microbiota present on the scalp and the means to modulate it. Different formulations of microbes need to be worked on to derive maximum benefit. There is almost no conclusive data on Pro and Prebiotic supplements that support the growth of beneficial microorganisms that can promote hair growth. Research should be focused on finding suitable pro and prebiotics to maintain a healthy microbiome both on the scalp and the gut.

Doctors are exploring Fecal Microbial Transplantations for gut-related disorders (Gupta et al. 2016). Similar attempts could be explored for treating scalp and skin infections as well. Transplantation of microbiome for healthy scalps could be one prospective treatment method to be explored.

The challenge is that hair-related disorders are not caused only due to imbalanced microbiota. Hair loss and scalp diseases are also a result of an individual's physiology, mental and emotional state, diet, lifestyle choices as well as genetic predispositions (Pinto et al. 2019; Rajput 2022). To find a singular solution to all microbiome-related problems would be tough to achieve. Which is why more research has to happen on a large number of subjects both healthy and affected to study the hair quality, microbiome pattern, and other physiological factors. Gut microbiome should also be considered while drawing conclusions on how scalp microbiota affects the hair quality since imbalances in gut microbiome has been previously implicated in skin conditions like psoriasis (Sikora et al. 2020).

There is dearth of research publications demonstrating the relationship between scalp microbiome and hair and scalp health, as well as the effect of hair care products on microbiome, which has led to the conclusion that while this is a major challenge it also stands out as an opportunity to do more in this field.

References

Alessandrini A, Piraccini B (2016) Essential of hair care cosmetics. Cosmetics 3(4):34. https://doi.org/10.3390/cosmetics3040034

Blume-Peytavi U, Vogt A (2011) Human hair follicle: reservoir function and selective targeting. Br J Dermatol 165:13–17. https://doi.org/10.1111/j.1365-2133.2011.10572.x

Bosch TCG, McFall-Ngai MJ (2011) Metaorganisms as the new frontier. Zoology 114(4):185–190. https://doi.org/10.1016/j.zool.2011.04.001

Byrd AL, Belkaid Y, Segre JA (2018) The human skin microbiome. Nat Rev Microbiol 16(3):143–155. https://doi.org/10.1038/nrmicro.2017.157

Cogen AL et al (2010) Staphylococcus epidermidis antimicrobial δ-toxin (phenol-soluble modulin-γ) cooperates with host antimicrobial peptides to kill group a streptococcus. PLoS One 5(1):e8557. https://doi.org/10.1371/journal.pone.0008557

Dias MFRG, de Almeida AM, Cecato PMR, Adriano AR, Pichler J (2014) The shampoo pH can affect the hair: myth or reality? Int J Trichology 6(3):95. https://doi.org/10.4103/0974-7753.139078

Dréno B et al (2016) Microbiome in healthy skin, update for dermatologists. J Eur Acad Dermatol Venereol 30(12):2038–2047. https://doi.org/10.1111/jdv.13965

Gavazzoni Dias MF (2015) Hair cosmetics: an overview. Int J Trichology 7(1):2. https://doi.org/10.4103/0974-7753.153450

Grimshaw SG, Smith AM, Arnold DS, Xu E, Hoptroff M, Murphy B (2019) The diversity and abundance of fungi and bacteria on the healthy and dandruff affected human scalp. PLoS One 14(12):e0225796. https://doi.org/10.1371/journal.pone.0225796

Gupta S, Allen-Vercoe E, Petrof EO (2016) Fecal microbiota transplantation: in perspective. Ther Adv Gastroenterol 9(2):229–239. https://doi.org/10.1177/1756283X15607414

Ho BS-Y et al (2019) Microbiome in the hair follicle of androgenetic alopecia patients. PLoS One 14(5):e0216330. https://doi.org/10.1371/journal.pone.0216330

Holland KT, Bojar RA (2002) Cosmetics. Am J Clin Dermatol 3(7):445–449. https://doi.org/10.2165/00128071-200203070-00001

Kerk SK, Lai HY, Sze SK, Ng KW, Schmidtchen A, Adav SS (2018) Bacteria display differential growth and adhesion characteristics on human hair shafts. Front Microbiol 9:2145. https://doi.org/10.3389/fmicb.2018.02145

Lazzaro BP, Zasloff M, Rolff J (1979) Antimicrobial peptides: application informed by evolution. Science 368(6490):2020. https://doi.org/10.1126/science.aau5480

Lousada MB et al (2021) Exploring the human hair follicle microbiome. Br J Dermatol 184(5):802–815. https://doi.org/10.1111/bjd.19461

Martel JL, Miao JH, Badri T (2023) Anatomy, hair follicle. StatPearls Publishing, Treasure Island, FL. https://www.ncbi.nlm.nih.gov/books/NBK470321/. Accessed 28 Jul 2023

Mayer W et al (2023) Biomolecules of fermented tropical fruits and fermenting microbes as regulators of human hair loss, hair quality, and scalp microbiota. Biomolecules 13(4):699. https://doi.org/10.3390/biom13040699

McGinley KJ, Leyden JJ, Marples RR, Path MRC, Kligman AM (1975) Quantitative microbiology of the scalp in non-dandruff, dandruff, and seborrheic dermatitis. J Invest Dermatol 64(6):401–405. https://doi.org/10.1111/1523-1747.ep12512335

Mikhal'chik EV et al (2014) Evaluation of ATP content in hair bulbs in human scalp. Bull Exp Biol Med 157(1):112–115. https://doi.org/10.1007/s10517-014-2504-2

Naik S et al (2012) Compartmentalized control of skin immunity by resident commensals. Science (1979) 337(6098):1115–1119. https://doi.org/10.1126/science.1225152

Oura H et al (2008) Adenosine increases anagen hair growth and thick hairs in Japanese women with female pattern hair loss: a pilot, double-blind, randomized, placebo-controlled trial. J Dermatol 35(12):763–767. https://doi.org/10.1111/j.1346-8138.2008.00564.x

Patel DP, Swink SM, Castelo-Soccio L (2017) A review of the use of biotin for hair loss. Skin Appendage Disord 3(3):166–169. https://doi.org/10.1159/000462981

Paus R, Ito N, Takigawa M, Ito T (2003) The hair follicle and immune privilege. J Investig Dermatol Symp Proc 8(2):188–194. https://doi.org/10.1046/j.1087-0024.2003.00807.x

Pinto D, Rinaldi F, Sorbellini E (2019) Microbiome and diet impact in scalp disorder: the example of alopecia areata. Int J Exper Biol 2(1):001–009

Polak-Witka K, Rudnicka L, Blume-Peytavi U, Vogt A (2020) The role of the microbiome in scalp hair follicle biology and disease. Exp Dermatol 29(3):286–294. https://doi.org/10.1111/exd.13935

Rajput R (2022) Influence of nutrition, food supplements and lifestyle in hair disorders. Indian Dermatol Online J 13(6):721. https://doi.org/10.4103/idoj.idoj_175_22

Sanford JA, Gallo RL (2013) Functions of the skin microbiota in health and disease. Semin Immunol 25(5):370–377. https://doi.org/10.1016/j.smim.2013.09.005

Saxena R et al (2018) Comparison of healthy and dandruff scalp microbiome reveals the role of commensals in scalp health. Front Cell Infect Microbiol 8:346. https://doi.org/10.3389/fcimb.2018.00346

Saxena R et al (2021) Longitudinal study of the scalp microbiome suggests coconut oil to enrich healthy scalp commensals. Sci Rep 11(1):7220. https://doi.org/10.1038/s41598-021-86454-1

Sikora M et al (2020) Gut microbiome in psoriasis: an updated review. Pathogens 9(6):463. https://doi.org/10.3390/pathogens9060463

Tsai W-H, Fang Y-T, Huang T-Y, Chiang Y-J, Lin C-G, Chang W-W (2023) Heat-killed Lacticaseibacillus paracasei GMNL-653 ameliorates human scalp health by regulating scalp microbiome. BMC Microbiol 23(1):121. https://doi.org/10.1186/s12866-023-02870-5

Turner GA, Hoptroff M, Harding CR (2012) Stratum corneum dysfunction in dandruff. Int J Cosmet Sci 34(4):298–306. https://doi.org/10.1111/j.1468-2494.2012.00723.x

Turner GA, Matheson JR, Li G-Z, Fei X-Q, Zhu D, Baines FL (2013) Enhanced efficacy and sensory properties of an anti-dandruff shampoo containing zinc pyrithione and climbazole. Int J Cosmet Sci 35(1):78–83. https://doi.org/10.1111/ics.12007

Wang G et al (2023) Commensal microbiome promotes hair follicle regeneration by inducing keratinocyte HIF-1α signaling and glutamine metabolism. Sci Adv 9(1):eabo7555. https://doi.org/10.1126/sciadv.abo7555

Yoon YC, Ahn BH, Min JW, Lee KR, Park SH, Kang HC (2022) Stimulatory effects of extracellular vesicles derived from Leuconostoc holzapfelii that exists in human scalp on hair growth in human follicle dermal papilla cells. Curr Issues Mol Biol 44(2):845–866. https://doi.org/10.3390/cimb44020058

9 Assessing Hair Care Product Safety and Compliance with Global Regulations Through Preclinical and Clinical Models

Ruchi Pandey, Abhishank Singh, Abhishek Singh, and Shakti Vinay Shukla

Abstract

Ensuring hair care product safety and regulatory compliance is crucial for consumer protection. This chapter explores the methodologies employed in preclinical and clinical models to assess the safety of hair care products and their adherence to global regulations. Preclinical models encompass in vitro and ex vivo studies. In vitro, cytotoxicity assays evaluate potential ingredient irritation or toxicity on cell cultures. Ex vivo scalp models assess product permeation and potential scalp irritation. Clinical models involve human volunteer studies. Testing identifies potential skin sensitization, while application studies assess product efficacy and safety in real-world use conditions. This chapter explores into the design and conduct of these studies, emphasizing the importance of adhering to international regulatory guidelines. It explores the regulatory landscape governing hair care products across critical markets like the European Union, the United States, and India. By integrating preclinical and clinical data, manufacturers can ensure the safety and compliance of their hair care products, fostering consumer confidence and market success.

Keywords

Regulatory compliance · Preclinical models · Clinical models · Toxicity

R. Pandey · S. V. Shukla (✉)
Fragrance and Flavour Development Centre, Kannauj, Uttar Pradesh, India

A. Singh · A. Singh
Growth Pharmaceutical Private Limited, Andhra Pradesh MedTech Zone (AMTZ), Visakhapatnam, Andhra Pradesh, India

A. Dwivedi et al. (eds.), *Hair Care Products*,
https://doi.org/10.1007/978-981-97-6740-3_9

9.1 Introduction

The cosmetic industry has undergone a significant transformation, witnessing a surge in demand for innovative cosmetic and hair care products. This surge is propelled by evolving consumer preferences and an increased awareness of personal grooming. The Federal Food, Drug, and Cosmetic Act (FD&C Act) delineates cosmetics, including hair care products, partially based on their intended purpose, defining them as "articles meant to be rubbed, poured, sprinkled, or sprayed on, introduced into, or otherwise applied to the human body, for cleansing, beautifying, promoting attractiveness, or altering the appearance" (FD&C Act, Section 201(i)) (Cavers 1939). The definition and regulatory scope of cosmetics are fundamental for ensuring consumer safety and product compliance. Section 3(a) of the Drugs and Cosmetics Act, 1940 (CDSCO 1940), offers a comprehensive definition of cosmetics, emphasizing their intended application on humans. This definition encompasses products aimed at cleansing, enhancing attractiveness, or altering appearance, including substances applied to the skin, hair, or other body parts for beautification or therapeutic purposes. Regulatory agencies such as the FDA and EMA echo similar definitions, underlining the intended use and mode of application of cosmetics. As consumers explore a plethora of cosmetics, including hair care options, the imperative to ensure the safety and compliance of these products with global regulatory standards becomes paramount (Gavazzoni Dias 2015). This chapter embarks on an exploration of the intricate process of assessing the safety of hair care products, employing a comprehensive approach that integrates clinical, preclinical, and ex-vivo hair follicle models. In recent years, the regulatory landscape governing cosmetics has experienced a paradigm shift, placing a heightened focus on ensuring the safety, efficacy, and transparency of products (Rocca et al. 2022). Having established the increasing demand for safe and compliant hair care products, it's crucial to examine how regulatory frameworks have evolved to meet these demands.

Regulatory bodies, including the U.S. Food and Drug Administration (FDA), the European Medicines Agency (EMA), and their global counterparts, have actively updated and reinforced guidelines to address the dynamic nature of the cosmetic industry. Compliance with these evolving standards is imperative for manufacturers, necessitating a sophisticated and multifaceted approach to safety assessments. Navigating through the intricacies of this evolving landscape, this chapter sheds light on the significance of employing diverse hair follicle models in assessing the safety of hair care products. From clinical studies involving human subjects to preclinical models and ex-vivo assays bridging the gap between in vitro and in vivo evaluations, each approach contributes unique insights into the safety profile of these products (Kaushik et al. 2023). In the contemporary cosmetic landscape, consumers are not merely seeking aesthetic enhancements but are increasingly concerned about the safety, sustainability, and ethical aspects of the products they use (Boz et al. 2020). Hair care products, ranging from shampoos and conditioners to styling aids, play a pivotal role in daily grooming routines (D'Souza and Rathi 2015).

Consequently, the safety of these products becomes paramount, prompting an in-depth exploration of assessment methodologies that go beyond traditional approaches. In response to the growing demand for safe and effective hair care solutions, regulatory bodies worldwide have refined their oversight mechanisms. In the pursuit of diverse and effective hair care solutions, there is an increasingly urgent need to ascertain the safety and compliance of these products with stringent global regulatory standards (Barthe et al. 2021). The regulatory framework governing cosmetics has witnessed a notable paradigm shift in recent years, with regulatory bodies such as the U.S. Food and Drug Administration (FDA), the European Medicines Agency (EMA), and other global counterparts actively engaging in updates and reinforcement of guidelines. This evolution underscores an enhanced focus on ensuring the safety, efficacy, and transparency of cosmetic products, including those within the hair care domain.

The Cosmetic Regulation in the European Union, for instance, has undergone substantial updates, including the prohibition of certain ingredients and the establishment of more stringent safety evaluation requirements. Simultaneously, the FDA continues to enhance its regulatory framework to ensure that cosmetic products, including those for hair care, meet the highest standards of safety and quality. Recognizing that the efficacy and safety of hair care products extend beyond conventional subjective assessments, this chapter explores clinical studies, preclinical models, and ex-vivo hair follicle models. Each contributes to a holistic understanding of product safety, with clinical trials offering real-world insights, preclinical models providing controlled environments for initial safety assessments, and ex-vivo models bridging the gap between the laboratory setting and the complex interactions within living organisms.

In light of these considerations, this chapter also aims to provide an up-to-date analysis of the methodologies employed in the assessment of hair care product safety. Delving into current reference standards, recent regulatory updates, and emerging trends, the goal is to offer a comprehensive resource that not only meets the demands of the present but also anticipates the evolving landscape of global regulations. As we navigate through the intricate web of safety assessments, the ultimate objective is to empower stakeholders with knowledge that transcends conventional boundaries, fostering a safer and more transparent future for the hair care industry (EC 2009; European Medicines Agency 2022; FDA 2022).

9.2 Adherence to Safety Standards

Manufacturers must ensure that their products meet the safety standards set by regulatory authorities in different markets. Compliance includes adherence to ingredient restrictions, labeling requirements, and overall safety protocols.

Understanding and complying with international standards and regulations (e.g., EU, US, the Association of Southeast Asian Nations; ASEAN) (Morel et al. 2023; ASEAN 2023) for cosmetic and personal care products are crucial for manufacturers who operate in a globalized market as summarized in Table 9.1. This may involve harmonizing formulations and testing methodologies to meet the diverse requirements of various regions. Globally while no single global framework exists,

Table 9.1 Regulatory framework for cosmetics: overview of key bodies and guidelines

Regulatory body	Regulation/guideline	Description	References
U.S. Food and Drug Administration (FDA)	Cosmetic Labeling Regulations	Regulations outlining requirements for labeling of cosmetic products, including ingredient listing and warning statements	FDA (2022)
European Medicines Agency (EMA)	Cosmetic Regulation	Legislation governing the safety and efficacy of cosmetic products marketed within the European Union (EU)	European Medicines Agency (2022)
Health Canada	Cosmetic Ingredient Hotlist	List of prohibited and restricted ingredients for use in cosmetic products sold in Canada	Health Canada (2022)
Therapeutic Goods Administration (TGA)	Standard for the Uniform Scheduling of Medicines and Poisons	Classification system for cosmetic ingredients based on potential health risks	Therapeutic Goods Administration (2022)
International Organization for Standardization (ISO)	ISO 22716: Cosmetics—Good Manufacturing Practices (GMP)	Guidelines for the production, control, storage, and shipment of cosmetic products	ISO (2022)
Cosmetic Ingredient Review (CIR)	Ingredient Safety Assessments	Independent expert panel that assesses the safety of cosmetic ingredients based on scientific data	CIR (2022)
Occupational Safety and Health Administration (OSHA)	Hazard Communication Standard (HCS)	Standardized approach to communicating the hazards of chemicals in the workplace, including those in cosmetics	OSHA (2022)
Environmental Protection Agency (EPA)	Federal Insecticide, Fungicide, and Rodenticide Act (FIFRA)	Regulatory framework for the registration, distribution, sale, and use of pesticides, including those used in hair care products	EPA (2022)
China	Cosmetic Supervision and Administration Regulation (CSAR)	Regulation governing the supervision and administration of cosmetics in China, effective from 2021	CSAR (2021)
Personal Care Products Council (PCPC)	Industry Guidelines and Testing Protocols	US-based industry association offering guidelines and testing protocols for safety and efficacy of personal care products	PCPC (2022)

the International Harmonization Cooperation (IHC) aims to align regulations across regions. Regional, major examples include, the European Union (EU), Regulation (EC) No 1223/2009 sets strict safety and labeling requirements for cosmetics sold within the EU. The United States (US), the Federal Food, Drug, and Cosmetic Act (FD&C Act), and Fair Packaging and Labeling Act (FPLA) are enforced by the Food and Drug Administration (FDA). China, the Cosmetic Supervision and Administration Regulation (CSAR) came into effect in 2021, outlining comprehensive registration, testing, and labeling requirements. In other regions, each country often has its own specific regulations, making navigating the global landscape complex. Industry Standards for Cosmetics, The International Organization for Standardization (ISO) offers voluntary standards for various aspects of the cosmetics industry, like quality management, good manufacturing practices, and ingredient definitions (e.g., ISO 16128 for natural and organic cosmetics). Personal Care Products Council (PCPC): US-based industry association offering guidelines and testing protocols for safety and efficacy. Other regional and national organizations: May have their own standards specific to their jurisdiction. The table below provides an overview of key regulations and guidelines from prominent regulatory bodies such as the U.S. Food and Drug Administration (FDA), European Medicines Agency (EMA), Health Canada, Therapeutic Goods Administration (TGA), International Organization for Standardization (ISO), and the Cosmetic Ingredient Review (CIR). By adhering to these regulations and guidelines, manufacturers can ensure the safety, efficacy, and compliance of hair care products, ultimately safeguarding consumer health and well-being.

With regulatory standards serving as the bedrock of safety assessments, it's imperative to delve into the methodologies employed overall. Regulatory bodies worldwide have established stringent guidelines to safeguard consumer health, prompting the need for comprehensive safety assessments.

9.3 Methodologies Employed in Evaluating the Safety of Hair Care Products

This chapter explores the multifaceted methodologies employed in evaluating the safety of hair care products, ranging from clinical testing to preclinical studies and ex-vivo hair follicle models.

9.4 Clinical Testing

Clinical trials represent the pinnacle of safety and efficacy assessment in the realm of medical interventions, including pharmaceuticals and other healthcare products. These trials, conducted on human subjects, meticulously evaluate the performance of interventions, such as new drugs or treatments, in controlled settings. However, they present formidable challenges due to their resource-intensive nature and ethical complexities. Essentially, clinical trials demand significant financial investment, encompassing expenses ranging from participant recruitment to regulatory compliance. This financial burden can impede the pace of research and delay the availability of potentially life-saving treatments. Moreover, the process, from designing the study to analyzing the final results, is time-consuming, often spanning several years.

Ethical considerations loom large in clinical trials, necessitating rigorous protocols to safeguard the well-being of participants. Obtaining informed consent is paramount, ensuring that individuals fully comprehend the risks and benefits before enrollment. Regulatory bodies closely oversee these trials to uphold ethical standards and maintain the integrity of the research. Clinical trials are widely recognized as the gold standard for evaluating the safety and efficacy of medical interventions, including pharmaceuticals and other healthcare products. These trials are conducted on human subjects under strict regulations and guidelines to ensure ethical standards and scientific rigor.

References and guidelines play a crucial role in shaping the conduct of clinical trials. One prominent reference is the Declaration of Helsinki, a set of ethical principles for medical research involving human subjects, developed by the World Medical Association (WMA 2014), the Declaration emphasizes the importance of voluntary informed consent, protection of participants' privacy and confidentiality, and minimizing harm and maximizing benefits. Regulatory norms also play a significant role in overseeing clinical trials. In the United States, the Food and Drug Administration (FDA) regulates clinical trials through its Good Clinical Practice (GCP) guidelines. GCP guidelines outline standards for designing, conducting, recording, and reporting clinical trials, ensuring the integrity and credibility of the data collected (Arango et al. 2016; Vijayananthan and Nawawi 2008). Additionally, the International Council for Harmonization of Technical Requirements for Pharmaceuticals for Human Use (ICH) develops guidelines that aim to harmonize regulatory requirements for pharmaceuticals across different regions. The ICH guidelines, such as ICH E6(R2) for Good Clinical Practice, provide internationally recognized standards for the conduct of clinical trials (ICH 2016).

Ethical review boards, also known as institutional review boards (IRBs) or ethics committees, play a critical role in evaluating the ethical aspects of clinical trials. These boards review research protocols to ensure that participant rights, safety, and well-being are protected (Grady 2015; Kim 2012). Therefore, adherence to references, guidelines, and regulatory norms is essential for conducting ethical and scientifically rigorous clinical trials. By upholding these standards, researchers can generate reliable data to inform medical decision-making and improve patient care.

9.5 Preclinical Testing

Preclinical testing in the context of hair care products refers to the phase of research that occurs before human/clinical trials. This phase involves laboratory and animal studies to assess the safety, efficacy, and mechanism of action of the product. Here's how preclinical testing typically unfolds in the development of hair care products. In vitro studies involve testing the hair care product's effects on isolated cells or tissues in a controlled laboratory environment. Researchers may examine the product's impact on cell viability, proliferation, differentiation, and other relevant factors. For example, the product interaction with hair follicle cells or scalp tissue cultures may be assessed. Various preclinical testing methods, including in vitro assays, animal studies, and computational modeling, enable researchers to evaluate the toxicity and efficacy of hair care ingredients. Preclinical testing also includes evaluation of the safety profile of hair care products and determines whether the product causes irritation, sensitization, or other adverse reactions. In vitro models may be used to reconstruct human epidermis (RHE) or 3D skin models, to simulate the skin's response to the product (Hofmann et al. 2023; Moon et al. 2021; Schmidt et al. 2020). Understanding of mechanism of action and how the hair care product works at a molecular level is crucial for optimizing its efficacy and preclinical studies help to understand the mechanism of action, including its effects on gene expression, protein synthesis, and signaling pathways involved in hair growth, hydration, or repair. Whereas animal studies serve as a bridge between in vitro experiments and human trials. Animal models, such as mice or rats, are being used to assess the product's effects on hair growth, follicle morphology, and overall hair health. These studies help to identify potential efficacy and safety concerns before testing the product in humans. Preclinical testing may also involve optimizing the formulation of the hair care product to enhance its stability, shelf life, and delivery to target tissues. Researchers may experiment with different ingredients, concentrations, and delivery systems to achieve the desired effects while minimizing potential side effects. Throughout the preclinical testing phase, adherence to regulatory guidelines and standards for cosmetic products is very important. These guidelines may vary depending on the region, but they generally require comprehensive safety and efficacy data to support the product's eventual marketing approval. Overall, preclinical testing plays a crucial role in the development of hair care products by providing valuable insights into their safety, efficacy, and mechanism of action. By conducting rigorous preclinical studies, researchers can identify promising candidates for further evaluation in human clinical trials and ultimately bring safe and effective products to market. Encompasses laboratory testing conducted on animals or in vitro models to evaluate the initial safety and efficacy of the products before human trials. This step is crucial in identifying potential risks and refining formulations.

9.6 Ex Vivo/In Vitro Hair Follicle Models

A diverse array of in-vitro/ex-vivo methods is currently accessible and regarded as effective for accurately interpreting safety data related to ingredients utilized in finalized cosmetic and personal care items. The ex-vivo hair follicle models utilize human or animal hair follicles outside the body to study the effects of hair care products. This helps to understand how products interact with hair at a cellular level and also helps to study hair biology and test products without relying on live animals. This addresses ethical concerns associated with animal testing and aligns with the growing emphasis on ethical treatment of animals in research. The use of

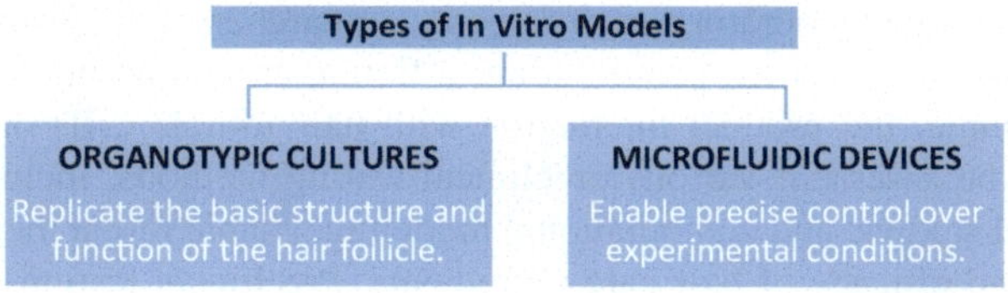

Fig. 9.1 Types of in vitro hair follicle model

ex-vivo models provides a humane alternative by eliminating the need to sacrifice animals for experimental purposes. This resonates with ethical considerations and contributes to the development of more responsible research practices. In-vitro models involve using human or animal hair follicle cells in 2D or 3D form, providing a more relevant and representative model of human hair biology (Fig. 9.1). This increases the likelihood of translating research findings to human applications accurately. In-vitro models offer controlled and reproducible experimental conditions, allowing to manipulate variables and study specific aspects of hair biology or product effects. This control enhances the reliability and consistency of experimental results. In vitro and ex-vivo models allow the study of hair growth, hair follicle cycling, and responses to various treatments. Researchers can investigate the impact of genetic factors, drugs, and environmental influences on hair follicle health without the complexity of in vivo systems. Hair care product development and drug testing can be efficiently carried out using ex-vivo hair follicle models. This approach enables the assessment of the safety and efficacy of products in a controlled setting before progressing to more complex and resource-intensive clinical trials. Ex-vivo models generally require fewer resources compared to in vivo studies.

9.7 In Vitro Models

9.7.1 Cell Monoculture Test Systems

In-vitro models often begin with simple cell monoculture test systems, where single types of cells are grown in a laboratory setting. These cells can be derived from various sources, such as human skin or hair follicles, and provide a basic understanding of how hair care products interact with individual cell types.

9.7.2 Three-Dimensional (3D) Reconstructed Tissue Models

As research progresses, more complex in vitro models like 3D reconstructed tissue models are employed. These models mimic the structure and function of human skin or hair follicles more accurately, providing a closer representation of real-world interactions. They offer a more comprehensive assessment of product penetration, absorption, and efficacy.

9.8 Ex Vivo Models

Explants: Ex vivo models involve the use of human or animal tissue samples obtained from donors. Hair follicle explants, for example, can be cultured in a laboratory environment to study hair growth, follicle health, and responses to various treatments. These models retain some aspects of the tissue's natural microenvironment, allowing researchers to observe how hair care products affect living tissue.

Organotypic Tissue Models: Similar to 3D reconstructed tissue models, organotypic models involve the cultivation of tissue in a way that mimics its natural architecture and function. These models may include multiple cell types arranged in a specific pattern to recreate the complexity of the skin or hair follicle environment. They offer valuable insights into product performance and safety.

Model type	Description
In-vitro models	
Cell monoculture test systems	Simple systems where single types of cells are grown in a laboratory setting
Three-dimensional (3D) reconstructed tissue models	Complex models that mimic the structure and function of human skin or hair follicles
Ex-vivo models	
Explants	Tissue samples were obtained from donors and cultured in a laboratory environment
Organotypic tissue models	Cultivated tissue that mimics natural architecture and function, including multiple cell types

This includes reduced time, cost, and personnel, making them a more practical option for certain types of research and product testing. Ex-vivo models can be developed using hair follicles derived from patients, allowing for the study of individual responses to treatments. This contributes to the emerging field of personalized medicine, where interventions are tailored to individual characteristics. Ex-vivo hair follicle models can be adapted for high-throughput screening, allowing to efficiently test multiple compounds or products simultaneously. This scalability enhances the speed and efficiency of experimental processes.

After discussing the intricacies of clinical trials, preclinical testing, and in vitro, ex-vivo hair follicle models individually, Table 9.2 provides a comprehensive comparison of these approaches. This comparative analysis highlights key aspects such as their purpose, participants, challenges, ethical considerations, regulatory oversight, experimental control, resource efficiency, applications, and contributions to personalized medicine. Understanding these differences underscores the significance of employing diverse methodologies in assessing the safety and efficacy of hair care products.

Table 9.2 Comparison of clinical trials, preclinical testing, and ex-vivo hair follicle models

Aspect	Clinical trials	Preclinical testing	In-vitro/ex-vivo hair follicle models
Purpose	Evaluate safety and efficacy of medical interventions	Assess safety, efficacy, and mechanism of action of hair care products	Study effects of hair care products at a cellular level
Participants	Human subjects	Laboratory models and animal subjects	Hair follicles outside the body
Challenges	Resource-intensive, time-consuming, ethical considerations	Resource-intensive, adherence to regulatory guidelines	Address ethical concerns, relevance to human biology
Ethical considerations	Informed consent, participant privacy, minimizing harm, maximizing benefit	Adherence to regulatory guidelines and minimizing harm	Eliminate need for live animal testing, ethical treatment of animals
Regulatory oversight	FDA, ICH, declaration of Helsinki, institutional review boards	Regulatory guidelines for cosmetics, region-specific requirements	Regulatory guidelines, ethical considerations
Experimental control	Controlled settings, rigorous protocols	Manipulation of variables, optimization of formulations	Controlled and reproducible conditions
Resource efficiency	Financial investment, time-consuming	Requires fewer resources compared to clinical trials	Fewer resources compared to in vivo studies
Applications	Evaluate medical interventions, inform patient care	Identify promising candidates for human trials, refine formulations	Study hair growth, responses to treatments, assess product safety
Personalized medicine	N/A	N/A	Study individual responses to treatments, contribute to precision medicine
High-throughput screening	N/A	N/A	Adapted for efficient testing of multiple compounds or products

9.9 Challenges and Future Directions

9.9.1 Applications in Safety and Compliance Testing

Ensuring the safety and compliance of hair care products is essential to protect consumer health and maintain regulatory compliance. Regulatory agencies provide guidelines for skin irritation testing to ensure the safety of consumer products (Fig. 9.2). Examples include the OECD (Organization for Economic Co-operation and Development) guidelines. Microscopic examination of skin tissue can provide insights into cellular changes associated with irritation. Regulatory agencies/systems such as the Globally Harmonized System (GHS) include criteria for

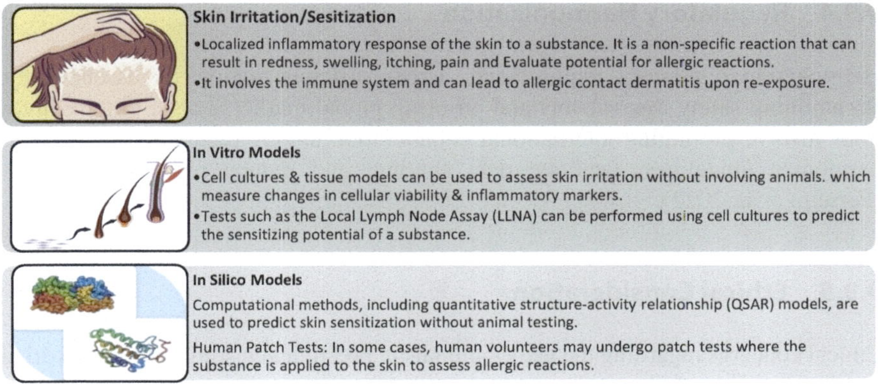

Fig. 9.2 Methods of pre-clinical and clinical hair assessment

classifying substances based on their sensitizing potential. Ex-vivo hair follicle models offer a promising alternative for assessing hair care product safety and compliance with global regulations. By overcoming challenges and fostering collaboration, these models can revolutionize the hair care industry, ensuring safe and effective products for consumers worldwide.

9.9.2 Advancements in Testing Methodologies

As the cosmetics industry continues to evolve, there is a growing need for more sophisticated testing methodologies to accurately assess the safety and efficacy of cosmetic products. Advanced techniques such as high-throughput screening, organ-on-a-chip models, and computational modeling are emerging as promising alternatives to traditional preclinical testing methods (Bhat et al. 2023; Pant et al. 2023; Wu et al. 2020).

9.9.3 Integration of Alternative Methods

Both preclinical and ex-vivo testing rely on alternative methods to traditional animal models. Future directions involve harnessing emerging technologies such as organoids, microfluidics, and computational modeling to enhance productivity and reduce reliance on animal testing (Balls et al. 2020; Bhat et al. 2023; Pandey and Poojan 2023).

9.9.4 Regulatory Harmonization

Harmonizing regulatory requirements across different regions is essential for streamlining safety assessments and ensuring global market access. Future directions involve promoting international collaboration among regulatory agencies to harmonize guidelines, facilitate data sharing, and reduce regulatory barriers (Ferreira et al. 2022).

9.9.5 Ethical Considerations

Ethical concerns regarding the use of animals in preclinical testing continue to drive the development of alternative methods and the adoption of principles such as the 3Rs (Replacement, Reduction, and Refinement). Future directions in preclinical testing should prioritize the refinement and validation of alternative methods to ensure ethical and humane treatment of animals (Kiani et al. 2022).

9.10 Integration of Omics Technologies

Omics technologies, including genomics, proteomics, and metabolomics, hold immense potential for enhancing preclinical testing by providing comprehensive insights into the molecular mechanisms underlying cosmetic product safety and efficacy. Integration of omics data with traditional preclinical testing approaches can lead to more predictive and personalized safety assessments (Dai and Shen 2022). Enhancing the predictive capabilities of safety assessments is crucial for identifying potential risks and ensuring consumer safety. Future efforts should focus on integrating multi-omics approaches, advanced imaging techniques, and computational modeling to improve predictivity across all testing methodologies (Buesen et al. 2017).

Overall, the assessment of hair care product safety and compliance with global regulations is a multifaceted endeavor that requires a comprehensive approach. From the stringent regulatory standards set forth by authorities such as the FDA, EMA, EPA, ISO, CIR, and others to the rigorous methodologies employed in clinical, preclinical, and ex-vivo testing, every step plays a crucial role in ensuring consumer safety and well-being. As the cosmetic industry continues to evolve and consumer preferences shift toward safer, more sustainable products, it becomes increasingly imperative for manufacturers to prioritize transparency, efficacy, and ethical considerations in their product development processes. By adhering to international standards, conducting thorough safety assessments, and embracing innovative testing methodologies, the hair care industry can foster trust and confidence among consumers while advancing toward a safer and more transparent future. Toward the quest of excellence, collaboration across stakeholders; from researchers and regulatory agencies to manufacturers and consumers remains essential. As a result, there is a strong desire to collaborate in order to progress the sector by sharing knowledge, ideas, and best practices while encouraging creativity and upholding the highest standards of safety and compliance. Ultimately, the goal is not just to create products that beautify and enhance but also to promote health, well-being, and sustainability. By embracing this holistic approach, one can ensure that every hair care product that graces the shelves meets not only the highest regulatory standards but also the expectations and trust of consumers worldwide.

References

Arango J, Chuck T, Ellenberg SS, Foltz B, Gorman C, Hinrichs H et al (2016) Good clinical practice training: identifying key elements and strategies for increasing training efficiency. Ther Innov Regul Sci 50(4):480–486

ASEAN (2023) Association of Southeast Asian Nations. Mutual Recognition Agreements (MRAs). https://asean.org/asean-sectoral-mras/. Accessed 4 Jan 2023

Barthe M, Bavoux C, Finot F, Mouche I, Cuceu-Petrenci C, Forreryd A et al (2021) Safety testing of cosmetic products: overview of established methods and new approach methodologies (NAMs). Cosmetics 8(2):50

Bhat D, Goyal S, Poojan S (2023) Human skin reconstructs model systems in mechanistic research, safety, and efficacy studies of cosmetics: pros and cons. In: Pant AB, Dwivedi A, Ray RS, Tripathi A, Upadhyay AK, Poojan S (eds) Skin 3-D models and cosmetics toxicity. Springer Nature Singapore, Singapore, pp 39–45

Boz Z, Korhonen V, Koelsch Sand C (2020) Consumer considerations for the implementation of sustainable packaging: a review. Sustain For 12(6):2192

Buesen R, Chorley BN, da Silva Lima B, Daston G, Deferme L, Ebbels T et al (2017) Applying 'omics technologies in chemicals risk assessment: report of an ECETOC workshop. Regul Toxicol Pharmacol 91:S3–S13. https://doi.org/10.1016/j.yrtph.2017.09.002

Cavers DF (1939) The food, drug, and cosmetic act of 1938: its legislative history and its substantive provisions. Law Contemp Probl 6(1):2–42. https://doi.org/10.2307/1189727

CDSCO (1940) Government of India, Drugs and Cosmetics Act. cdsco.gov.in

CIR (2022) Cosmetic ingredient review, ingredient safety assessments: independent expert panel that assesses the safety of cosmetic ingredients based on scientific data

CSAR (2021) Cosmetic supervision and administration regulation (CSAR), China

D'Souza P, Rathi SK (2015) Shampoo and conditioners: what a dermatologist should know? [Review]. Indian J Dermatol 60(3):248–254

Dai X, Shen L (2022) Advances and trends in omics technology development. [Review]. Front Med 9:911861. https://doi.org/10.3389/fmed.2022.911861

EC (2009) European Commission, Regulation (EC) No 1223/2009 of the European Parliament and of the Council of 30 November 2009 on cosmetic products. Off J Eur Union. https://eur-lex.europa.eu/legal-content/EN/ALL/?uri=celex%3A32009R1223

EPA (2022) Environmental protection agency, summary of the federal insecticide, fungicide, and rodenticide act

European Medicines Agency (2022). https://www.ema.europa.eu/en/homepage

FDA (2022) US Food and Drug Administration. https://www.fda.gov/cosmetics

Ferreira M, Matos A, Couras A, Marto J, Ribeiro H (2022) Overview of cosmetic regulatory frameworks around the world. Cosmetics 9(4):72

Gavazzoni Dias MF (2015) Hair cosmetics: an overview. [Review]. Int J Trichol 7(1):2–15

Grady C (2015) Institutional review boards: purpose and challenges. [Research Support, N I H, Intramural Review]. Chest 148(5):1148–1155

Health Canada (2022) Cosmetic ingredient hotlist: list of prohibited and restricted ingredients for use in cosmetic products sold in Canada

Hofmann E, Schwarz A, Fink J, Kamolz LP, Kotzbeck P (2023) Modelling the complexity of human skin in vitro. [Review]. Biomedicines 11(3):794
ICH (2016) International council for harmonisation, harmonised guideline integrated addendum to ICH E6(R1): guideline for good clinical practice E6(R2) Current Step 4 version
ISO (2022) International organization for standardization, 22716: cosmetics—good manufacturing practices (GMP): Guidelines for the production, control, storage, and shipment of cosmetic products
Kaushik M, Farooq U, Ali MS, Ansari MJ, Iqbal Z, Mirza MA (2023) Safety concern and regulatory status of chemicals used in cosmetics and personal care products. Dermato 3(2):131–157
Kiani AK, Pheby D, Henehan G, Brown R, Sieving P, Sykora P et al (2022) Ethical considerations regarding animal experimentation. [Review]. J Prev Med Hyg 63(2 Suppl 3):E255–E266
Kim WO (2012) Institutional review board (IRB) and ethical issues in clinical research. Korean J Anesthesiol 62(1):3–12
Moon S, Kim DH, Shin JU (2021) In vitro models mimicking immune response in the skin. [Review]. Yonsei Med J 62(11):969–980
Morel S, Sapino S, Peira E, Chirio D, Gallarate M (2023) Regulatory requirements for exporting cosmetic products to extra-EU countries. Cosmetics 10(2):62
OSHA (2022) Occupational safety and health administration. Hazard communication standard (HCS)
Pandey R, Poojan S (2023) Artificial skin models for animal-free testing: 3D skin reconstruct approach, a journey in the past two decades. In: Pant AB, Dwivedi A, Ray RS, Tripathi A, Upadhyay AK, Poojan S (eds) Skin 3-D models and cosmetics toxicity. Springer Nature Singapore, Singapore, pp 1–17
Pant AB, Dwivedi A, Ray RS, Tripathi A, Upadhyay AK, Poojan S (2023) Skin 3-D models and cosmetics toxicity. Springer Nature Singapore
PCPC (2022) Personal care products council, Industry Guidelines and Testing Protocols
Rocca R, Acerbi F, Fumagalli L, Taisch M (2022) Sustainability paradigm in the cosmetics industry: state of the art. Cleaner Waste Syst 3:100057. https://doi.org/10.1016/j.clwas.2022.100057
Schmidt FF, Nowakowski S, Kluger PJ (2020) Improvement of a three-layered in vitro skin model for topical application of irritating substances. [Original Research]. Front Bioeng Biotechnol 8:388. https://doi.org/10.3389/fbioe.2020.00388
Therapeutic Goods Administration (2022) Standard for the Uniform Scheduling of Medicines and Poisons: Classification system for cosmetic ingredients based on potential health risks
Vijayananthan A, Nawawi O (2008) The importance of good clinical practice guidelines and its role in clinical trials. Biomed Imaging Interv J 4(1):1
WMA (2014) World medical association declaration of Helsinki: ethical principles for medical research involving human subjects. [Practice Guideline]. J Am Coll Dent 81(3):14–18
Wu F, Zhou Y, Li L, Shen X, Chen G, Wang X et al (2020) Computational approaches in preclinical studies on drug discovery and development. [Review]. Front Chem 8. https://doi.org/10.3389/fchem.2020.00726

The Potential of Artificial Intelligence in Hair Product Toxicity Prediction: A Global Perspective

10

Neeraj Kumar Tiwari, Ajay Prasad, Prachi Kapoor, and Prabhat Ranjan

Abstract

This chapter explores the potential of artificial intelligence (AI) to revolutionize safety assessment in the cosmetics industry, with a specific focus on hair care products. By drawing parallels to its success in predicting the dangers of novel substances in medicine, the chapter argues for AI's ability to ensure more comprehensive and efficient safety standards. The focus on hair care products highlights the importance of safety not just for physical health but also for consumer confidence and the overall well-being of the cosmetics industry.

Keywords

Artificial intelligence (AI) · Safety assessment · Cosmetics industry · Hair care products · Consumer safety · Toxicity prediction · Regulatory compliance · Product development

10.1 Introduction

Artificial intelligence (AI) has been used to increase the accuracy and efficiency of approaches for predicting the possible danger of novel substances before they are evaluated in human clinical trials, which saves time and money (Tran et al. 2023).

N. K. Tiwari (✉)
Babasaheb Bhimrao Ambedkar University (Satellite Centre), Lucknow, Uttar Pradesh, India

A. Prasad · P. Kapoor
University of Petroleum and Energy Studies, Dehradun, Uttrakhand, India

P. Ranjan
Division of Cardiovascular Disease, Department of Medicine, The University of Alabama at Birmingham, Birmingham, AL, USA

A. Dwivedi et al. (eds.), *Hair Care Products*,
https://doi.org/10.1007/978-981-97-6740-3_10

This offers a more comprehensive and efficient approach to ensuring that the products we use daily, from cosmetics to medicines, meet rigorous safety standards. In this rapidly evolving field, the potential for AI to transform how we assess toxicity is boundless, promising a safer and more sustainable future. Safe hair products are vital for health, confidence, and the cosmetics industry. They protect against harm, boost self-esteem, and ensure consumer trust. Hair care involves prioritizing all the safety benefits.

10.1.1 Brief Overview of the Importance of Safe Hair Products

Safe hair products are of paramount importance in the daily lives of individuals across the globe. This significance stems from various factors that encompass health and well-being, as well as aesthetics and confidence.

First and foremost, safe hair products are essential to safeguarding the health of consumers. Hair, being a prominent part of one's body, is susceptible to the adverse effects of harmful chemicals and toxins found in some hair care products. These substances can lead to a range of health issues, including skin irritation, allergies, and even more severe conditions such as hair loss. Thus, ensuring the safety of hair products is integral to protecting individuals from potential harm.

Furthermore, safe hair products play a pivotal role in enhancing self-esteem and confidence. Hair is often regarded as a symbol of beauty and personal identity. Healthy, well-maintained hair contributes significantly to one's appearance and self-assurance. Conversely, the use of unsafe products that lead to hair damage or adverse reactions can have detrimental effects on an individual's self-esteem and overall well-being. In addition to personal well-being, safe hair products have broader societal implications. The beauty and cosmetics industry is a global economic powerhouse, and the demand for hair care products is ever-growing. Ensuring the safety of these products not only protects consumers but also contributes to the credibility and trustworthiness of the industry itself.

In conclusion, the importance of safe hair products is multifaceted. It encompasses the protection of individual health, the promotion of self-confidence, and the overall integrity of the beauty and cosmetics industry. Recognizing the significance of safe hair products is not only a matter of personal well-being but also a crucial aspect of consumer rights and industry responsibility. As such, the development and use of safe hair products should be a priority for both manufacturers and consumers alike.

10.1.2 Introduction to the Use of Artificial Intelligence (AI) in Toxicity Prediction

The introduction of artificial intelligence (AI) in toxicity prediction represents a groundbreaking paradigm shift in the field of toxicology and safety assessment. AI, with its machine learning and deep learning capabilities, has revolutionized the way we analyze and predict the potential harm posed by various substances, including those found in cosmetics, pharmaceuticals, and chemicals (Saikia et al. 2022).

Traditionally, toxicity assessment relied on time-consuming and costly animal testing and empirical studies. AI has ushered in a new era where predictive models are trained on vast datasets, allowing for more accurate, efficient, and humane methods of evaluating toxicity (Jeong and Choi 2022). This introduction of AI offers several key advantages:

1. **Precision:** AI models can analyze a wide array of chemical properties and biological data to predict toxicity with exceptional precision, identifying subtle patterns that might be missed by traditional methods.
2. **Speed:** AI can rapidly process enormous datasets, significantly reducing the time required for toxicity assessments and expediting product development.
3. **Cost-Efficiency:** By reducing the need for extensive animal testing and empirical studies, AI-driven toxicity prediction can save both time and resources.
4. **Safety:** Consumers benefit from the enhanced safety of products, as AI helps identify potential hazards more comprehensively, ensuring that only safe products reach the market.
5. **Environmental Impact:** AI can aid in assessing the environmental impact of chemicals, promoting sustainability and responsible manufacturing.

10.1.3 The Significance of a Global Perspective in Hair Product Safety

A global perspective in hair product safety is of paramount significance in our interconnected world. Hair products traverse international borders, and consumers worldwide rely on them for personal care. Understanding the global context ensures that safety standards are not only consistent but also adaptive to diverse needs and preferences. Today's medical practice does require knowledge of hair cosmetics, aesthetic techniques, hair shaft structure, and physical behavior (Bolduc and Shapiro 2001).

The dermatologist's awareness of hair care products, how to use them, and any potential side effects can extend to their knowledge of cosmetic resources and aid them in treating hair and scalp disorders more effectively in light of the variety of hair types and ethnicities (Gavazzoni Dias 2015).

First, globalization has led to the exchange of beauty and grooming products across continents. A product deemed safe in one region may have different effects on individuals elsewhere due to varying genetics, environments, and usage patterns. A global perspective allows for a more comprehensive assessment of these products' safety on a diverse population.Second, it encourages collaboration among nations to harmonize safety regulations and standards. Shared research and data enable quicker identification of potentially harmful ingredients, leading to more effective consumer protection.

In conclusion, a global perspective in hair product safety transcends geographical boundaries and fosters safer products, international cooperation, and consumer well-being. It's a crucial component of modern safety assessments in an increasingly interconnected world.

10.2 Understanding Hair Product Toxicity

Understanding hair product toxicity is essential as these products have a direct impact on personal health. Hair care items can contain various chemicals that, when used incorrectly or excessively, may lead to skin irritations, allergies, or even long-term health issues. Toxicity knowledge involves recognizing harmful ingredients, assessing potential side effects, and understanding cumulative exposure risks. It's vital for consumers to make informed choices, manufacturers to prioritize safety, and regulators to establish stringent standards to ensure that hair products do not compromise health and well-being. In essence, comprehending hair product toxicity safeguards both individual users and public health.

10.2.1 Types of Hair Products and Common Ingredients

Hair products come in a diverse array, each designed to cater to specific hair care needs and styling preferences. Common types include shampoos, conditioners, hair serums, hair sprays, gels, and hair dyes (Shapiro and Maddin 1996). These products often share common ingredients, including water, surfactants (cleansing agents), emollients (softeners), proteins, and preservatives. For example, shampoos contain cleansing agents like sodium lauryl sulfate, while conditioners feature emollients like silicone to provide smoothness and shine. Styling products may contain polymers for hold, and hair dyes use pigments and oxidizing agents. Understanding these ingredients helps consumers select products tailored to their hair type and needs, ensuring effective and safe hair care (La Torre and Bhushan 2006).

10.2.2 Risks Associated with Toxic Hair Products

Risks associated with toxic hair products are a significant concern for consumer safety. Toxic hair products may contain harmful ingredients such as parabens, formaldehyde, phthalates, and sulfates, which can lead to various adverse effects (He et al. 2022). These risks include skin and scalp irritations, allergic reactions, hair damage, and, in severe cases, even long-term health problems like hormone disruption and cancer. Additionally, overuse or misuse of certain hair products can exacerbate these risks. The lack of standardized regulations in the beauty industry poses challenges in identifying and avoiding toxic ingredients. Hence, consumers must educate themselves and prioritize safe, nontoxic hair care products to protect their health and well-being.

10.2.3 Regulatory Challenges in Ensuring Hair Product Safety

Regulatory challenges in ensuring hair product safety persist due to several factors. These challenges include the rapid introduction of new ingredients, limited oversight, and varying international standards. Cosmetic regulations often lag behind scientific advancements, making it difficult to assess and regulate emerging substances. Furthermore, the burden of proving safety primarily falls on manufacturers, and inadequate transparency can hinder oversight. International harmonization of safety standards remains an ongoing challenge, making it essential for consumers to stay informed and for regulatory bodies to adapt swiftly to emerging hair product safety concerns.

10.3 The Role of AI in Toxicity Prediction

The role of artificial intelligence (AI) in toxicity prediction is transformative. AI leverages machine learning and deep learning algorithms to analyze vast datasets of chemical and biological information. It excels in recognizing complex patterns and relationships, enabling accurate toxicity assessments. AI-driven models can swiftly evaluate the safety of substances, replacing time-consuming and costly traditional methods. This technology enhances efficiency, reduces animal testing, and contributes to more precise safety evaluations, ultimately leading to the development of safer products for consumers across various industries, including cosmetics and pharmaceuticals. Preclinical safety investigations involve a variety of animal species, including rats, rabbits, monkeys, and horses, which require a huge number of animals and take a lot of time. Consequently, employing artificial intelligence (AI) to foresee medication toxicity throughout the early stages of research could be a possible technique to solve the aforementioned problem (Malani et al. 2023).

10.3.1 Overview of AI and Its Capabilities

Artificial intelligence (AI) is a cutting-edge field of computer science that endows machines with the ability to mimic human intelligence and perform tasks that typically require human cognition. Its capabilities encompass a wide spectrum, including machine learning, natural language processing, computer vision, and more. AI systems excel at data analysis, recognizing patterns, and making predictions or decisions based on data. These capabilities enable AI to automate complex tasks, optimize processes, and enhance decision-making across various domains, from healthcare to finance and beyond. AI's potential is boundless, driving innovation, efficiency, and transformative advancements in nearly every facet of modern life (Mikalef and Gupta 2021).

10.3.2 Historical Context of AI in Toxicology

The historical context of AI in toxicology dates back to the late twentieth century. Early applications primarily focused on expert systems, which used rule-based approaches to predict toxicity based on known chemical properties. As computing power advanced, machine learning techniques gained prominence. In recent decades, AI has revolutionized toxicology by analyzing extensive datasets and identifying intricate chemical interactions (Mintz and Brodie 2019). This shift has accelerated toxicity prediction, replacing time-consuming animal testing with rapid, accurate computational models. Today, AI-driven tools are integral in assessing the safety of chemicals, drugs, and consumer products, marking a significant transition in toxicological science toward more efficient and humane methods.

10.3.3 Advantages of Using AI for Hair Product Toxicity Prediction

Using AI for hair product toxicity prediction offers numerous advantages. First, AI enables the rapid assessment of a wide range of ingredients, saving time and resources compared to traditional testing methods (Vatiwutipong et al. 2023). It enhances the precision of toxicity predictions by analyzing vast datasets, recognizing subtle patterns, and identifying potential risks that may go unnoticed through manual assessments.

Second, AI-driven models are adaptable and can continuously improve through machine learning (Tetko et al. 2022). They can incorporate new data and research findings, staying up-to-date with the evolving landscape of hair product ingredients and safety standards. Furthermore, AI reduces reliance on animal testing, aligning with ethical considerations and promoting more humane approaches to safety assessment. Additionally, it enhances transparency by providing clear and data-driven insights into toxicity assessments. AI empowers the beauty and cosmetics industry to develop safer products, protects consumers from potential harm, and fosters responsible innovation in the field of hair product safety.

10.3.4 Challenges and Limitations of AI in Hair Product Toxicity Prediction

Despite its potential, AI in the context of hair product toxicity prediction faces several challenges and limitations. One significant challenge is the availability of high-quality, comprehensive data, as accurate models rely on extensive and diverse datasets. Interpretability remains an issue, as complex AI models can be challenging to understand fully. Ethical concerns regarding data privacy and the potential misuse of AI in this field must be addressed. Moreover, AI models may not capture rare or novel toxicity patterns, highlighting the importance of continued validation and human expertise. Despite these challenges, AI holds immense promise, and ongoing research aims to address these limitations for more robust and effective toxicity prediction (Bhhatarai et al. 2019).

10.4 Data Collection and Management

Data collection and management are pivotal aspects of harnessing AI for hair toxicity prediction. AI systems require vast and diverse datasets encompassing chemical properties, biological interactions, and toxicity outcomes. Data quality is paramount; therefore, rigorous data preprocessing, cleansing, and standardization are essential. Advanced data management tools and databases help organize and maintain this wealth of information.

Furthermore, AI continuously learns from updated data, necessitating effective strategies for data storage and retrieval. Collaboration and data sharing among research and regulatory bodies are crucial for building comprehensive databases, which, in turn, enhance the accuracy and reliability of AI models. Ultimately, meticulous data management ensures AI's effectiveness in predicting hair product toxicity accurately and efficiently.

10.4.1 Importance of Data in AI-Driven Toxicity Prediction

Data is the lifeblood of AI-driven toxicity prediction. It forms the foundation for machine learning models to recognize patterns, relationships, and potential hazards in chemicals. Comprehensive and diverse datasets enable AI to identify subtle toxicity trends that may elude traditional methods. The quality of data, including accuracy and relevance, directly impacts prediction accuracy. Moreover, ongoing data updates ensure AI remains effective in a continually evolving chemical landscape. Without robust and well-managed data, AI's potential to revolutionize toxicity prediction and enhance product safety in various industries, including cosmetics and pharmaceuticals, would be severely compromised.

10.4.2 Sources of Toxicity Data for Hair Products

Toxicity data for hair products can be sourced from various channels (Mayr et al. 2016). These include:

1. **Laboratory Studies:** Controlled experiments and lab tests on specific hair product ingredients.
2. **Clinical Trials:** Data from trials assessing the safety of hair products on human volunteers.
3. **Cosmetic Industry:** Internal safety assessments and data collected by manufacturers.

4. **Regulatory Agencies:** Publicly available reports and assessments from health and safety agencies.
5. **Research Institutions:** Studies conducted by independent research institutions and universities.
6. **Consumer Complaints:** Reports and feedback from consumers who experienced adverse reactions.
7. **Chemical Databases:** Comprehensive databases containing toxicity information on various chemicals used in hair products.
8. **Literature Review:** Scientific publications and research papers on hair product safety and toxicity.

10.4.3 Data Quality and Preprocessing

In the context of hair toxicity prediction, data quality and preprocessing are critical. Ensuring high-quality data is essential because AI models rely on accurate information to make reliable predictions. This involves cleaning data to remove errors, inconsistencies, or outliers that could skew results. Standardization is vital to ensure that data from various sources are compatible and can be effectively analyzed. Moreover, missing data must be handled appropriately to avoid bias. Proper preprocessing, including feature selection and extraction, enhances the model's ability to discern relevant patterns. In essence, data quality and preprocessing are fundamental steps in harnessing AI for accurate and actionable hair toxicity prediction.

10.5 AI Models and Algorithms

AI Models and Algorithms are the cornerstone of accurate hair product toxicity prediction. Understanding machine learning and deep learning principles is fundamental. Exploring specific AI models, their strengths, and their application in toxicity assessment provides a robust foundation. Customizing these models to the unique challenges of the cosmetics industry ensures precise, industry-specific predictions. In essence, AI Models and Algorithms are the engine that drives effective and tailored toxicity assessments for hair products.

10.5.1 Introduction to Machine Learning and Deep Learning

Introduction to machine learning and deep learning is fundamental in harnessing AI for hair product toxicity prediction. Machine learning is the field of AI where algorithms learn patterns and make predictions from data. Deep learning, a subset of machine learning, involves artificial neural networks inspired by the human brain. Deep learning excels in handling complex, unstructured data, making it particularly valuable for deciphering intricate chemical interactions and predicting toxicity accurately. Understanding these concepts lays the groundwork for leveraging AI effectively in ensuring hair product safety (Mintz and Brodie 2019).

10.5.2 Specific AI Models Used for Toxicity Prediction

Several specific AI models are employed in toxicity prediction for hair products. These include Random Forests, Support Vector Machines (SVM), Gradient Boosting Machines, and Neural Networks. Random Forests excel in handling diverse datasets, SVMs are adept at classifying complex chemical structures, Gradient Boosting Machines refine predictions iteratively, and Neural Networks, particularly Convolutional Neural Networks (CNNs), are powerful for image-based assessments. These models provide diverse tools for accurate and comprehensive hair product toxicity predictions (LeCun et al. 2015).

10.5.3 Customizing AI Models for Hair Product Toxicity

Customizing AI models for hair product toxicity is a crucial step in ensuring accurate and industry-specific predictions. This process involves tailoring existing models to accommodate the unique characteristics of cosmetic ingredients, chemical interactions, and safety standards. Customization may include feature engineering, data augmentation, or the creation of domain-specific datasets. By fine-tuning AI models to address the specific challenges of the cosmetics industry, such as assessing the effects of various formulations on human hair and skin, these models can provide precise and relevant toxicity assessments.

10.6 Global Data Integration

Global data integration is pivotal in the context of hair product safety. It involves harmonizing data from various regions to create a comprehensive and diverse dataset (Guo et al. 2021). This approach ensures that toxicity predictions consider the impact on a global population, taking into account genetic and environmental variations. By synthesizing data from multiple sources and regions, it fosters a more accurate and holistic understanding of safety implications. Global data integration not only enhances product safety assessments but also promotes international collaboration and a shared commitment to consumer well-being.

10.6.1 The Need for a Global Perspective in Data Collection

A global perspective in data collection is imperative in today's interconnected world. It acknowledges that products, like hair care items, reach consumers across borders. Regional variations in genetics, climates, and lifestyle factors can influence how individuals react to these products. Collecting data from diverse geographic areas ensures that safety assessments are inclusive and reflective of a global consumer base. This approach fosters a more comprehensive understanding of potential risks and benefits. Moreover, it promotes responsible product development and regulatory decisions that prioritize the well-being of consumers worldwide.

10.6.2 Challenges in Integrating Data from Various Regions

Integrating data from various regions presents significant challenges in ensuring a comprehensive perspective on hair product safety. First, variations in data quality and reporting standards across regions can hinder seamless integration. Additionally, regulatory disparities and language barriers can complicate the harmonization of information. Cultural differences may also influence data collection practices and reporting bias. Moreover, privacy concerns and legal constraints may impede the sharing of data across borders (Martínez-García and Hernández-Lemus 2022). Overcoming these challenges requires collaborative efforts, standardized protocols, and a commitment to transparency to create a truly global perspective on hair product safety.

10.6.3 Benefits of a Diverse Dataset

A diverse dataset holds several key benefits in the context of hair product safety assessment. First, it captures a wide range of genetic and environmental factors, reflecting real-world conditions, which leads to more accurate and robust toxicity predictions. Second, it helps identify rare or unexpected patterns and interactions that may be missed in homogeneous datasets (Helgheim et al. 2019). Third, a diverse dataset enhances model generalization, enabling reliable predictions across different populations. Additionally, it promotes inclusivity and fairness by ensuring that safety assessments cater to a broad spectrum of users, ultimately leading to safer, more effective, and globally relevant hair care products.

10.7 Case Studies

10.7.1 Highlighting Successful AI Applications in Hair Product Toxicity Prediction

Artificial intelligence (AI) has emerged as a game-changer in the realm of hair product toxicity prediction, revolutionizing the way we evaluate product safety. Let's delve into some remarkable case studies that underscore the success of AI applications in this critical field.

10.7.1.1 Case Study 1: Deep Learning for Skin Sensitization

One compelling case study involves the use of deep learning techniques to predict skin sensitization potential in hair care products. Historically, identifying skin irritants required extensive animal testing. However, a pioneering AI-driven model, developed by a research team in collaboration with a cosmetic company, has proven its mettle in predicting skin sensitization accurately. By analyzing a vast dataset of chemical properties and known skin sensitization outcomes, the model achieved remarkable accuracy (Wilm et al. 2021). This breakthrough has not only accelerated product safety assessments but also led to a substantial reduction in animal testing, aligning with ethical concerns.

10.7.1.2 Case Study 2: Global Data Integration for Ingredient Safety

In another case study, AI was employed to integrate global data on cosmetic ingredients, providing a comprehensive safety assessment. Hair product manufacturers often use different formulations for various regions, which can pose challenges in assessing ingredient safety on a global scale. AI-driven data integration systems helped harmonize data from diverse regions. By considering genetic diversity, environmental factors, and local preferences, these models offer a more holistic perspective on ingredient safety. This not only streamlines regulatory compliance but also enhances consumer safety by accounting for regional variations in response to hair products.

10.7.2 Real-World Examples of AI-Driven Global Toxicity Assessments

Example 1: Assessing the Impact of Hair Dyes

Hair dyes have been a focal point in hair product safety due to potential chemical exposures (Patel et al. 2013). AI-driven global toxicity assessments, conducted by regulatory agencies and independent researchers, have revealed critical insights. These assessments analyze large-scale databases containing information on dye ingredients, consumer usage patterns, and reported adverse reactions. The AI models consider genetic variations among different populations, helping to identify individuals at higher risk of allergies or sensitivities to specific dye components. This information informs labeling requirements and encourages the development of safer dye formulations.

Example 2: Evaluating the Safety of Keratin Treatments

Keratin treatments, popular for hair straightening and smoothing, have faced scrutiny for their chemical content (Burnett et al. 2021). AI-driven assessments have examined the safety of various formulations by analyzing ingredient toxicity data and adverse event reports from consumers worldwide. These assessments revealed that certain keratin treatments contained formaldehyde at levels exceeding safety thresholds. This information prompted regulatory action to limit formaldehyde content and improve consumer awareness of potential risks. The global nature of these assessments ensures that consumers worldwide benefit from safer keratin treatments.

10.7.3 Impact on the Hair Product Industry and Consumer Safety

The impact of AI-driven global toxicity assessments on the hair product industry and consumer safety cannot be overstated.

10.7.3.1 Advancing Product Safety

AI applications have catalyzed advancements in product safety by offering rapid and precise toxicity predictions. Manufacturers now have access to sophisticated tools that help them identify potentially harmful ingredients during product development. This leads to the reformulation of products to meet stringent safety standards, reducing the risk of adverse effects for consumers.

10.7.3.2 Streamlining Regulatory Compliance

Regulatory bodies benefit from AI-driven assessments as well. They can leverage these models to evaluate safety data more efficiently, enabling faster decision-making. AI also aids in the harmonization of safety regulations across regions, reducing regulatory hurdles for international hair product manufacturers.

10.7.3.3 Enhancing Consumer Awareness

Consumers stand to gain significantly from AI-driven assessments. They receive safer products and benefit from increased transparency. Clear labeling of potential allergens and warnings related to specific ingredients enhance consumer awareness, allowing individuals to make informed choices about the products they use.

10.7.3.4 Reducing Animal Testing

Perhaps one of the most significant ethical achievements is the substantial reduction in animal testing. AI models are increasingly capable of replacing animal trials in toxicity assessments, contributing to a more humane approach to product safety (Tran et al. 2023).

AI-driven global toxicity assessments have ushered in a new era of hair product safety. These assessments provide in-depth insights into product safety, streamline regulatory processes, and prioritize consumer well-being. As technology continues to evolve, AI will play an even more pivotal role in ensuring the safety and efficacy of hair care products on a global scale, marking a remarkable step forward for both the industry and consumers alike.

10.8 Ethical and Regulatory Considerations

10.8.1 Ethical Issues Surrounding AI in Toxicity Prediction

As the integration of AI in toxicity prediction becomes more prevalent, a range of ethical concerns emerges, demanding careful consideration.Data Privacy and Consent: AI models rely heavily on data, often including personal information. Ethical dilemmas arise regarding the collection, storage, and use of this data. It's crucial to ensure informed consent, data anonymization, and adherence to data protection regulations to safeguard individual privacy.Algorithmic Bias: AI models are susceptible to bias, reflecting existing disparities in historical data. This bias can result in unequal assessments and decisions, disproportionately affecting certain demographic groups. Ethical AI requires efforts to identify and mitigate bias, fostering fairness and inclusivity.Transparency and Interpretability: Complex AI algorithms can be challenging to interpret, raising concerns about transparency and accountability. Ethical practices involve efforts to make AI systems more explainable, enabling users to understand how predictions are made and ensuring that unjust decisions can be identified and rectified.

10.8.2 Regulatory Frameworks for AI-Driven Safety Assessments

The ethical concerns surrounding AI in toxicity prediction have prompted the development of regulatory frameworks to ensure responsible and transparent use of these technologies.

FDA Guidance for AI in Medical Devices: The U.S. Food and Drug Administration (FDA) has issued guidelines for the development and deployment of AI-driven medical devices, which include safety assessments. These guidelines emphasize the need for robust testing, validation, and transparency in AI applications to ensure patient safety.

EU AI Regulation: The European Union (EU) has proposed comprehensive regulations for AI systems, addressing safety and ethical considerations. These regulations require high-risk AI systems, including those used in safety assessments, to undergo rigorous conformity assessments and adhere to transparency and accountability standards.

International Collaboration: Various international bodies, including the World Health Organization (WHO) and the United Nations (UN), are working to establish global AI governance frameworks. These frameworks aim to harmonize AI regulations, promote ethical practices, and ensure the responsible use of AI technologies in areas like toxicity prediction.

10.8.3 Ensuring Transparency and Accountability

Transparency and accountability are paramount in the ethical application of AI in toxicity prediction.

Model Explainability: AI models should be designed with explainability in mind, allowing users to understand how predictions are generated. Techniques such as feature importance analysis and model-agnostic explanations enhance transparency.

Regulatory Audits: Regular audits and assessments by regulatory bodies help ensure that AI-driven toxicity predictions comply with established safety standards and ethical guidelines. These audits can uncover biases or transparency issues that require correction.

Public Awareness and Education: Transparency extends to the end-users, consumers, and stakeholders. Raising awareness about AI applications in toxicity prediction, potential risks, and ethical considerations is vital for informed decision-making.

Ethical Guidelines: The development and adoption of ethical guidelines within the AI community and industry can provide a roadmap for responsible AI practices. These guidelines often encompass principles like fairness, accountability, and transparency.

The ethical and regulatory considerations surrounding AI in toxicity prediction are essential to ensure the responsible and equitable use of these technologies. Balancing innovation with ethical concerns requires ongoing collaboration among governments, regulatory bodies, industry stakeholders, and the AI community. Transparency, accountability, and adherence to ethical principles should remain at the forefront of AI-driven safety assessments, ultimately benefiting consumer safety and fostering trust in these groundbreaking technologies.

10.9 Future Directions and Challenges

10.9.1 Emerging Trends in AI and Toxicology

The future of AI in toxicology holds exciting possibilities and innovative trends that promise to reshape the landscape of hair product safety assessments.Advanced Deep Learning Architectures: Deep learning continues to evolve with novel architectures like Transformers and Graph Neural Networks. These advancements enhance the ability to capture complex chemical interactions and predict toxicity with unprecedented accuracy (https://www.frontiersin.org/articles/10.3389/frai.2023.1269932/full).

Explainable AI (XAI): The demand for transparent AI models is driving the development of XAI techniques. Explainable AI allows users to understand how AI systems arrive at predictions, promoting trust and accountability in toxicity assessments.

Personalized Safety Assessments: AI's potential to incorporate individual genetics, lifestyle factors, and environmental influences is on the horizon. This opens doors to personalized safety assessments, where products are tailored to an individual's unique profile, minimizing risks.

10.9.2 Addressing Ongoing Challenges and Limitations

Challenges and limitations persist in the integration of AI into hair product safety assessments, necessitating ongoing research and innovation.

Data Quality and Availability: Ensuring high-quality data and overcoming data limitations, such as sparse or biased datasets, remains a challenge. Efforts to improve data collection and curation are critical.

Interpretable Models: Enhancing model interpretability is an ongoing challenge, especially in complex deep-learning models. Developing effective techniques for explaining AI-driven predictions will be crucial for regulatory compliance and user trust.

Ethical Concerns: As AI applications expand, ethical dilemmas may intensify. Ensuring privacy, fairness, and accountability will require continued vigilance and evolving ethical frameworks.

Regulatory Adaptation: Regulatory bodies must adapt swiftly to the evolving AI landscape, ensuring that safety assessments keep pace with technological advancements while maintaining rigorous standards.

10.9.3 The Potential for AI to Revolutionize Hair Product Safety Globally

AI holds the potential to usher in a global revolution in hair product safety, benefiting consumers, manufacturers, and regulators alike.Accelerated Product Development: AI streamlines product development by identifying potential hazards early in the formulation process. This accelerates time-to-market for safer hair products, meeting consumer demands efficiently.Reduced Animal Testing: AI's capacity to replace animal testing with in silico assessments marks a significant step toward ethical and humane safety evaluations, aligning with modern sensibilities.

Enhanced Consumer Safety: By delivering more accurate toxicity predictions and personalized assessments, AI ensures that consumers receive safer, tailored products, reducing the risk of adverse reactions.

Global Collaboration: The global perspective afforded by AI-driven assessments fosters international collaboration, harmonizes safety standards, and promotes the responsible development of hair products on a worldwide scale.

Future of AI in hair product safety assessment is promising, with emerging trends, ongoing challenges, and transformative potential. As AI continues to advance, it holds the power to revolutionize the industry by enhancing product development, safeguarding consumer health, and reducing the environmental and ethical footprint of safety assessments. Addressing challenges and embracing AI's potential ethically and responsibly will be key to realizing the full benefits of this technology in the realm of hair product safety.

10.10 Conclusion

10.10.1 Summary of Key Takeaways

In this comprehensive exploration of AI in hair product toxicity prediction, several key takeaways emerge:

1. **AI Transforms Safety Assessment:** Artificial Intelligence has revolutionized the way we evaluate the safety of hair products. It enables rapid and accurate toxicity predictions, reducing reliance on animal testing and expediting product development.
2. **Global Perspective is Essential:** Recognizing the global nature of hair product consumption is crucial. A diverse dataset that considers regional variations ensures comprehensive safety assessments.
3. **Ethical Considerations Matter:** Addressing ethical concerns, such as data privacy, algorithmic bias, and transparency, is vital for responsible AI implementation.
4. **Regulatory Frameworks Evolve:** Regulatory bodies worldwide are adapting to the AI era, developing frameworks to ensure the safe and ethical use of these technologies in safety assessments.
5. **Personalization is on the Horizon:** AI's potential to incorporate individual factors, like genetics, promises personalized safety assessments, enhancing consumer well-being.

10.10.2 The Future Outlook for AI in Hair Product Toxicity Prediction

The future outlook for AI in hair product toxicity prediction is bright and dynamic. Continued advancements in deep learning, explainable AI, and data quality will further refine toxicity assessments. Personalized safety evaluations will become more prevalent, tailoring products to individual profiles. As AI evolves, it will continue to reduce animal testing and enhance transparency, earning consumer trust.

Furthermore, AI will play an integral role in global harmonization efforts. Regulatory bodies, industry stakeholders, and researchers will collaborate to establish consistent standards, fostering safer and more responsible product development. Ethical considerations will remain at the forefront, ensuring privacy, fairness, and accountability in AI-driven assessments.

10.10.3 Encouraging Global Collaboration for Safer Hair Products

Encouraging global collaboration is paramount for ensuring safer hair products worldwide. To achieve this, stakeholders must work together to:

1. **Share Data:** Industry and regulatory bodies should collaborate to share safety data and adverse event reports, creating a comprehensive knowledge base that benefits all.
2. **Harmonize Regulations:** Efforts to harmonize safety regulations across regions will reduce barriers to international product development and enhance consumer safety.
3. **Ethical Guidelines:** Develop and adhere to ethical guidelines that prioritize consumer privacy, fairness, and accountability in AI applications.
4. **Public Awareness:** Raise public awareness about the benefits of AI-driven safety assessments and the importance of reading product labels and reporting adverse events.
5. **Research Collaboration:** Foster collaboration among researchers, industry experts, and regulatory authorities to advance AI techniques and refine safety assessment methodologies.

In conclusion, AI is poised to continue its transformative role in ensuring the safety of hair products globally. By embracing ethical practices, advancing regulatory frameworks, and promoting global collaboration, we can usher in a future where hair products are not only effective but also safe, ethical, and tailored to individual needs. This collective effort will contribute to the well-being of consumers and the sustainability of the beauty and cosmetics industry.

References

Bhhatarai B, Walters WP, Hop CECA, Lanza G, Ekins S (2019) Opportunities and challenges using artificial intelligence in ADME/Tox. Nat Mater 18(5):418–422. https://doi.org/10.1038/s41563-019-0332-5. PMID: 31000801; PMCID: PMC6594826

Bolduc C, Shapiro J (2001) Hair care products: waving, straightening, conditioning, and coloring. Clin Dermatol 19:431–436

Burnett CL, Bergfeld WF, Belsito DV et al (2021) Safety assessment of keratin and keratin-derived ingredients as used in cosmetics. Int J Toxicol 40(2_suppl):36S–51S. https://doi.org/10.1177/10915818211013019

Gavazzoni Dias MF (2015) Hair cosmetics: an overview. Int J Trichology 7(1):2–15. https://doi.org/10.4103/0974-7753.153450. PMID: 25878443; PMCID: PMC4387693

Guo Y, Zhang Y, Lyu T, Prosperi M, Wang F, Xu H, Bian J (2021) The application of artificial intelligence and data integration in COVID-19 studies: a scoping review. J Am Med Inform Assoc 28(9):2050–2067. https://doi.org/10.1093/jamia/ocab098. PMID: 34151987; PMCID: PMC8344463

He L, Michailidou F, Gahlon HL, Zeng W (2022) Hair dye ingredients and potential health risks from exposure to hair dyeing. Chem Res Toxicol 35(6):901–915. https://doi.org/10.1021/acs.chemrestox.1c00427

Helgheim BI, Maia R, Ferreira JC, Martins AL (2019) Merging data diversity of clinical medical records to improve effectiveness. Int J Environ Res Public Health 16(5):769. https://doi.org/10.3390/ijerph16050769. PMID: 30832447; PMCID: PMC6427263

https://www.frontiersin.org/articles/10.3389/frai.2023.1269932/full

Jeong J, Choi J (2022) Artificial intelligence-based toxicity prediction of environmental chemicals: future directions for chemical management applications. Environ Sci Technol 56(12):7532–7543. https://doi.org/10.1021/acs.est.1c07413

La Torre C, Bhushan B (2006) Nanotribological effects of silicone type, silicone deposition level, and surfactant type on human hair using atomic force microscopy. J Cosmet Sci 57:37–56

LeCun Y, Bengio Y, Hinton GE (2015) Deep learning. Nature 521:436–444. https://doi.org/10.1038/nature14539

Malani M, Kasturi A, Moinul M, Gayen S, Hota C, Nirmal J (2023) Role of artificial intelligence in the toxicity prediction of drugs. In: Mohanan PV, Kappalli S (eds) Biomedical applications and toxicity of nanomaterials. Springer, Singapore. https://doi.org/10.1007/978-981-19-7834-0_22

Martínez-García M, Hernández-Lemus E (2022) Data integration challenges for machine learning in precision medicine. Front Med (Lausanne) 8:784455. https://doi.org/10.3389/fmed.2021.784455. PMID: 35145977; PMCID: PMC8821900

Mayr A, Klambauer G, Unterthiner T, Hochreiter S (2016) DeepTox: toxicity prediction using deep learning. Front Environ Sci 3:80. https://doi.org/10.3389/fenvs.2015.00080

Mikalef P, Gupta M (2021) Artificial intelligence capability: conceptualization, measurement calibration, and empirical study on its impact on organizational creativity and firm performance. Inform Manage 58(3):103434. ISSN 0378-7206. https://doi.org/10.1016/j.im.2021.103434

Mintz Y, Brodie R (2019) Introduction to artificial intelligence in medicine. Minim Invasive Ther Allied Technol 28(2):73–81. https://doi.org/10.1080/13645706.2019.1575882

Patel D, Narayana S, Krishnaswamy B (2013) Trends in use of hair dye: a cross-sectional study. Int J Trichology 5(3):140–143. https://doi.org/10.4103/0974-7753.125610. PMID: 24574693; PMCID: PMC3927172

Saikia S, Prajapati JB, Prajapati BG, Padma VV, Pathak YV (2022) The role of artificial intelligence in therapeutic drug monitoring and clinical toxicity. In: Amponsah SK, Pathak YV (eds) Recent advances in therapeutic drug monitoring and clinical toxicology. Springer, Cham. https://doi.org/10.1007/978-3-031-12398-6_5

Shapiro J, Maddin S (1996) Medicated shampoos. Clin Dermatol 14:123–128

Tetko IV, Klambauer G, Clevert D-A, Shah I, Benfenati E (2022) Artificial intelligence meets toxicology. Chem Res Toxicol 35(8):1289–1290. https://doi.org/10.1021/acs.chemrestox.2c00196

Tran TTV, Surya Wibowo A, Tayara H, Chong KT (2023) Artificial intelligence in drug toxicity prediction: recent advances, challenges, and future perspectives. J Chem Inf Model 63(9):2628–2643. https://doi.org/10.1021/acs.jcim.3c00200. Epub 2023 Apr 26

Vatiwutipong P, Vachmanus S, Noraset T, Tuarob S. (2023). Artificial intelligence in cosmetic dermatology: a systematic literature review. IEEE Access, pp 1–1. https://doi.org/10.1109/ACCESS.2023.3295001

Wilm A, Garcia de Lomana M, Stork C, Mathai N, Hirte S, Norinder U, Kühnl J, Kirchmair J (2021) Predicting the skin sensitization potential of small molecules with machine learning models trained on biologically meaningful descriptors. Pharmaceuticals 14(8):790. https://doi.org/10.3390/ph14080790

11 Omics Approaches for the Assessment of Hair Care Product Safety, Efficacy and Toxicity: Current Techniques and Future Advancements

Tanmayee Basu, Rashi Chugh, Ranjit Singh Gujjar, Dibyalochan Mohanty, and Atul Kumar Upadhyay

Abstract

The cosmetic business has experienced a significant increase in the desire for hair care products that are both innovative and safe in recent years. As a result, researchers are now investigating sophisticated approaches to thoroughly assess the safety, effectiveness and potential harmful effects of the products. Omics technologies, including genomics, proteomics, metabolomics and transcriptomics, have become valuable tools for understanding the intricate molecular mechanisms that contribute to the effects of hair care products on the hair and scalp. This study concisely summarises the current status of omics methodologies used to evaluate hair care products. It presents a complete survey of the techniques utilised in genomics, proteomics, metabolomics and transcriptomics. Genomic research has facilitated a more profound comprehension of the hereditary elements that impact individual reactions to hair care treatments, providing personalised product suggestions. Proteomics investigations have shown the complex protein interactions occurring inside the hair follicles and the scalp, offering valuable information about the molecular pathways influenced by various formulations. Metabolomics provides valuable insights into product metabolism and potential systemic effects by analysing the tiny chemicals present in hair and the surrounding tissues. Transcriptomics investigations have revealed

T. Basu · R. Chugh · A. K. Upadhyay (✉)
Department of Biotechnology, Thapar Institute of Engineering and Technology, Patiala, Punjab, India
e-mail: atul.upadhyay@thapar.edu

R. S. Gujjar
Division of Crop Improvement, Indian Institute of Sugarcane Research, Lucknow, Uttar Pradesh, India

D. Mohanty
Department of Pharmaceutics, Anurag University, Hyderabad, Telangana, India

A. Dwivedi et al. (eds.), *Hair Care Products*,
https://doi.org/10.1007/978-981-97-6740-3_11

the ever-changing gene expression patterns linked to different hair care treatments, exposing the molecular indicators of effectiveness and possible harm.

Moreover, this abstract explores imminent progress in omics technologies, emphasising the possible incorporation of artificial intelligence (AI) and machine learning (ML) for enhanced data analysis and predictive modelling. These advanced procedures have great potential to transform the development and evaluation of hair care products, guaranteeing their safety and effectiveness in a personalised way. This review provides a clear and precise guide for researchers, industry professionals and regulators, leading them towards a more knowledgeable and progressive period of evaluating hair care products.

Keywords

Omics technologies · Genomics · Proteomics · Metabolomics · Transcriptomics · Personalised hair care · Hair follicle biology · Product safety and effectiveness · Artificial intelligence · Machine learning · Predictive modelling

11.1 Introduction

When it is time to choose the right hair care product, there is a lot to ponder about. How well is it going to work? Is it good for my hair type? However, the most important question we all need to be asking is whether the product contains any toxic ingredients because hair care products that have a long-lasting shelf life, good aroma, create form and shine easily often contain chemicals that have carcinogenic properties and are endocrine disrupters, which interfere with the body's endocrine systems and lead to several adverse effects in the reproductive, nervous and immune systems of the body.

In the modern era, people are more interested in using natural and botanical ingredients that are free from claims like (sulphates, silicones, parabens, formaldehyde and sodium laurel sulphate, etc.) which are responsible for adverse effects. With the rise in demand for personalised hair care products, it has become mandatory to implement cutting-edge techniques to examine the quality and effectiveness of these products. This chapter aims to shed light on the current advances and technologies in hair toxicology assessment.

11.1.1 Importance of Omics Approaches in Toxicology Research

Elevations in the field of hair care products have led to the growth of various omics approaches for analysing safety, efficacy and toxicity. Omics approaches include collective technologies like genomics, transcriptomics, metabolomics and proteomics that measure the role, action and relationship of certain molecules such as proteins, genes, etc. Through the integration of such approaches, researchers can gain valuable knowledge about the mechanism of action, the effects of these products and potential markers of toxicity in the product. For example, genomics approaches have potential applications in analysing gene expression profiles, providing insights into the mechanism and potential risks associated with the use of hair care products. In hair sample metabolites, protein quantification and identification can be done through proteomics and metabolomics techniques to look over product efficacy and to identify the toxicological effects. One of the techniques in the omics approaches, transcriptomics, plays a vital role in the assessment of hair care products. Transcriptomics involves the study of total RNA within an organism. This technique can give a valuable understanding of the expression and regulation of genes in the hair cycle, hair growth and hair follicle health. For instance, gene expression using RNA sequencing (RNA-Seq) technology or microarray can help in analysing differential expression of genes in response to the use of hair care products. Omics-based approaches revolutionised the traditional method of toxicity assessment and the use of animal models to test carcinogenic and toxicity potencies, as shown in Fig. 11.1.

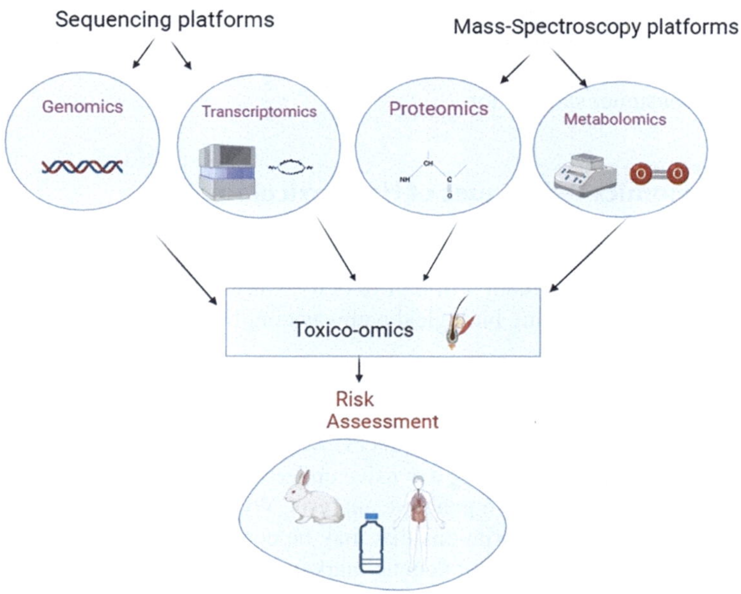

Fig. 11.1 Representing important branches of "Omics" technology

11.2 Genomic-Based Approaches for Hair Toxicity Assessment

Genomics is the study of the complete genetic material of an organism, and it has transformed the branch of toxicology by giving unprecedented insights into the molecular mechanisms involved in toxicology responses. Toxico-genomics, the use of genomics in toxicology, has led to the identification of particular genes and toxicity-related pathways, as well as the identification of putative biomarkers to monitor and recognise harmful consequences on time.

11.2.1 Use of Genomics to Study Genetic Markers in Hair Follicles

Finding genetic markers has become a viable method in recent years for assessing the performance and safety of hair care products. Genetic markers refer to distinct DNA sequences associated with particular qualities or characteristics. These markers are optimal for evaluating an individual's reaction to diverse hair care products. By examining the genetic markers, scholars can acquire significant knowledge regarding the potential responses of individuals to particular chemicals or formulations. By examining gene expression patterns about exposure to hair care products, genomics can yield valuable insights into the possible negative consequences and the fundamental biological mechanisms implicated in the growth and upkeep of hair. For instance, a genome-wide association study (GWAS) is employed to ascertain genetic markers linked to the effectiveness of hair growth treatments. Similarly, the utilisation of genetic markers for the evaluation of the safety and potential toxicity of hair dyes. The findings, as mentioned earlier, underscore the potential of genetic markers in enhancing the process of selecting hair care products and improving overall consumer satisfaction.

11.2.2 Genomics Techniques of Hair Toxicology

The utilisation of genomics-based methodologies holds the promise of fundamentally transforming the evaluation of hair toxicity, since it enables a thorough comprehension of the underlying biological pathways implicated in this process.

11.2.2.1 Whole Genome Sequencing

Among the omics techniques most frequently employed in the evaluation of hair toxicity is whole-genome sequencing (WGS). WGS entails sequencing every DNA molecule in an individual, offering extensive understanding of their genetic composition. Through comprehensive genomic analysis, WGS makes it possible to identify genetic mutations and variations that may be connected to hair poisoning and will be helpful in identification of genetic markers linked to drug metabolism.

11.2.2.2 Gene Expression Profiling

Thousands of genes and their expression levels can be measured simultaneously because of a potent omics-based method called gene expression profiling. It offers insightful information about the intricate cellular systems that regulate responses to toxins. Gene expression profiling was utilised in a study by Brima et al. (2006) to find differentially expressed genes of hyperkeratotic skin lesions from inner Mongolians chronically exposed to arsenic. This method made the identification of new biomarkers and clarification of the molecular mechanisms behind arsenic poisoning possible. Such results emphasise the significance of omics-based approaches and the possibility of gene expression profiling in toxicity assessment.

11.2.2.3 Single Nucleotide Polymorphism Analysis

Recently, omics-based study for hair toxicity assessment has given significant emphasis to single nucleotide polymorphism (SNP) analysis. SNPs represent the most prevalent form of genetic variation among individuals. This examination permits the discovery of particular genetic markers that could be connected to hair toxicity. For instance, research has revealed that SNPs in the genes responsible for drug metabolism, like the cytochrome P450 enzymes, may affect a person's vulnerability to toxicity from hair. Consequently, SNP analysis offers insightful information about the hereditary elements that lead to hair toxicity and may be employed in customised toxicity evaluations.

11.3 Transcriptomics-Based Approaches for Hair Toxicity Assessment

Transcriptomics refers to the scientific investigation of the entirety of RNA molecules transcribed by cells or tissues inside an organism, encompassing various physiological circumstances and stages of development. The primary emphasis of transcriptome research lies in investigating mRNA since its expression can be a proxy for the corresponding protein of interest (Nohynek et al. 2004). Transcriptomics, a specialised field within omics-based research, enables comprehensive analysis of gene expression patterns. Transcriptome profiling enables scientists to identify specific genes and pathways that demonstrate differential expression in response to harmful stimuli. In a recent study conducted by Zhang et al. (2022), the researchers examined the impact of cadmium exposure on hair follicle cells through the application of transcriptomics. This investigation revealed notable gene alterations associated with DNA repair mechanisms and oxidative stress. The findings highlight the significant insights gained regarding cellular reactions and potential indicators of hair toxicity.

11.3.1 Identification of Potential Toxicological Effects Through Transcriptomic Analysis

Research on transcriptomes can show the entire range of gene activities and structures as well as the molecular mechanisms underlying certain biological processes in disease. Nowadays, high-throughput RNA-Seq and gene expression microarrays are frequently used in medicinal, biological and clinical research. By evaluating the expression pattern of genes in response to toxic components, transcriptomes allow for the finding of specific gene markers related to hair toxicity.

Through the examination of the complete transcriptome, scientists are able to detect changes in the expression of genes that serve as indicators of cellular stress, disturbance or toxicity. Numerous investigations have effectively employed transcriptome analysis to examine the safety and toxicity of diverse cosmetic and personal care goods, encompassing hair care products.

11.3.1.1 RNA Sequencing

With an broad application spectrum and high sensitivity, NGS is used in RNA-Seq to identify and measure RNA in biological samples, ranging from targeted sequencing to full transcriptome analysis. While mRNA-Seq uses poly(A) selection to sequence protein-coding gene expression, total RNA or whole-transcriptome sequencing is capable of simultaneously analysing coding and non-coding RNAs. Widely used NGS methods rely on the concept of short-read sequencing (less than 200 nucleotides per fragment), which can yield reliable and high-throughput results but has limited capacity to identify transcript isoforms. RNA-Seq technique greatly contributes to drug development and discovery, increasing our understanding of toxicity mechanisms. Evaluating transcriptomics data results in the identification of the differentially expressed genes, forming substructure to discover biomarkers and to analyse drug-related mechanisms (Yang et al. 2020).

11.3.1.2 Microarray Technology

Microarray technology has been known by various other names such as Gene chips, Biochip and DNA/RNA chips. It was invented in the 1990s, and since then it has been known as microarray (Khelurkar et al. 2017). The fundamental steps involved in microarray technology are: obtaining mRNA from suitable biological samples, labelling RNA or cDNA copies with fluorescence, hybridising the labelled RNA or cDNA with microarray (up to millions of probes) for a while before washing off the excess, scanning the microarray under a laser and analysing data using the appropriate software (Chavan et al. 2006). This technique can provide a large sample of data information with high accuracy in a shorter duration of time. Through microarray analysis, this study aims to examine the alterations in gene expression inside human hair follicles subjected to various hair care products.

The results can indicate notable modifications in various crucial genes associated with hair development, control of the hair cycle and inflammation-related pathways. These insights can assist in detecting potential negative consequences or give evidence supporting the effectiveness of hair care products in enhancing hair health and encouraging hair growth.

11.4 Proteomics-Based Approaches for Hair Toxicity Assessment

The utilisation of proteomics-based methodologies has developed as a robust means of evaluating the toxicity of hair. The aforementioned methodologies encompass a thorough examination of proteins that are present in hair samples, so yielding significant knowledge regarding their possible harmful effects and the fundamental molecular processes at play (Rice 2011; Adeola et al. 2017). Proteomics has been widely utilised in the field of toxicology to gain insight into the molecular pathways that underlie the hazardous effects of different drugs. Proteomics has yielded significant insights into the proteins and signalling pathways implicated in toxicological responses through the examination of alterations in protein expression, post-translational modifications and protein–protein interactions (Richards et al. 2021). In addition, using proteomics alongside other omics methodologies augments our comprehension of the intricate interplay of diverse molecular components and toxic compounds within the hair. An integrative approach exhibits significant potential in the advancement of hair toxicity assessment and the facilitation of personalised toxicity monitoring strategies. In addition, utilising proteomics methodologies has facilitated the discovery of possible toxicity-associated biomarkers, enhancing the advancement of precise diagnostic instruments for evaluating toxicological impacts (Gao et al. 2009; Adeola et al. 2017).

11.4.1 Evaluation of Protein Expression Changes in Hair Follicles After Product Application

Proteomics approaches are essential tools to assess protein expression abnormalities in hair follicles following the application of a treatment. By employing proteomic analysis, scientists can detect and measure alterations in protein expression levels, thereby offering significant knowledge regarding the impacts of hair care products on hair follicles. Methods such as two-dimensional gel electrophoresis (2D-GE) and mass spectrometry (MS)-based techniques, such as liquid chromatography–mass spectrometry (LC–MS), have been extensively employed to examine the proteome of hair follicles and to quantify differentially expressed proteins associated with hair toxicity (Lee et al. 2006; Doll and Burlingame 2014). These methodologies allow the detection of distinct protein markers and the evaluation of changes in protein expression, enhancing our comprehensive understanding of the safety, effectiveness and adverse effects of hair care products.

11.4.2 Identification of Protein Biomarkers for Hair Care Product Safety and Efficacy

Protein biomarkers are of the utmost significance in evaluating the safety and effectiveness of hair care products. These biomarkers can ascertain potential dangers and assess the efficacy of those products. One example of the application of proteome analysis is its utilisation in identifying alterations in peptide biomarkers linked to cosmetic treatments, which enables researchers to ascertain whether the product has detrimental effects on the scalp or hair follicles (Sinclair et al. 2012; Goecker et al. 2021). In addition, conducting a comparative analysis of protein biomarkers prior to and subsequent to the application of the product can yield valuable insights regarding the effectiveness of the product. By quantifying specific proteins, such as keratins or structural proteins, it becomes feasible to assess the efficacy of the product in enhancing hair strength and quality (Lee et al. 2006; Adav et al. 2018). Therefore, employing proteomics to identify protein biomarkers is imperative, as this is crucial for conducting a thorough evaluation of the safety and effectiveness of hair care products.

11.4.3 Proteomics-Based Techniques for Hair Toxicology

The proteomics-based methodologies employed encompass a thorough examination of proteins that are present in hair samples, with the objective of identifying potential indicators of toxicity. One example of the application of MS-based proteomics is the analysis of the proteome of hair samples to detect and characterise proteins that exhibit differential expression patterns in response to exposure to certain toxins (Lee et al. 2006). Furthermore, the utilisation of protein microarray technology can facilitate the concurrent identification and quantification of numerous proteins within hair samples. This advancement has greatly aided in the discovery of previously unidentified protein biomarkers associated with toxicity (Goud et al. 2013; Tosto and Reitz 2016). Proteomics methodologies exhibit significant potential in enhancing the precision and sensitivity of hair toxicity evaluation, hence facilitating the implementation of individualised preventive strategies and therapies (Tosto and Reitz 2016).

11.4.3.1 MS-Based Proteomics

The utilisation of MS-based proteomics has become a prominent method in the field of hair toxicity evaluation, as it enables the detection and measurement of proteins that are found within hair samples. In a study conducted by Tang et al. (2016), the proteome of hair samples obtained from individuals who dyed their hair frequently was analysed using liquid chromatography–tandem mass spectrometry (LC–MS/MS). The researchers effectively discovered increase in hydroxyl radicals and metal ions such as copper and iron, thus, increasing photo-oxidative damage. In a separate investigation conducted utilising matrix-assisted laser desorption/time of flight (MALDI–TOF/TOF) to identify 95 differentially expressed proteins in a mouse model to investigate the molecular mechanisms underlying hair follicle cycling Wang et al. (2016), discovered 44 of these proteins to have a strong correlation with hair follicle cycling.

11.4.3.2 Protein Microarray Analysis

The utilisation of protein microarray analysis represents a robust method within the field of omics, allowing for the concurrent identification and measurement of a multitude of target proteins in a manner that is efficient and scalable. Protein microarrays have demonstrated efficacy in diverse domains, encompassing biomarker exploration, identification of therapeutic targets and investigations in systems biology (Goud et al. 2013; Zhu and Qian 2012). A study conducted by Lueking et al. (2005), utilised alopecia areata as a model for autoimmune dermatological diseases employing protein microarrays to identify particular biomarkers for scalp lesions associated with autoimmunity, employing cDNA expression library systems.

11.5 Metabolomics-Based Approaches for Hair Toxicity Assessment

Metabolomics encompasses the thorough examination of metabolites within a biological sample, offering a momentary depiction of an organism's dynamic metabolic reaction. It has the potential to provide significant insights into the biochemical alterations that arise as a result of toxic exposures within the framework of hair toxicity evaluation (Araújo et al. 2021). For instance, a research endeavour using metabolomics techniques to discern metabolic changes in hair samples of persons subjected to heavy-metal exposure (Hamid Akash et al. 2023), or another investigation aimed to examine the metabolic profiles of human urine samples from individuals with prolonged phthalate exposure (Hsu et al. 2019). These studies emphasise the potential of metabolomics in revealing the fundamental molecular pathways of toxicity and its utilisation in assessing hair toxicity. Metabolomics techniques, such as nuclear magnetic resonance (NMR) spectroscopy and MS, have been widely utilised in the identification and quantification of modified metabolic pathways resulting from hazardous exposure (Ren et al. 2018).

11.5.1 Detection of Metabolic Changes Associated with Hair Care Product Use

Metabolomics has the potential to offer significant insights into the metabolic alterations linked to the utilisation of hair care products. For example, the utilisation of metabolomic profiling can provide insights into changes observed in the concentrations of particular metabolites, such as lipids, amino acids and carbohydrates, which are known to have significant impacts on the overall well-being and functioning of hair (Jung et al. 2011; Araújo et al. 2021). These alterations may serve as indications of the impact of hair care products on the metabolic pathways associated with hair development, maintenance and repair.

In addition, metabolomics has the potential to contribute to the discovery of metabolic biomarkers that can be utilised as indicators of the safety, effectiveness or toxicity of a product (Jang et al. 2019). Metabolomic analysis can be employed to investigate the impact of hair dyes and bleaching agents on the hair metabolome, offering valuable insights into the potential hazards and toxicity associated with using these products.

11.5.2 Analysis of Small-Molecule Metabolites in Hair Samples

This technique offers significant insights into the inherent metabolic response of hair to various hair care treatments. Moreover, there has been a notable increase in examining small-molecule metabolites in hair samples recently. Multiple pathways can facilitate the incorporation of small compounds, such as drugs, toxins and metabolites, into hair strands (Kempson and Lombi 2011). Metabolites provide significant insights into an individual's historical orientation and can assist in evaluating the safety and effectiveness of hair care products.

11.5.3 Metabolomics-Based Techniques for Hair Toxicology

The utilisation of metabolomics-based methodologies can become prominent in the field of hair toxicity evaluation, owing to its capacity to offer a thorough and quantitative examination of small-molecule metabolites found in biological specimens. The combination of NMR spectroscopy and MS with chromatographic techniques, such as gas chromatography (GC) or LC (Ren et al. 2018), has become the favoured approach for detecting and measuring small-molecule metabolites in hair samples. Metabolomics enables the identification and quantification of particular metabolites that function as biomarkers for toxic exposure (Jang et al. 2019). In addition, the progress in high-resolution MS instruments and sophisticated data analysis tools has significantly enhanced the sensitivity and specificity of these techniques (Hsu et al. 2019). These techniques provide valuable insights into the mechanisms that contribute to hair toxicity and have the potential to facilitate the advancement of non-invasive biomonitoring technologies (Behringer and Deschner 2017).

11.5.3.1 Nuclear Magnetic Resonance Spectroscopy

NMR spectroscopy is a very effective analytical method frequently employed in metabolomics-based investigations for the purpose of characterising the structure, dynamics and interactions of molecules. The utilisation of this non-destructive methodology yields comprehensive data regarding the chemical composition and spatial organisation of molecules, hence presenting significant elucidation into the mechanisms and routes of bio-molecular processes (Emwas et al. 2019). Furthermore, NMR spectroscopy can be utilised for the evaluation of hair toxicity by detecting and quantifying harmful chemicals such as formaldehyde and other metabolites added in various hair styling products (Monakhova et al. 2013).

11.5.3.2 Liquid Chromatography-Mass Spectrometry

LC–MS is a highly effective analytical technique that is often employed in metabolomics-based studies for the evaluation of hair toxicity, as it can demonstrate potential in offering a comprehensive understanding of people's exposure history and prospective health impacts, hence contributing to the progress of toxicological research.

In order to analyse endogenous metabolites and chemicals exposed throughout time in the same hair sample, an untargeted metabolomics process was developed utilising ultra-high-performance liquid chromatography–high-resolution mass spectrometry (UHPLC–HRMS)-based metabolomics to perform a thorough examination of their metabolic profiles on a universal level (Chen et al. 2021).

11.6 Future Advancements in Omics Approaches for Hair Care Product Assessment

11.6.1 Combined Analysis of Genomics, Transcriptomics, Proteomics and Metabolomics Data

Research that integrates numerous omics techniques is critical because it provides a thorough and integrated understanding of the impacts of hair care products on different biological levels. Through the simultaneous analysis of several omics data, researchers can acquire and examine the diverse layers of molecular information. This approach allows for a more comprehensive understanding of biological mechanisms and their errors, leading to deeper insights in the field. Genomic analysis provides for detecting genetic differences that can impact the individual's response to hair care products. Transcriptomic analysis enables the investigation of alterations in gene expression after the application of a product, thereby offering valuable insights into the fundamental molecular mechanisms involved. The utilisation of proteomic analysis can provide additional insights into the influence of hair care products on the levels of and changes in proteins, thereby enhancing our understanding of their effects at the protein level. Finally, applying metabolomic analysis enables the detection and measurement of alterations in metabolism resulting from the utilisation of hair care products. Furthermore, omics approaches can also facilitate the identification of key biomarkers that serve as indicators of safety and efficacy, streamlining the evaluation process and reducing overall costs. Therefore, the integration of omics techniques (illustrated in Fig. 11.2) offers a promising solution to enhance the efficiency and cost-effectiveness of safety and efficacy evaluation in the hair care industry.

Despite the considerable potential of omics-based methodologies in evaluating hair toxicity, there exist numerous challenges and limits that necessitate attention and resolution.

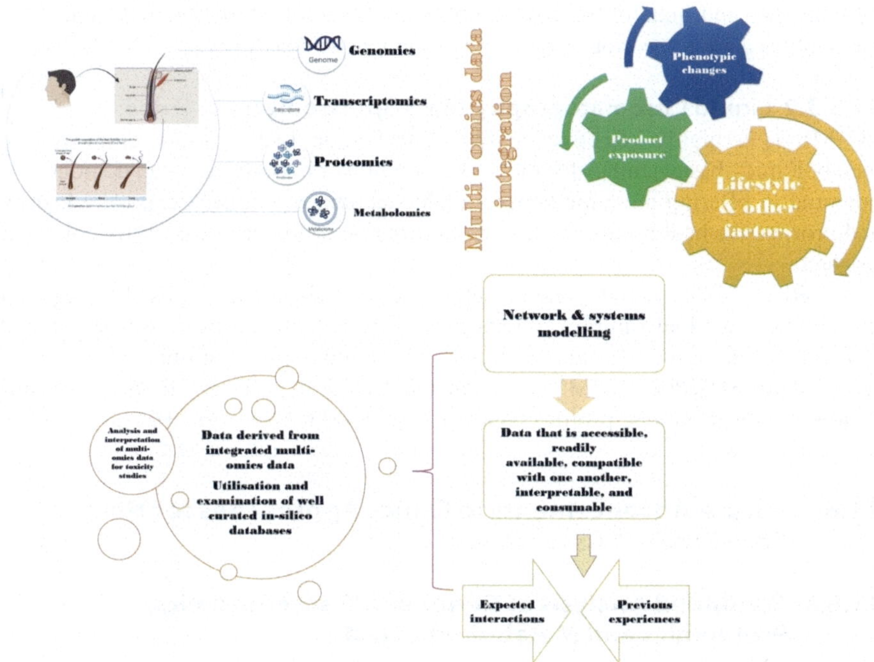

Fig. 11.2 Integration of "Omics" technology for combined analysis

- The analysis and integration of data generated from various omics platforms present a substantial computational and bioinformatics challenge.
- Furthermore, the absence of standardised methods and quality control mechanisms across various omics technologies introduces intricacy to the process of integrating and comparing data.
- Moreover, the process of identifying dependable biomarkers from omics data necessitates rigorous validation and replication in separate cohorts, a task that can be both time-consuming and labour-intensive.

Therefore, it is imperative to address these problems in order to effectively utilise the advantages of integrating omics data in hair toxicity evaluation.

11.6.2 Comprehensive Understanding of Hair Care Product Effects on Various Biological Levels Through High-Throughput Screening Methods

High-throughput screening (HTS) approaches have become increasingly influential in efficiently assessing hair care products. HTS facilitates the expeditious evaluation of extensive arrays of substances or combinations, enabling the detection of prospective safety issues, effectiveness and toxicity (Zhu et al. 2014). Advances in technology have led to the expansion of HTS methodologies, encompassing a range of tests, including cell viability, gene expression, proteomics and metabolomics. This broadens the scope of the assessment of product effects on diverse biological levels. For example, HTS techniques can be employed to assess the potential toxicity of hair care products by quantifying their impact on cellular viability and utilising cell-based assays (Borrel et al. 2020).

Furthermore, HTS enables the evaluation of alterations in gene expression that are caused by these particular products. This process provides valuable insights into the potential effects they may have on hair growth or the occurrence of inflammatory reactions (Borrel et al. 2020). Utilising HTS, the evaluation process will be enhanced in terms of efficiency, cost-effectiveness and the ability to detect potential safety issues and optimise formulations of hair care products.

11.6.3 Development of Automated Platforms for Rapid Assessment of Multiple Hair Care Products

The creation of automated platforms for quick assessment of multiple hair care products is a promising area of research with potential implications for evaluating their safety, efficacy and toxicity in the future. These platforms will employ omics methodologies to produce extensive data on the diverse constituents and their interactions within the products. Through the utilisation of high-throughput technologies, such as microarrays and next-generation sequencing, these platforms possess the capability to efficiently evaluate numerous hair care products, encompassing shampoos, conditioners and treatments, simultaneously. This facilitates the acquisition of a comprehensive understanding regarding the impacts of these products on both the hair and scalp.

11.6.4 Advanced Data Analysis Techniques—Utilisation of Artificial Intelligence and Machine Learning Algorithms

The application of artificial intelligence (AI) and machine learning (ML) algorithms has become a groundbreaking method in multiple domains, such as evaluating the safety, effectiveness and toxicity of hair care products. AI and ML algorithms provide notable benefits in the examination of extensive data sets, the detection of patterns and the extraction of valuable insights (Lin and Chou 2022; Tetko et al. 2022; Hartung 2023). Hair care products possess inherent hazards, hence requiring the implementation of thorough testing and evaluation procedures. Researchers can enhance the accuracy and efficiency of evaluating the safety and effectiveness of hair care products using these methods. These models also provide the ability to analyse extensive quantities of data, encompassing chemical composition, formulation and biological characteristics, to forecast the possible toxicity and efficacy of a given product. Furthermore, the utilisation of AI and ML algorithms can be instrumental in the identification of intricate relationships among variables. This capability empowers researchers to unveil novel perspectives and expedite the advancement of forthcoming hair care items.

11.6.5 Improved Prediction Models for Hair Care Product Safety and Efficacy

The need for more advanced prediction models that can evaluate the performance and safety of hair care products has grown in recent years. Prediction models play a crucial role for both consumers and manufacturers since they provide a more precise assessment of the prospective advantages and disadvantages associated with the utilisation of certain commodities. Presently, the evaluation of the safety and effectiveness of hair care products predominantly relies on in vitro and animal experimentation, which may not consistently mirror human reactions. Nevertheless, the progress made in omics technologies presents excellent opportunities for creating more dependable prediction models. By examining molecular profiles and pathways associated with hair growth and maintenance, omics techniques offer significant contributions to understanding the potential effectiveness and safety of hair care products. The development of safer and more effective hair care products, as well as a more comprehensive understanding of their impact on human health, is an area where these breakthroughs offer considerable promise for the future of hair care product development and regulation.

11.6.6 Personalised Hair Care Product Assessment—Individualised Omic Profiling for Personalised Product Recommendations and Tailored Hair Care Solutions Based on Genetic and Metabolic Characteristics

The notion of personalised medicine has garnered considerable attention within the healthcare sector in recent years. Personalised hair care product assessment can be developed as a novel and unique strategy to effectively cater to the diverse differences in hair features and individual needs. The recent progress in omics technologies has facilitated the customisation of hair care products by taking into account an individual's genetic composition, protein expression profile and metabolic pathways, and their utilisation has significant potential in elucidating the diverse responses individuals exhibit towards hair care products. Consequently, this knowledge can be leveraged to provide tailored suggestions to consumers. The efficacy and safety of hair care products are optimised by the implementation of a customised approach that takes into account the distinctive attributes and needs of an individual's hair.

The industry is increasingly focusing on the concept of personalised hair care products that take into account an individual's phenotypic traits, such as the brand "Ravel," which was pitched in one of the seasons of Shark Tank India. This

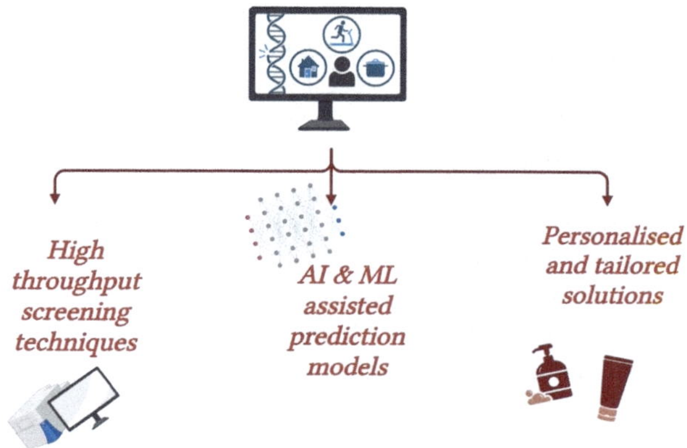

Fig. 11.3 New-age techniques for the toxicity studies of hair care products

methodology acknowledges the distinct requirements of each hair type and endeavours to offer tailored remedies to get the most favourable outcomes. For instance, a correlation exists between genetic variants in the P2RY5 gene and the characteristics of hair thickness and texture (Pasternack et al. 2008), enabling the creation of tailored treatments that specifically address these particular attributes. In addition, the analysis of individuals' metabolic profiles offers valuable information regarding shortages or imbalances in nutrients that could impact the health of hair. Implementing customised hair care solutions that consider genetic and metabolic factors holds the promise of significantly transforming the industry, as it offers personalised and optimised experience for individuals (illustrated in Fig. 11.3).

11.7 Conclusions

Utilisation of omics-based methodologies has significantly broadened our comprehension of hair toxicity assessment. It can be achieved by applying integrated omics methodologies to identify critical biomarkers and their respective functions in toxicity pathways. In addition, using advanced methods in omics-based investigations has facilitated the identification of several minute constituents within hair samples. This has enhanced the precision and reliability of evaluating potential health effects. The application of omics-based methodologies has resulted in substantial advancements in our understanding of the molecular mechanisms behind toxicity. One illustrative instance involves utilising proteomics techniques to evaluate specific protein indicators associated with hair damage. At the same time, transcriptome research elucidates gene expression alterations resulting from applying hair care products (illustrated in Fig. 11.3). In addition, the field of metabolomics has provided insights into alterations in metabolic pathways induced by hazardous constituents. Omics-based methodologies offer a comprehensive understanding of hair toxicity, facilitating the development of targeted and efficacious intervention strategies. Although the current methods have contributed to expanding knowledge in this field, additional research is still necessary. A call for action is required to address the existing deficiencies and challenges within this field. For instance, the need for standardised protocols for sample preparation, data processing and interpretation introduces challenges in comparing outcomes.

In comparison to other areas of study, the field in question exhibits a relative scarcity of evaluation practices. Therefore, professionals from various industries and researchers must collaborate and allocate resources towards advancing standardised protocols and extensive research. This will enable the seamless incorporation of omics-based approaches as a regular tool for assessing the safety, effectiveness and potential harmful effects of hair care products.

References

Adav SS, Subbaiaih RS, Kerk SK, Lee AY, Lai HY, Ng KW, Sze SK, Schmidtchen A (2018) Studies on the proteome of human hair—identification of histones and deamidated keratins. Sci Rep 8(1):1599. https://doi.org/10.1038/s41598-018-20041-9

Adeola HA, Van Wyk JC, Arowolo A, Ngwanya RM, Mkentane K, Khumalo NP (2017) Emerging diagnostic and therapeutic potentials of human hair proteomics. Proteomics Clin Appl 12(2). https://doi.org/10.1002/prca.201700048

Araújo AM, Carvalho F, Guedes de Pinho P, Carvalho M (2021) Toxicometabolomics: small molecules to answer big toxicological questions. Metabolites 11(10):692. https://doi.org/10.3390/metabo11100692

Behringer V, Deschner T (2017) Non-invasive monitoring of physiological markers in primates. Horm Behav 91:3–18. https://doi.org/10.1016/j.yhbeh.2017.02.001

Borrel A, Huang R, Sakamuru S, Xia M, Simeonov A, Mansouri K, Houck KA, Judson RS, Kleinstreuer NC (2020) High-throughput screening to predict chemical-assay interference. Sci Rep 10(1):3986. https://doi.org/10.1038/s41598-020-60747-3

Brima EI, Haris PI, Jenkins RO, Polya DA, Gault AG, Harrington CF (2006) Understanding arsenic metabolism through a comparative study of arsenic levels in the urine, hair and fingernails of healthy volunteers from three unexposed ethnic groups in the United Kingdom. Toxicol Appl Pharmacol 216(1):122–130. https://doi.org/10.1016/j.taap.2006.04.004

Chavan P, Joshi K, Patwardhan B (2006) DNA microarrays in herbal drug research. Evid Based Complement Alternat Med 3(4):447–457. https://doi.org/10.1093/ecam/nel075

Chen Y, Guo J, Xing S, Yu H, Huan T (2021) Global-scale metabolomic profiling of human hair for simultaneous monitoring of endogenous metabolome, short- and long-term exposome. Front Chem 9:674265. https://doi.org/10.3389/fchem.2021.674265

Doll S, Burlingame AL (2014) Mass spectrometry-based detection and assignment of protein post-translational modifications. ACS Chem Biol 10(1):63–71. https://doi.org/10.1021/cb500904b

Emwas A-H, Roy R, McKay RT, Tenori L, Saccenti E, Gowda GA, Raftery D, Alahmari F, Jaremko L, Jaremko M, Wishart DS (2019) NMR spectroscopy for metabolomics research. Metabolites 9(7):123. https://doi.org/10.3390/metabo9070123

Gao Y, Holland RD, Yu L-R (2009) Quantitative proteomics for drug toxicity. Brief Funct Genomic Proteomic 8(2):158–166. https://doi.org/10.1093/bfgp/elp006

Goecker ZC, Legg KM, Salemi MR, Herren AW, Phinney BS, McKiernan HE, Parker GJ (2021) Alternative LC–MS/MS platforms and data acquisition strategies for proteomic genotyping of human hair shafts. J Proteome Res 20(10):4655–4666. https://doi.org/10.1021/acs.jproteome.1c00209

Goud S, Espina V, Liotta LA (2013) Protein microarrays. In: Brenner's encyclopedia genetics, pp 478–481. https://doi.org/10.1016/b978-0-12-374984-0.00939-6

Hamid Akash MS, Yaqoob A, Rehman K, Imran M, Assiri MA, Al-Rashed F, Al-Mulla F, Ahmad R, Sindhu S (2023) Metabolomics: a promising tool for deciphering metabolic impairment in heavy metal toxicities. Front Mol Biosci 10:1218497. https://doi.org/10.3389/fmolb.2023.1218497

Hartung T (2023) Toxaicology—the evolving role of artificial intelligence in advancing toxicology and modernizing regulatory science. ALTEX 40:559–570. https://doi.org/10.14573/altex.2309191

Hsu J-F, Tien C-P, Shih C-L, Liao P-M, Wong HI, Liao P-C (2019) Using a high-resolution mass spectrometry-based metabolomics strategy for comprehensively screening and identifying biomarkers of phthalate exposure: method development and application. Environ Int 128:261–270. https://doi.org/10.1016/j.envint.2019.04.041

Jang W-J, Choi JY, Park B, Seo JH, Seo YH, Lee S, Jeong C-H, Lee S (2019) Hair metabolomics in animal studies and clinical settings. Molecules 24(12):2195. https://doi.org/10.3390/molecules24122195

Jung H, Kim SJ, Lee W, Chung BC, Choi MH (2011) Gas chromatography/mass spectrometry based hair steroid profiling may reveal pathogenesis in hair follicles of the scalp. Rapid Commun Mass Spectrom 25(9):1184–1192. https://doi.org/10.1002/rcm.4975

Kempson IM, Lombi E (2011) Hair analysis as a biomonitor for toxicology, disease and health status. Chem Soc Rev 40(7):3915. https://doi.org/10.1039/c1cs15021a

Khelurkar VC, Ingle KP, Padole DA (2017) DNA microarray: basic principle and it's applications. Trends Bioscie 10(2):488–490

Lee YJ, Rice RH, Lee YM (2006) Proteome analysis of human hair shaft. Mol Cell Proteomics 5(5):789–800. https://doi.org/10.1074/mcp.m500278-mcp200

Lin Z, Chou W-C (2022) Machine learning and artificial intelligence in toxicological sciences. Toxicol Sci 189(1):7–19. https://doi.org/10.1093/toxsci/kfac075

Lueking A, Huber O, Wirths C, Schulte K, Stieler KM, Blume-Peytavi U, Kowald A, Hensel-Wiegel K, Tauber R, Lehrach H, Meyer HE, Cahill DJ (2005) Profiling of alopecia areata autoantigens based on protein microarray technology. Mol Cell Proteomics 4(9):1382–1390. https://doi.org/10.1074/mcp.t500004-mcp200

Monakhova YB, Kuballa T, Mildau G, Kratz E, Keck-Wilhelm A, Tschiersch C, Lachenmeier DW (2013) Formaldehyde in hair straightening products: rapid ^{1}H NMR determination and risk assessment. Int J Cosmet Sci 35(2):201–206. https://doi.org/10.1111/ics.12027

Nohynek GJ, Fautz R, Benech-Kieffer F, Toutain H (2004) Toxicity and human health risk of hair dyes. Food Chem Toxicol 42(4):517–543. https://doi.org/10.1016/j.fct.2003.11.003

Pasternack SM, von Kügelgen I, Aboud KA, Lee Y-A, Rüschendorf F, Voss K, Hillmer AM, Molderings GJ, Franz T, Ramirez A, Nürnberg P, Nöthen MM, Betz RC (2008) G protein–coupled receptor P2Y5 and its ligand LPA are involved in maintenance of human hair growth. Nat Genet 40(3):329–334. https://doi.org/10.1038/ng.84

Ren J-L, Zhang A-H, Kong L, Wang X-J (2018) Advances in mass spectrometry-based metabolomics for investigation of metabolites. RSC Adv 8(40):22335–22350. https://doi.org/10.1039/c8ra01574k

Rice RH (2011) Proteomic analysis of hair shaft and nail plate. J Cosmet Sci 62(2):229–236

Richards AL, Eckhardt M, Krogan NJ (2021) Mass spectrometry-based protein—protein interaction networks for the study of human diseases. Mol Syst Biol 17(1):e8792. https://doi.org/10.15252/msb.20188792

Sinclair R, Flagler MJ, Jones L, Rufaut N, Davis MG (2012) The proteomic profile of hair damage. Br J Dermatol 166:27–32. https://doi.org/10.1111/j.1365-2133.2012.10862.x

Tang Y, Dyer JM, Deb-Choudhury S, Li Q (2016) Trace metal ions in hair from frequent hair dyers in China and the associated effects on photo-oxidative damage. J Photochem Photobiol B Biol 156:35–40. https://doi.org/10.1016/j.jphotobiol.2016.01.010

Tetko IV, Klambauer G, Clevert D-A, Shah I, Benfenati E (2022) Artificial intelligence meets toxicology. Chem Res Toxicol 35(8):1289–1290. https://doi.org/10.1021/acs.chemrestox.2c00196

Tosto G, Reitz C (2016) Use of "OMICS" technologies to dissect neurologic disease. Neuroepidemiology 138:91–106. https://doi.org/10.1016/b978-0-12-802973-2.00006-9

Wang L, Xu W, Cao L, Tian T, Yang M, Li Z, Ping F, Fan W (2016) Differential expression of proteins associated with the hair follicle cycle—proteomics and bioinformatics analyses. PLoS One 11(1):e0146791. https://doi.org/10.1371/journal.pone.0146791

Yang X, Kui L, Tang M, Li D, Wei K, Chen W, Miao J, Dong Y (2020) High-throughput transcriptome profiling in drug and biomarker discovery. Front Genet 11:19. https://doi.org/10.3389/fgene.2020.00019

Zhang Y, Hu B, Qian X, Xu G, Jin X, Chen D, Tang J, Xu L (2022) Transcriptomics-based analysis of co-exposure of cadmium (CD) and 2,2′,4,4′-tetrabromodiphenyl ether (BDE-47) indicates mitochondrial dysfunction induces NLRP3 inflammasome and inflammatory cell death in renal tubular epithelial cells. Ecotoxicol Environ Saf 241:113790. https://doi.org/10.1016/j.ecoenv.2022.113790

Zhu H, Qian J (2012) Applications of functional protein microarrays in basic and clinical research. Adv Genet 79:123–155. https://doi.org/10.1016/b978-0-12-394395-8.00004-9

Zhu H, Zhang J, Kim MT, Boison A, Sedykh A, Moran K (2014) Big data in chemical toxicity research: the use of high-throughput screening assays to identify potential toxicants. Chem Res Toxicol 27(10):1643–1651. https://doi.org/10.1021/tx500145h

Next Generation of Hair Care Products: Innovations for Safe and Healthy Consumer Choices

12

Akshita Puri, Shishir Bobate, and Abhay Bajaj

Abstract

In recent times, there is an increasing demand of hair care products or cosmetics for the maintenance and styling of hair and treatment of hair-related problems. Hair care and styling products have taken the market by storm. These products are often considered as the luxury of a common man and are significantly related to health and beauty of a person. Although there are a variety of chemical-based products available in the market, they cause adverse effects and damage to the health as well the environment. The industry is evolving toward innovations for safe and healthy alternatives to these chemical-based products. The present chapter outlines some of the traditionally used hair care products and the new bio-based ingredients of microbial and plant origin which are safe for both human health and environment during their entire life cycle. The growing awareness and availability of bio-based products have led to a significant shift in consumer choice toward green cosmetics. The development of such innovation-backed ecologically friendly products is the need of the hour to work toward achieving sustainable goals at the global level through involvement of the masses.

Keywords

Hair care · Green cosmetics · Sustainable · Safety · Efficacy

A. Puri (✉)
PGTD Zoology, RTM Nagpur University, Nagpur, Maharashtra, India

S. Bobate · A. Bajaj
Environmental Toxicology Group, FEST Division, CSIR-Indian Institute of Toxicology Research (CSIR-IITR), Lucknow, India

A. Dwivedi et al. (eds.), *Hair Care Products*,
https://doi.org/10.1007/978-981-97-6740-3_12

12.1 Introduction

Hair, the thread-like structures found in vertebrates, are the vital components that are derived from the ectoderm and are considered as protective appendages present on the body. The texture of the hair, their color and style are considered as a sign of physical appearance and health, complementing the beauty of a person. The variation in the appearance of hair is caused due to the type of hair growth by follicles and the variety of hair care products or cosmetics used by individuals. Hair is composed of keratinized epithelium cells and follows a regulated hair growth cycle, which is divided into three phases that correspond to growth, regression, and resting phase (Padule et al. 2022). The growth cycle is regulated by certain factors whose alteration affects the cell cycle, thereby leading to apoptosis and inhibition of hair growth, hair loss, and other hair disease conditions. Changing lifestyles, stress, hormonal imbalance and environmental factors cause hair damage or other hair problems. To overcome hair damage or hair problems and for the maintenance of hair, there are several cosmetic or hair care products available in the market. These products work by efficient cleansing, nourishing, deep conditioning, providing strength, controlling dandruff, and reducing hair loss. Some of these products may even modulate the hair growth cycle to regain normal hair growth and texture. Depending upon their chemical properties, the products either alter the hair with temporary effects or may alter the hair shaft with permanent effects.

Due to high consumer demand, there is a huge variety of hair care products available for the enhancement and maintenance of both hair and scalp. However, these products derived from petrochemical source or chemical additives pose side effects and damage to health as well as environment. Therefore, there is a global need to evolve with alternatives to these chemical-based products and come up with eco-sustainable products which do not have adverse health effects and are not harmful to the environment during their entire life cycle. The research and development strategies involve developing bio-based formulations of hair care products or "bio-cosmetics", following the guidance laid down by regulatory authorities. The major sources for such bio-cosmetics are microorganisms and plants. The present chapter highlights the commonly used traditional hair care products and the bio-based formulations that are innovated to replace those chemical ingredients for safety and efficacy. The growing public awareness for health and environmental concerns has led to the emergence of several bio-based formulations, leading to a huge shift in consumer choice toward green cosmetics (Fig. 12.1).

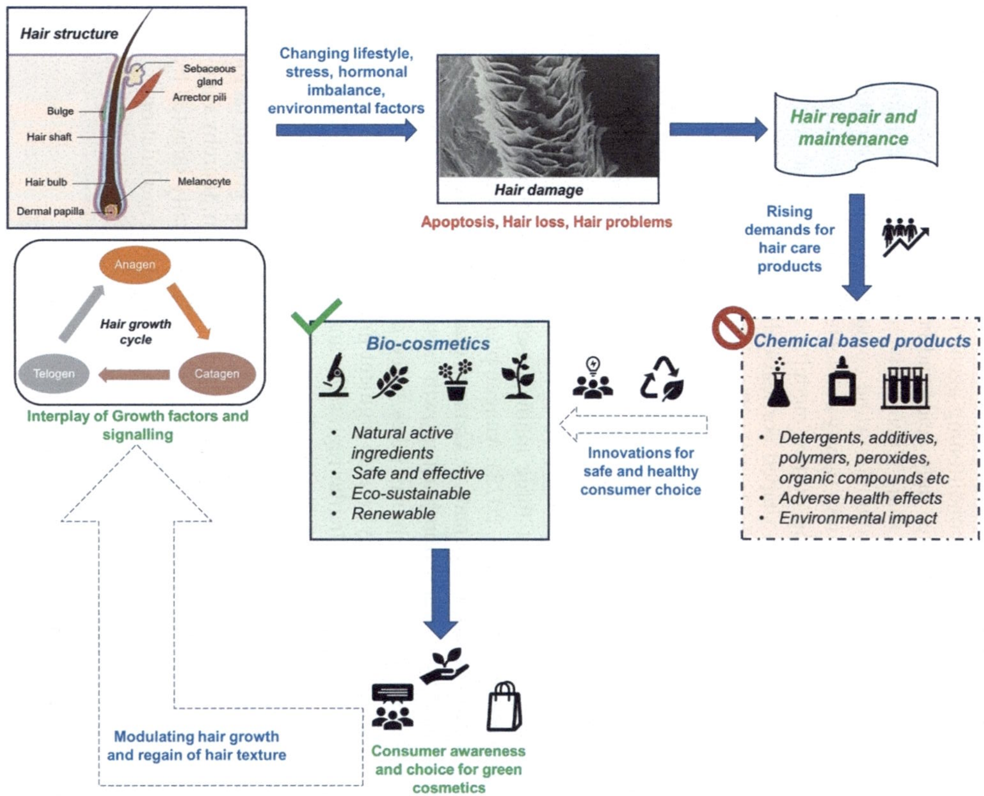

Fig. 12.1 Innovations in hair care products for safe and healthy consumer choices. Hair, an epidermal derivative present as protective appendage follows a regulated growth cycle. Conditions like changing lifestyle, stress, hormonal imbalance, or environmental factors cause hair cycle modulation and apoptosis, ultimately leading to hair damage and other hair problems. For enhancement and maintenance of hair, many chemical-based *hair care products are available in the market which fulfil the purpose but cause adverse effects to the health and environmental. As an alternative, research and development strategies have come up with bio-based ingre*dients replacing chemical ingredients, thereby providing eco-sustainable products. The rising consumer awareness for health and environmental effects led to a shift in choice for green cosmetics

12.2 Hair Structure

The hair is an epidermal derivative formed due to keratinization of the ectoderm-derived germinative cells. It is divided into two major parts: hair follicle (under the skin) and hair shaft (visible on body surface). Externally, it is composed of fully keratinized epithelium cells and internally, cylinder-like epithelial grow into dermis and subcutaneous fat, leading into hair bulb formation at the base surrounding the dermal papilla. A naturally secreted conditioning agent from the sebaceous gland called sebum (composed of mainly free fatty acids and neutral fats) is responsible for the maintenance of the grown hair shaft. Chemically, hair is composed of amorphous matrix rich in keratin, and insoluble cystine containing protein complexes cross-linking with other proteins, forming tightly connected fibers. Structurally, the hair shaft is composed of three regions: cortex, medulla, and cuticle. The cortex is constituted by tightly packed spindle-shaped cortical cells separated by a narrow region rich in proteinaceous intercellular lamella. This region provides strength and mechanical properties to the hair shaft i.e., form, stability, and texture. The inner layer, medulla, is made up of translucent cells and air gaps. This layer is rich in proteins trichohyalin, glycogen, and melanosomes. The cuticle is made up of keratinized scales that form a barrier protecting the cortex from environmental damage. Hair conditioning products are used to enhance and restore this cuticle layer (Gubitosa et al. 2019; Trüeb and Lee 2014).

The hair growth cycle consists of three main phases: anagen phase, catagen phase, and telogen phase (Padule et al. 2022). Anagen phase involves follicle regeneration, active hair growth, and pigmentation. Catagen phase, also known as regressive phase, is induced by catagen fibroblast growth factor-5 (FGF-5). It involves decrease in mitotic activity, and apoptosis takes place due to bcl-2 suppression. Also, there is contraction of melanocytes and pigment production ceases. Telogen phase is the resting phase between follicular regression and beginning of the anagen phase. There are certain factors, like insulin-like growth factor (IGF), vascular endothelial growth factor (VEGF), FGF, epidermal growth factor (EGF), wingless-related integration site (WNT), prostaglandin (PGD), transforming growth factor beta (TGF-β), whose stimulation/inhibition affects gene expression, thereby regulating the growth cycle (Padule et al. 2022). These factors are usually targeted by therapeutic agents to alter signaling pathways, leading to hair cycle modulation.

12.3 Hair Damage and Repair

Hair structure and the organization of its constituents is crucial for imparting physical properties to hairs. Any mechanical or chemical trauma altering the structural organization of hair, hair cycle modulation, or hormonal imbalance may lead to hair damage, hair loss (alopecia), gray hair, or weathering condition. An intact cuticle represents strong, shiny, smooth, and manageable hair and is easy to detangle due to overlapping scales of the cuticle. Environmental factors, like air pollution, sunlight, dirt, chlorinated water, or hair care products used for cleansing and styling may lead to cuticular loss, known as weathering. Other than these external factors, some inherited abnormalities like androgenetic alopecia (male pattern baldness) may also lead to intrinsically weak hair (Dias 2015). In androgenic alopecia, increased level of dihydrotestosterone binds to the androgen receptor and mediates gene activation, leading to the transformation of a normal follicle into a miniaturized one. This causes the hair follicle size to reduce, a diminished anagen and a prolonged telogen phase, thus leading to hair loss. Another condition is alopecia areta, an autoimmune disease, characterized by the presence of antigen in the follicle (Padule et al. 2022). Hair care products or hair cosmetics can help in controlling the hair damage by cleansing, increasing strength, and better conditioning. Treatment of scalp, hair loss or decreased hair density, hair cycle modulation is the likely purpose of these hair care products.

12.4 Hair Care Products

Hair care products are formulations for hair or scalp maintenance, especially for the purpose of cleansing, better appearance, and protection. They are well described as "preparations intended for placing in contact with the hair and scalp, with the purpose of cleansing, promoting attractiveness, altering appearance, and/or protecting them in order to maintain them in good condition" (Trüeb 2005). These products are broadly categorized into two groups: hair-altering products with temporary effects and hair shaft-altering products with permanent effects.

12.4.1 Products with Temporary Effects

12.4.1.1 Shampoos

Shampoos have clearly replaced the use of soaps for hair cleansing. They are not just cleansing agents but are also expected to possess conditioning benefits, so that hair are not dried out, are nonirritant to the scalp, and are cost-friendly. These shampoos are expected to provide wide benefits to a broader target population of different age groups, various hair types, and different lifestyles. Thus, to meet consumer needs, shampoos are made up of a specialized formulation consisting of several ingredients: detergents or surfactants (cleansing agents like sodium lauryl sulfate (SLS), polyethoxylated analogs, etc.) for removal of sebum; conditioning agents (like silicones, guar hydroxypropyltrimonium chloride, etc.) for shine and softness; buffering agents and additives like parabens, EDTA, alkanolamides, polypropylene glycol, etc., for stability; fragrances (like perfume oils) for product attraction; active ingredients or plant extracts (special care agents like selenium disulfide) for treatment of specific conditions like dandruff, hair loss, greying (Trüeb 2007). Each ingredient is tested, approved, and labeled as per cosmetics regulatory norms laid down by regulatory bodies across the globe.

12.4.1.2 Conditioners

Though shampoos contain conditioning agents, a separate conditioner is usually applied for more gloss and to minimize frizzing in hair. The common ingredients of conditioners are cationic surfactants (like propyltrimonium), polymers (like mono- and polypeptides), emollients (oily compounds, esters, or wax), active oils like silicones, salts like sodium chloride, fatty alcohols like cetyl alcohol, and emulsifiers like polysorbate-80 (D'Souza and Rathi 2015). Conditioners work by neutralizing negative electric charge, greasing the cuticle, thereby reducing the hydrophilicity and also help in lowering down raised cuticles (Dias 2015). Conditioners are usually applied immediately after cleansing the hair with shampoo, left over for a few minutes for penetration, and then washed off. Another variety of conditioners are oil-free, contain proteins, and can be left on the hair, thus coating the hair shaft and imparting smoothness (D'Souza and Rathi 2015).

12.4.1.3 Temporary Colors or Hair Dyes

Earlier, hair dyes were used to blacken/brown grey hairs caused due to aging. Nowadays, it is more of a lifestyle trend to color hair differently within a few minutes, and these hair dyes are used equally by both men and women. Temporary colors were often used, and the most common example is natural henna (obtained from the plant *Lawsonia alba*), which was earlier the preferred method for coloring grey hairs and for conditioning. Other temporary color products (comprising of water-soluble acidic and basic dyes with azo−/anthraquinone groups) are also available, which can be easily applied in the form of a lotion but are retained for a short duration, usually a wash or two, as they are simply deposited on the cuticle of the hair. The process of hair coloring is nonoxidative (He et al. 2022). On the contrary, there are several other chemical dyes (commonly composed of nitroaromatic amines, azo−/anthraquinones, triphenylmethanes) which are usually retained for a longer duration and wash cycles (6–10 washes or more). They are also categorized as semi-permanent and permanent, based on the chemical reaction and penetration of coloring molecules (Dias 2015). The ingredients of semi-permanent dyes ingredients do not oxidize and hence do not bind to hair proteins but penetrate the middle layer of the hair cuticle, mediated by ionic interactions or van der Waals forces. Since they are water-soluble, they are easily washed off after 6–10 cycles (Neuser and Schlatter 2010). Though studies have reported their impact on the skin and eye as an irritant and sensitizer with some ingredients showing mutagenic effects too, they were the most widely used products due to their cost-effectiveness and ease of application. Furthermore, prolonged use of such dyes is linked with developmental and other life-threatening adverse effects like carcinogenesis (He et al. 2022).

12.4.1.4 Other Products

Other than above-listed products applied for the purpose of maintenance and fashion, there are styling products like hair styling gel, hair spray, hair wax, or mousse. They are synthetic polymer-based products usually applied on towel-dried hair to increase hold to hair away from scalp or sprayed on finished hair style to maintain hair in desired style and look (Trüeb and Lee 2014). The most common functional groups for these polymers are vinyl esters, acrylate esters, vinylpyrrolidone, amides with variations like addition of acidic groups or neutralizers based on function and application (Musa and Tallon 2013). Though the preferred ingredients are used in formulations within prescribed limits, inadvertent use may lead to irritation and sensitizing conditions.

12.4.2 Products with Permanent Effects

12.4.2.1 Permanent Styling

In temporary methods, the effect is usually hair-altering, seen after topical application of a cosmetic product, and is lost in shorter duration. In permanent methods, where the effect is hair shaft-altering, hair internal structure is modified by a chemical process, thereby changing the shape of the hair. The two major permanent styling methods are permanent waves and hair straightening (Dias 2015). In permanent waves, the chemicals break and reform the disulfide bonds in the hair shaft, resulting in permanent hair curls. The commonly used waving lotion is ammonium thioglycolate, which penetrates deep inside the cortex to break the disulfide bonds, and then oxidizing agents like hydrogen peroxide are used for rebuilding the disulfide bonds in new order, giving curly hair. On the contrary, in hair straightening methods, chemical relaxers are applied, which penetrate up to the endocuticle, react with keratin, and cause rearrangement of disulfide bonds, making the keratin molecules soft and stretched. The commonly used chemicals are ammonium thioglycolate and hydroxides (sodium, lithium, or calcium either individually or in combination). Hydrogen peroxide is used for oxidizing, and a hot iron is also used during the hair straightening process. Thus, the process sometimes causes scalp burns, hair breakage, dandruff, hair loss, or weakening of hair (Dias 2015).

12.4.2.2 Permanent Colors or Dyes

Permanent dyes are mainly composed of nitroaromatic amines (precursor and coupling agents) that undergo oxidation reaction (oxidizing agent hydrogen peroxide) in the presence of ammonia and thus penetrate deep into the pores of the hair shaft (Dias 2015; He et al. 2022). The coloring molecules are further transmitted to the keratin of cuticle and cortex, thereby mimicking natural coloring molecules and giving permanent color change to the hair that is resistant to wash for 4–6 weeks. As they are chemical-based dyes, some of them contain hydrogen peroxide and resorcinol, and some intermediates (like derivates of p-phenylenendiamine) are formed during the coloring reaction which can lead to allergic reactions and other disease conditions upon prolonged use. Although the allowed percentages of such ingredients are within the respective permissible limits, as directed by the regulatory authorities, there are a few reports on the occurrence of dermatitis despite using concentrations lower than what is allowed (6% in hair dyes) (Søsted et al. 2004). This condition mainly arises due to hyper-sensitization and allergic induction response (He et al. 2022).

12.5 Global Need: Innovation Backed with Sustainability

Strong, shiny, and thick scalp hair are considered to be a symbol of health, beauty, and youth. Hence, there is a huge consumer demand for hair cosmetics for the enhancement and maintenance of both hair and scalp. There are many chemical-based hair cosmetics available that fulfil their purpose to a great extent but also pose threat and damage to health. Therefore, as an alternative, there is a growing consumer demand and need for natural and bio-based products with least adverse effects. The innovation for safer and healthy consumer choices reflected by their increased demand are mainly based on ingredients i.e., products comprising mostly natural and botanical ingredients and free from parabens, sulfates and other harmful components. Furthermore, there is a rising demand for organic products, which is expected to reach $44.77 billion by 2030 as per Grand View Research report (https://www.grandviewresearch.com/press-release/global-organic-personal-care-market). The safety and efficacy of such natural or organic products are regulated by regulatory authorities in most countries. For instance, in India, cosmetic products are regulated by the Bureau of Indian Standards (BIS) under the Drug and Cosmetics Act 1940 and Rules 1945 (https://cdsco.gov.in/opencms/opencms/en/Acts-Rules). As per the rules, there are 11 categories, which include hair care products, and specific guidelines for manufacturing, registration, and labeling are laid down to ensure safety and transparency to the consumer. Also, there are certain chemicals like arsenic and mercury, which are restricted. In the United States, hair care products are regulated under the Food, Drug, and Cosmetic Act (1938) by the Food and Drug Administration (FDA) (https://www.fda.gov/about-fda/histories-product-regulation/1938-food-drug-and-cosmetic-act). The rules laid down under the Act provide a mandate for the manufacturer to provide safety of ingredients used in the cosmetics. In the European Union (EU), the harmonized EU cosmetics directive (established in 1976 and reevaluated in 2009) provides guidelines for the placement of safe products in the EU market without any adverse effects on human health (https://eur-lex.europa.eu/legal-content/EN/LSU/?uri=CELEX:31976L0768). Also, there is a list of substances which are prohibited to be used in cosmetic products or are restricted owing to maximum allowed permissible limit in the formulations based on the safety assessment, as they might lead to adverse effects upon usage. Likewise, other countries have also adopted or are adopting such rules to ensure the safety aspects of a cosmetic product usage. The use of these alternative bio-based products not only reduces adverse effects on human health but also protects the environment when they come in direct contact or during disposal of these cosmetic products. Hence, it is the need of the hour to shift toward such sustainable and efficient products.

12.6 Bio-Cosmetics for Hair Care

12.6.1 Biosurfactants

Healthy and beautiful hair improve physical appearance as well as play a role in self-perception. Throughout history, hair care has been an important practice for both men and women. Shampooing is the universal routine for hair cleansing. A shampoo typically contains four categories of ingredients—cleansing agents, additives for enhanced stability, conditioning agents, and special care ingredients (Trüeb 2001). Surfactants, in particular anionic surfactants, are employed as cleansing agents as they possess excellent foaming and help in removing soil/dirt and other greasy substances from negatively charged hair fiber. Other types of surfactants, namely cationic, nonionic, and amphoteric surfactants, are used as co-surfactants or additives due to their poor lather formation and cleansing activities. Biosurfactants (BSs) are surface-active molecules produced by microorganisms, which have garnered attention for use in various industries, including the personal care and cosmetics sector, as a substitute to synthetic surfactants. In hair care cosmetics like shampoos, conditioners, and other related products, BSs offer three distinct advantages over traditional synthetic surfactants. First, BSs are generally milder on scalp and hair, compared to anionic surfactants, which cause skin irritations and allergies, interact with proteins, remove lipids from epidermal surface, and also harm living cells in the skin (Vecino et al. 2017). Second, many of the BSs possess potent antimicrobial activity, which is helpful in treating specific hair-related disorders caused by microorganisms. For example, it has been well established that *Malassezia furfur* causes dandruff (Trüeb 2001; Rincón-Fontán et al. 2020). A lipopeptide BS from *Bacillus subtilis* showed inhibition of growth and biofilm formation as well as removal of preformed biofilms of *M. furfur* clinical strains (Da Silva et al. 2021). The inhibition of *M. furfur* growth by the BS was comparable to the commercial antifungal agent fluconazole. Moreover, in toxicity assessment on murine fibroblast cells, the BS showed no toxicity, highlighting its potential for use in personal care products, in particular hair cosmetics (Da Silva et al. 2021). The third advantage of this BS over its synthetic counterparts is its environment-friendly nature. Considering the large quantities of hair care products used and correspondingly large quantities of surfactants ultimately discharged in water bodies, the hazardous effects of poorly biodegradable synthetic surfactants have been of a major concern. Additionally, synthetic surfactants, being of petrochemical origin, are the farthest from being sustainable. BSs, with their moieties originating from basic cellular processes like lipid, amino acid, or carbohydrate metabolism, are completely biodegradable as well as sustainable due to the use of renewable substrates in their production by fermentation.

In addition to the above-mentioned advantages, BSs can potentially reduce the number and quantity of ingredients required in the formulation of hair care products. BSs have excellent surface tension reduction, solubilization, lathering, cleansing, emulsification, and emulsion stabilization activities, on par with commonly used anionic surfactants like sodium laureth sulphate (SLES) and SLS. These anionic surfactants are harsh, and cause irritation on hair and skin (Sen et al. 2021; Leoty-Okombi et al. 2021), to compensate which certain additives need to be included in the formulation. BSs can substitute both the primary surfactant and co-surfactant, as well as reduce the quantities of additives in the formulation owing to being milder on the skin. Fernández-Peña et al. (2019) demonstrated a successful increase in the efficiency of a conditioner formulation by the substitution of SLES with rhamnolipid. A conditioner functions by the deposition of a polymer and surfactant on damaged hair fiber. The rhamnolipid–polymer mixture showed significantly higher deposition on the materials mimicking hair fibers as compared to the conventional SLES–polymer mixture (Fernández-Peña et al. 2019). In another study assessing the adsorption of lipopeptide BS extracted from corn steep liquor on hair, a rapid adsorption (within a few min) of BS on hair fiber with a maximum capacity of 3679 mg/g was observed (Rincón-Fontán et al. 2016). In special formulations for the treatment of microorganism-associated hair disorders, like antidandruff shampoos, antimicrobial agents such as zinc pyrithione and ketoconazole are included in the formulations (Rincón-Fontán et al. 2020). Due to well-established antimicrobial activity of BSs against various pathogens (Chopra et al. 2020; Gharaie et al. 2023) and even against *M. furfur*, the causative agent of dandruff (Da Silva et al. 2021; Choudhary et al. 2022), BSs can reduce or completely substitute the antimicrobial agents in the formulation. Another interesting role that BSs can play in hair care formulations is the emulsification or solubilization of commonly used hydrophobic ingredients like zinc pyrithione, which are generally solubilized by addition of synthetic surfactants at very high concentrations. Zinc pyrithione, present as a fine powder, exhibits a strong antimicrobial activity and slows down the rapid turnover of epidermal cells (Rincón-Fontán et al. 2020). In a study by Rincón-Fontán et al. (Rincón-Fontán et al. 2020), a lipopeptide BS-based formulation greatly enhanced the solubility of zinc pyrithione to 2948 from 20 mg/L, underscoring the potential of the BS as an emulsifier. Taken together, due to various properties of BSs, they can serve diverse functions in hair care formulations with the potential to partially or completely substitute synthetic surfactants.

Hair care formulations using BSs in various roles have been patented. A patent (US8183198B2) described the application of BSs in treating baldness, in which concentrated rhamnolipid formulations, when applied on scalp with no hair present, resulted in hair growth. The same patent also demonstrated the long-lasting antimicrobial effect of rhamnolipid on scalp when used in shampoo formulation. Another patent (JP2011026277A) claimed the hair cosmetic formulations prepared with mannosyl erythritol lipid (MEL) BS produced by *Pseudozyma* sp. protected hair from physical and chemical irritation in addition to rendering the hair moist and smooth. Several other patents, including WO2013098066A3, JP2011026276A (MEL-based, provide protection from physical and chemical stimulations as well as moistness, resilience, and strength to hair), CN108699096B (rhamnolipid-based; hair fragrance retention), describe similar BS-based formulations for hair care.

Although BSs have numerous advantages over petrochemical-derived synthetic surfactants for use in hair care products and cosmetics, the picture is not completely green upon closer inspection. The major challenge, as with any practical application of BSs, is the production cost of BSs. To make BS-based hair care products more economical, the BS should be available at prices comparable to synthetic surfactants. The factors contributing to high production costs of BSs include high cost of substrates, low yields obtained in the fermentation, and high downstream purification costs (Mohanty et al. 2021; Schultz and Rosado 2020; Bertrand et al. 2018). For applications like use in cosmetics and personal care products, BSs of high purity are required. Research has been ongoing to combat these key limitations and make BSs available at a commercial scale. Glycolipids, especially rhamnolipid, appear promising in terms of commercialization potential. Currently, Evonik is at the forefront in the commercial production of rhamnolipid and sophorolipid from sustainably sourced substrates. Green surfactants have been made available to other brands like Unilever through partnerships. Evonik launched REWOFERM® RL 100, a commercially available rhamnolipid produced sustainably using renewable feedstocks, at the end of 2022 (Guzmán et al. 2023). A few other chemical and biotechnological companies, including AGAE Technologies, Allied Carbon Solutions Ltd., Henkel, Lion Corporation, LLC, Paradigm Biomedical Inc., and TeeGene Biotech, are also involved in rhamnolipid production (Guzmán et al. 2023). Still, similar exploration is highly required for other types of BSs, including lipopeptides, due to their diverse properties and to facilitate their application in the consumer market.

12.6.2 Bio-Actives

A mature hair follicle forms by the differentiation of matrix progenitors promoted by dermal papilla cells (Mesler et al. 2017). The proliferation and elongation of hair shaft is brought about by VEGF and TGF-β secreted by dermal papilla cells (Inoue et al. 2009; Yano et al. 2001). These cells also express an enzyme responsible for the conversion of testosterone to dihydrotestosterone (DHT)—5α-reductase (3-oxo-5α-steroid-delta 4-dehydrogenase). The overexpression of 5α-reductase disrupts the natural hair cycle by increased conversion of testosterone to DHT in hair follicles, increasing hair fall and leading to the development of androgenic alopecia (Escamilla-Cruz et al. 2023). Thus, hair growth is extensively dependent on two growth factors, VEGF and TGF-β, and the activity of 5α-reductase.

Bacillus coagulans MTCC 5856 is a bacterium with Generally Recognized as Safe (GRAS) status, as reviewed by the US FDA. A bioactive extracellular metabolite preparation from *B. coagulans* culture supernatant has been patented for skin and hair application. The patent (US20180344781) assessed the impact and mechanism of action of the extracellular metabolite preparation from a strain of *B. coagulans* on hair proliferation. The study noted that the preparation increased the proliferation of dermal papilla cells as well as the expression of VEGF by 30% and TGF-β2 by 70.3%, promoting hair growth. The preparation (0.5% v/v) also inhibited the 5α-reductase activity by 16.2% and reduced the DHT levels.

Recently, a cosmetic comprising *B. coagulans* culture supernatant-based postbiotic was launched that has antimicrobial, antioxidant, UV protection properties and promotes hair growth (https://sabinsa.com/press-release-2021/1237-sabinsa-s-postbiotic-lactosporin-wins-1st-place-in-nie-awards). LactoSporin also showed reduction in facial wrinkles with no adverse effects on skin in a randomized, double-blinded, placebo-controlled study (Majeed et al. 2023). The study demonstrates the use of *B. coagulans*-derived bioactive, highlighting the unexplored potential of microbial bioactive compounds for skin and hair cosmetics.

12.6.3 Essential Oils

Essential oils (EOs) are concentrated essence of the plants usually obtained by distillation and do not contain the whole plant constituents. They are a common renewable bio-resource of natural ingredients, with many beneficial properties like anti-inflammatory, antioxidant, antidandruff, anti-hair loss, and antimicrobial and have a very pleasant aroma. These are oily hydrophobic substances extracted from various parts of plants, i.e., flowers, fruits, seeds, roots, etc., and can be well exploited to produce eco-friendly products. They are a complex mixture of several bioactive compounds like terpenes, ketones, phenols, esters, etc., with around 30,000 known structures (Moghaddam and Mehdizadeh 2017). They differ in properties and activities mainly due to varying cyclic structure, method of extraction, isolation species, harvest, and climate conditions. Among the terpenes, the majority of EOs are comprised of monoterpenes. Their lipophilicity and low irritation power accounts for the usage in cosmetic products. Thus, EOs are becoming a common interest in hair care products due to their broad biological significance. EOs not only penetrate the scalp quickly but also nourish the deep hair follicles, and provide nutrition, thereby stimulating follicle growth and strengthening the hairs. Furthermore, they also help in unblocking the pores by effectively clearing the unwanted metabolites.

EOs are majorly extracted from the angiosperm family of plants, secreted usually as part of secondary metabolism, or signaling and defense. Of note, species compositional richness of Eos, depicted by the existence of specific biosynthetic pathways in plants along with different methods of isolation, gives possibility of obtaining extracts with different compositions. The recognized effectiveness of individual EO components provides different benefits, and is used accordingly in designing cosmetic formulations. There are several studies reporting the use of EOs in hair care products with different functions based on their properties (Abelan et al. 2022). Rosemary oil (*Rosmarinus officinalis*) and chamomile oil (*Matricaria chamomilla L.*) stimulate hair growth, prevent hair loss, and help in conditioning by nourishing the hair follicles and moisturizing the hair roots. Rosemary oil is composed of 1,8-cineole, camphor, α-pinene, verbenone, and borneol while chamomile oil is a mixture of sesquiterpene, farnesene, polyphenols, and α-bisabolol (Celiktas et al. 2007). Additionally, rosemary oil is reported to have skin tonic properties, a calming effect, vasodilating properties, and promote increased blood circulation in hair follicle and scalp stimulation, thereby becoming a choice of treatment of both dandruff and oily hair. Rosemary oil was reported to inhibit 5α-reductase activity, thereby enhancing vascularity of the scalp, leading to hair follicle regeneration (Abelan et al. 2022). Lavender oil (*Lavandula angustifolia*) is used as a repellent for lice in hair. It comprises of numerous monoterpenes, organic acids, coumarins, mineral salts, and tannins. Tea tree oil (*Melaleuca alternifolia*), thyme oil (*Thymus vulgaris*), and bergamot oil (*Citrus bergamia*) possess antidandruff properties and are commonly used to control dandruff. Tea tree oil is composed of terpinen-4-ol, delta-terpinene, alpha-terpinene, 1,8-cineole, p-cymene, and terpinolene. Thyme oil is a mixture of p-cymene, γ-terpinene, thymol, geraniol, linalool, gamma-terpineol, and carvacrol. Bergamot oil comprises both volatile and nonvolatile fractions, which mainly include monoterpene limonene, linalool, linalyl acetate, γ-terpinene, and β-pinene. Bergamot oil is also reported for its role in hair growth. Also, phenolic compounds from thyme have been reported for their antioxidant property of protecting biological systems from oxidative stress-promoting substances. Peppermint oil (*Mentha piperita L.*) is also reported to promote hair growth, protect dermal papillae vascularization, and early-stage anagen induction. Peppermint oil is a mixture of menthol, menthone, menthyl acetate, neomenthol, 1,8-cineole, isomenthone, and menthofuran. Thus, it is effectively used as a therapeutic agent in preventing hair loss (Yano et al. 2001; Majeed et al. 2023). Besides, EOs are also linked with the control of specific gene expression associated with hair growth regulation. Furthermore, a mixture of EOs was found to enhance the efficacy of product (shampoo) formulation without any adverse effects as compared to conventional chemical-based shampoos. Another mixture of EOs (thyme, rosemary, lavender, cedar wood, and grape seed) was tested and found to be safe and effective for the treatment of alopecia. Due to the antimicrobial activity of some EO components, they are also reported for their role in the treatment of pathological conditions. Seborrhea, a condition associated with oiliness of the hair fibers and causing dysfunction of capillary physiology, can be partially mitigated by EOs (Guzmán and Lucia 2021).

Although EOs have so many advantages for use in green cosmetic products several drawbacks like high volatile nature, poor water solubility, thermal and chemical labilities pose challenges to be overcome while manufacturing, storing, and packaging EO-based products. Also, allergic and toxicity assessment is important while preparing EO-based formulations to avoid reaction due to any of the components and to ensure safe production of these eco-sustainable products.

12.6.4 Other Herbal/Natural Products

With the rising trend and consumer demand for alternatives to chemical-based products, the cosmetic industry is evolving and coming up with a variety of natural products as per consumer choice. Herbs and herbal extracts have been used since ancient times for cleansing and beauty purposes. Herbs are simply the plant or its parts like root, stem, leaf, or flower in freshly cut or dried form. Now, the cosmetic products are formulated with such natural products as a response to growing trend and awareness for maximum efficacy without side effects (Gubitosa et al. 2019; Penkar et al. 2023). Neem (*Azadirachta indica*) leaves are rich in polyphenolic flavonoids (quercetin and beta-sitosterol), nimbin, nimbanene, 6-desacetylnimbinene, nimbandiol, etc., and are commonly used to treat hair problems. Fenugreek (*Trigonella foenum graecum*), which is rich in micronutrients like vitamin B complex, minerals like potassium and iron, saponins, alkaloids, flavonoids, folic acid, and vitamins, has several properties like preventing premature greying of hairs, strengthening roots, stimulating hair regrowth, repelling lice, and controlling dandruff. Shikakai (*Acacia concinna*), which is a common cleansing soap for hair used since ancient times, is rich in saponins, alkaloid, tannin, anthraquinone glycosides, sugar, and flavonoids. It also helps in controlling dandruff and strengthening hairs. Ginger (*Zingiber officinale*) contains gingerol, which enhances blood circulation to the hair follicles by relaxing the blood vessels (Li et al. 2022). Furthermore, it helps in controlling dandruff, itchy scalp, and prevents hair thinning. Bhringraj (*Eclipta alba* (L.) Hassk), rich in flavonoids and isoflavonoids, is a good source of minerals and vitamins; hence, it prevents thinning of hair; treats hair greying, hair density reduction, itchy scalp, and dandruff; and promotes scalp rejuvenation. Coffee (*Coffea arabica*), rich in caffeine, sterols, diterpenes, triacylglycerol, phospholipids, and keratin, is known to promote hair growth and provides shine and elasticity. It inhibits apoptosis and promotes proliferation by downregulating the TGF-β2 factor while upregulating the IGF factor in hair follicles (Padule et al. 2022). *Solanum nigrum* berries contains linoleic acid, steroidal glycoalkaloids, palmitic, elaidic, and stearic acids promote hair growth and is also used for treatment of alopecia. They act by blocking 5α-reductase activity and also downregulating the AKT signaling pathway. Amla (*Phyllanthus embelica*), rich in vitamins, minerals, tannins, and phyllemblin, is used as a hair tonic for enhancing growth and hair pigmentation. Also, it strengthens the roots and prevents dryness and dandruff. Likewise, there are so many herbal products (reetha, brahmi, marigold, etc.), which can be used in hair product formulations due to their hair growth, cleansing, maintenance, or treatment properties. These herbal products are very efficient and safe for long-term use. However, the selection and formulation of these natural ingredients pose a major challenge for an effective cosmetic product. One of the effective formulations consisting of amla, reetha, and shikakai is used for washing of hair and was reported as a good cleanser along with nourishing hair, promoting growth, removing dryness, and dandruff, retaining shine and natural oil, thus restoring natural hair texture (Gubitosa et al. 2019). Likewise, there are other formulations of herbal products which are evaluated and reported for synergistic and beneficial properties for hair (hair proliferation, anagen phase induction, hair cycle modulation, and treatment of several hair problems), thereby providing safe and effective products without side effects.

Interestingly, these herbal formulations are now advanced with nanotechnology, which provides advantage of small particle size, thus allowing effective penetration through the skin barrier, i.e., stratum corneum. A patent (patent number US 2010/0104646a1) described a nanoparticle herbal hair care composition (lecithin capsulated nanoparticles) for promoting hair growth and preventing hair loss due to antioxidant effects. Another patent (patent number Wo2017057881a1) described formulation from ginseng derived exosome like vesicles for hair growth promotion and prevention of hair loss. There are more such reports of herbal nano-formulations; however, separate rules and regulations for their safety and efficacy check are yet to be provided by regulatory authorities (Padule et al. 2022).

12.7 Consumer Choice: Green Cosmetics

Taking good care of hair is very important, especially in this changing lifestyle and polluted environmental conditions. Growing awareness of adverse health effects and environmental concerns has led to a rise in consumer choice for eco-sustainable products with natural/bio-based ingredients instead of chemical additives or ingredients from the petrochemical industry. This is further strengthened by international regulations which have put up restrictions on the use of harmful chemicals having adverse effects. Thus, it has become necessary for the industry to evolve toward eco-friendly product formulations to meet the growing consumer demand. Due to growing consumer interest in sustainability, research and development strategies are seeking to reformulate products with green ingredients and with the highest efficacy. The most promising natural sources include plants and microorganisms, providing a wide range of active ingredients with potential biological properties which can be exploited for health and beauty. Furthermore, oral products like hair gummies containing natural ingredients like grape seed extract, multivitamins, biotin, etc., have also come up, which strengthen the hair, promote hair growth, and help in overall hair maintenance. This path towards green cosmetics is of high concern for manufacturer, not only for the choice of ingredients in response to consumer's demand but ultimately for end of product lifecycle without creating waste epidemic. Thus, along with the selection of natural ingredients, the manufacturing process, packaging, and end of product life cycle are equally important challenges to be taken care of when it comes to innovation for sustainable and green cosmetic products.

References

Abelan US et al (2022) Potential use of essential oils in cosmetic and dermatological hair products: a review. J Cosmet Dermatol 21(4):1407–1418. https://doi.org/10.1111/JOCD.14286

Bertrand B, Martínez-Morales F, Rosas-Galván NS, Morales-Guzmán D, Trejo-Hernández MR (2018) Statistical design, a powerful tool for optimizing biosurfactant production: a review. Colloids Interfaces 2(3):1–18. https://doi.org/10.3390/colloids2030036

Celiktas OY, Kocabas EEH, Bedir E, Sukan FV, Ozek T, Baser KHC (2007) Antimicrobial activities of methanol extracts and essential oils of Rosmarinus officinalis, depending on location and seasonal variations. Food Chem 100(2):553–559. https://doi.org/10.1016/J.FOODCHEM.2005.10.011

Chopra A, Bobate S, Rahi P, Banpurkar A, Mazumder PB, Satpute S (2020) Pseudomonas aeruginosa RTE4: a tea rhizobacterium with potential for plant growth promotion and biosurfactant production. Front Bioeng Biotechnol 8:1–14. https://doi.org/10.3389/fbioe.2020.00861

Choudhary M, Verma V, Saran R, Bhagyawant SS, Srivastava N (2022) Natural biosurfactant as antimicrobial agent: strategy to action against fungal and bacterial activities. Cell Biochem Biophys 80(1):245–259. https://doi.org/10.1007/s12013-021-01045-1

D'Souza P, Rathi SK (2015) Shampoo and conditioners: what a dermatologist should know? Indian J Dermatol 60(3):248. https://doi.org/10.4103/0019-5154.156355

Da Silva GO et al (2021) Effects of lipopeptide biosurfactants on clinical strains of Malassezia furfur growth and biofilm formation. Med Mycol 59(12):1191–1201. https://doi.org/10.1093/mmy/myab051

Dias MFRG (2015) Hair cosmetics: an overview. Int J Trichology 7(1):2. https://doi.org/10.4103/0974-7753.153450

Escamilla-Cruz M, Magaña M, Escandón-Perez S, Bello-Chavolla OY (2023) Use of 5-alpha reductase inhibitors in dermatology: a narrative review. Dermatol Ther (Heidelb) 13(8):1721–1731. https://doi.org/10.1007/s13555-023-00974-4

Fernández-Peña L et al (2020) Effect of molecular structure of eco-friendly glycolipid biosurfactants on the adsorption of hair-care conditioning polymers. Colloids Surf B Biointerfaces 185:110578. https://doi.org/10.1016/j.colsurfb.2019.110578

Gharaie S, Ohadi M, Hassanshahian M, Shakibaie M, Shahriary P, Forootanfar H (2023) Glycolipopeptide biosurfactant from Bacillus pumilus SG: physicochemical characterization, optimization, antibiofilm and antimicrobial activity evaluation. 3 Biotech 13(10):321. https://doi.org/10.1007/s13205-023-03728-3

Gubitosa J, Rizzi V, Fini P, Cosma P (2019) Hair care cosmetics: from traditional shampoo to solid clay and herbal shampoo, a review. Cosmet 6(1):13. https://doi.org/10.3390/COSMETICS6010013

Guzmán E, Lucia A (2021) Essential oils and their individual components in cosmetic products. Cosmet 8(4):114. https://doi.org/10.3390/COSMETICS8040114

Guzmán E, Ortega F, Rubio RG (2023) Exploring the world of rhamnolipids: a critical review of their production, interfacial properties, and potential application. Curr Opin Colloid Interface Sci 69:101780. https://doi.org/10.1016/j.cocis.2023.101780

He L, Michailidou F, Gahlon HL, Zeng W (2022) Hair dye ingredients and potential health risks from exposure to hair dyeing. Chem Res Toxicol 35(6):901–915. https://doi.org/10.1021/acs.chemrestox.1c00427

Inoue K et al (2009) TGF-β 2 is specifically expressed in human dermal papilla cells and modulates hair folliculogenesis. J Cell Mol Med 13(11–12):4643–4656. https://doi.org/10.1111/j.1582-4934.2009.00739.x

Leoty-Okombi S et al (2021) Effect of sodium lauryl sulfate (Sls) applied as a patch on human skin physiology and its microbiota. Cosmetics 8(1):1–12. https://doi.org/10.3390/COSMETICS8010006

Li Y et al (2022) Hair-growth promoting effect and anti-inflammatory mechanism of Ginkgo biloba polysaccharides. Carbohydr Polym 278:118811. https://doi.org/10.1016/J.CARBPOL.2021.118811

Majeed M, Nagabhushanam K, Paulose S, Rajalakshmi HR, Mundkur L (2023) A randomized double-blind, placebo-controlled study to evaluate the anti-skin-aging effect of LactoSporin—the extracellular metabolite from Bacillus coagulans (Weizmannia coagulans) MTCC 5856 in healthy female volunteers. Clin Cosmet Investig Dermatol 16:769–782. https://doi.org/10.2147/CCID.S403418

Mesler AL, Veniaminova NA, Lull MV, Wong SY (2017) Hair follicle terminal differentiation is orchestrated by distinct early and late matrix progenitors. Cell Rep 19(4):809–821. https://doi.org/10.1016/j.celrep.2017.03.077

Moghaddam M, Mehdizadeh L (2017) Chemistry of essential oils and factors influencing their constituents. In: Soft chemistry and food fermentation. Academic Press, pp 379–419

Mohanty SS et al (2021) A critical review on various feedstocks as sustainable substrates for biosurfactants production: a way towards cleaner production. Microb Cell Fact 20(1):1–13. https://doi.org/10.1186/S12934-021-01613-3

Musa OM, Tallon MA (2013) Hair care polymers for styling and conditioning. In: Polymers for personal care and cosmetics., 1148, pp 233–284

Neuser F, Schlatter H (2010) Hair dyes. In: Cosmetic dermatology: products and procedures. John Wiley & Sons, Ltd., pp 227–235

Padule K, Shinde S, Chitlange S, Giram P, Nagore D (2022) The advancement of herbal-based nanomedicine for hair. Cosmet 9(6):118. https://doi.org/10.3390/COSMETICS9060118

Penkar GM et al (2023) An overview on Indian herbs in hair care therapy. Res J Pharmacogn Phytochem 15(2):161–166. https://doi.org/10.52711/0975-4385.2023.00025

Rincón-Fontán M, Rodríguez-López L, Vecino X, Cruz JM, Moldes AB (2016) Adsorption of natural surface active compounds obtained from corn on human hair. RSC Adv 6(67):63064–63070. https://doi.org/10.1039/c6ra13823c

Rincón-Fontán M, Rodríguez-López L, Vecino X, Cruz JM, Moldes AB (2020) Novel multifunctional biosurfactant obtained from corn as a stabilizing agent for antidandruff formulations based on Zn pyrithione powder. ACS Omega 5(11):5704–5712. https://doi.org/10.1021/acsomega.9b03679

Schultz J, Rosado AS (2020) Extreme environments: a source of biosurfactants for biotechnological applications. Extremophiles 24(2):189–206. https://doi.org/10.1007/S00792-019-01151-2/FIGURES/3

Sen S, Borah SN, Deka S (2021) Biotechnologically derived bioactive molecules for skin and haircare application. In: Biosurfactants for a sustainable future. Wiley, pp 443–463

Søsted H, Rastogi SC, Andersen KE, Johansen JD, Menné T (2004) Hair dye contact allergy: quantitative exposure assessment of selected products and clinical cases. Contact Derm 50(6):344–348. https://doi.org/10.1111/J.0105-1873.2004.00362.X

Trüeb RM (2001) The value of hair cosmetics and pharmaceuticals. Dermatology 202(4):275–282. https://doi.org/10.1159/000051658

Trüeb RM (2005) Dermocosmetic aspects of hair and scalp. J Investig Dermatology Symp Proc 10(3):289–292. https://doi.org/10.1111/J.1087-0024.2005.10118.X

Trüeb RM (2007) Shampoos: ingredients, efficacy and adverse effects. J Dtsch Dermatol Ges 5(5):356–365. https://doi.org/10.1111/J.1610-0387.2007.06304.X

Trüeb RM, Lee W-S (2014) Hair care and cosmetics. In: Male alopecia. Springer, Cham, pp 239–261

Vecino X, Cruz JM, Moldes AB, Rodrigues LR (2017) Biosurfactants in cosmetic formulations: trends and challenges. Crit Rev Biotechnol 37(7):911–923. https://doi.org/10.1080/07388551.2016.1269053

Yano K, Brown LF, Detmar M (2001) Control of hair growth and follicle size by VEGF-mediated angiogenesis. J Clin Invest 107(4):409–417. https://doi.org/10.1172/JCI11317

Global Regulatory Frameworks for Hair Care Industry: An Overview of Key Agencies and Guidelines

13

Akshay Anand M.

Abstract

The hair care industry is a rapidly growing sector that is subject to various regulations and guidelines. This chapter provides an overview of the key agencies and guidelines that regulate the hair care industry globally. It covers the regulatory frameworks of the United States, the European Union, and India. It outlines the regulatory requirements and pathways for hair care products under the categories of cosmetics and drugs. This chapter also explores how regulatory variations affect the safety and availability of these products in the specified regions.

Keywords

Hair care products · Hair care industry · Regulatory frameworks · Guidelines · Regulatory requirements · Regulatory bodies

13.1 Introduction

The hair care industry is a significant part of the beauty and personal care sector, covering hygiene and beauty treatments for hair on the scalp, face, and body. Haircare practices vary widely based on cultural backgrounds and hair types. According to a report by Fortune Business Insights, the global hair care market is expected to grow from $99.53 billion in 2023 to $147.49 billion by 2030 (Fortune Business Insights 2030). In India, the hair care market was valued at $3.4 billion in 2023 and is projected to grow at a Compound Annual Growth Rate (CAGR) of

A. Anand M. (✉)
Regulatory Affairs, ITC Limited—ITC Life Sciences and Technology Centre, Bengaluru, Karnataka, India
e-mail: akshay.anand@itc.in

A. Dwivedi et al. (eds.), *Hair Care Products*,
https://doi.org/10.1007/978-981-97-6740-3_13

11.4% during the same period, as per the TechSci Research (TechsciResearch Pvt Ltd. 2029).

Given the rapid growth and increasing demand for hair care products, it is crucial to ensure these products are safe and meet the regulatory standards. Regulatory frameworks are essential in maintaining these standards. This chapter will provide an overview of the key regulatory agencies and guidelines governing the hair care industry globally. It will cover safety standards, labeling requirements, and testing procedures. By the end of this chapter, the reader will have a better understanding of the global regulatory frameworks and their impact on the hair care industry.

13.2 Hair Care Products

Broadly, hair care products can be categorized into two distinct categories based on the application. The first is personal care or cosmetic applications while the second is therapeutic or medicated applications.

Under cosmetic or personal care use cases, the term "hair care" may be widely applied to a number of activities involved in maintaining hygiene and beautification of hair. The hair on a human body can undergo various grooming processes, such as coloring, trimming, cutting, shaving, plucking, or removal through methods like waxing, sugaring, and threading. The assortment of products used in these processes, along with the equipment and techniques employed, collectively constitutes the "hair care" category. This category encompasses a wide range of items from shampoos and conditioners to styling tools and hair removal devices, reflecting the diverse approaches to hair maintenance and styling influenced by personal preferences and cultural practices.

Cleaning and maintaining hair will employ cosmetic products such as soap, shampoo, conditioner, oil, serum, tonic wax, gel, spray, pomade, color, etc., and styling requires equipment such as hair dryer, comb, hairbrush, straightener, iron, roller, curler, clipper, scissors, ribbons, clips, pins, etc.

There are different types of cosmetic hair care products available on the global market. Some examples have been listed below. All these products come in different types and formulations.

1. **Hair Soap:** Hair soap is a type of soap that is used to clean hair and scalp. Some soaps may have a high alkaline pH, which causes hair tangling and roughness and are not usually preferred.
2. **Hair Shampoo:** A hair care product that is used to clean hair and scalp. It helps to remove dirt, oil, and product buildup from hair.
3. **Hair Conditioner:** A hair care product that is used to moisturize and de-tangle hair. It helps to make hair soft, smooth, and shiny.
4. **Hair Oil:** A hair care product that is used to nourish and strengthen hair. It helps to prevent breakage and split ends.
5. **Hair Serum:** A hair care product that is used to add shine and smoothness to hair. It helps to tame frizz and flyaways.
6. **Hair Tonic:** A liquid hair moisturizer that typically includes light oils or liquid petroleum; it is used to moisturize the hair and scalp.

7. **Hair Mask:** A hair care product that is used to deep-condition and repair damaged hair. It helps to restore moisture and improve the overall health of the hair.
8. **Hair Spray:** A hair care product that is used to hold hairstyles in place. It helps to add volume and texture to the hair.
9. **Hair Gel:** A hair care product that is used to style hair. It helps to create a variety of hairstyles, from slicked-back looks to spiky styles.
10. **Hair Wax/Pomade:** A thick hairstyling product containing wax, used to assist with holding the hair.
11. **Hair Color:** A range of products that are used to change the color of hair. These products include hair dyes, hair toners, and hair color removers.

While cosmetic hair care products suffice for the needs of an individual in terms of maintaining and beautifying their hair, there are products that have therapeutic properties and are used to treat hair-related ailments.

The therapeutic/medicinal/drug products for hair care are usually prescribed by a dermatologist or a trichologist, or are available as over-the-counter (OTC) drugs and are used for treating disorders such as

1. Allergic reactions to chemical ingredients in cosmetic hair care products.
2. Infestation of mites or lice, infections of follicles or fungus on the scalp.
3. Dandruff, psoriasis, eczema, seborrheic dermatitis.
4. Hair loss (alopecia), baldness, breakage, other damages to hair.

The hair care products used for therapeutic indications typically contain one or more active pharmaceutical ingredients. Some examples are as follows (Stylecraze 2023; Muinos 2022; Drugs.com n.d.):

1. **Cortisone:** A steroid and potent hair re-growth medicine and has been shown effective when taken directly on the scalp as injections.
2. **Minoxidil:** A popular non-prescription hair-loss remedy, promotes hair growth; available as liquids, shampoos, foams, etc.
3. **Finasteride:** A key drug that promotes hair growth in the crown region; helps in preventing a receding hairline in men. Available as oral and topical formulations.
4. **Dutasteride:** Used to treat alopecia in men (off-label); similar to Finasteride in action but more effective. Available as oral formulations.
5. **Spironolactone:** Used (off-label) as a treatment for hair loss in women. Available as oral formulations.
6. **Diphenylcyclopropenone:** A prescription only drug used to promote hair growth and treat alopecia; available as liquid and applied topically.
7. **Selenium sulfide:** Non-prescription, OTC drug used to treat dandruff, scalp seborrheic dermatitis. Available as topical preparations (shampoo).
8. **Zinc pyrithione:** Non-prescription, OTC drug used to treat dandruff, scalp seborrheic dermatitis. Available as topical preparations (shampoo).

9. **Ciclopirox:** Anti-fungal drug used to treat dandruff, scalp seborrheic dermatitis. Available as topical preparations (shampoo).
10. **Ketoconazole:** Anti-fungal drug used to treat dandruff, scalp seborrheic dermatitis. Available as topical preparations (shampoo).
11. **Betamethasone:** Corticosteroid drug used to treat dandruff, scalp seborrheic dermatitis. Available as topical preparations (shampoo).
12. **Clobetasol:** Corticosteroid drug used to treat dandruff, scalp seborrheic dermatitis. Available as topical preparations (shampoo).
13. **Fluocinolone:** Corticosteroid drug used to treat dandruff, scalp seborrheic dermatitis. Available as topical preparations (solution and shampoo).

Such products will need to go through a more stringent regulatory approval process before being placed on the market. Those hair care products that are categorized as cosmetics/personal care products undergo a comparatively less complex regulatory pathway to be commercialized.

13.3 Regulatory Framework

The regulatory requirements differ for products that are considered cosmetic and products categorized as drugs. While safety is the primary concern for cosmetics, efficacy along with safety needs to be proven for drug products.

The cosmetic industry, including hair care, is known for its innovation and rapid development. It is crucial to regulate this segment to ensure product safety and quality, and to prevent any adverse health effects on consumers. However, regulatory frameworks vary widely across different markets, both locally and globally. This lack of harmonization makes it challenging to sell the same product worldwide (Cosmetics Europe et al. 2021; Ferreira et al. 2022).

The regulatory framework for cosmetic hair care products varies by country and region and is relatively less complex. On the other hand, the products categorized as drugs will require pre-market approval in most of the global markets and go through a stringent regulatory pathway. The companies that manufacture hair care products must ensure that their products meet these standards before they enter the market.

Most regulatory agencies enforce regulations related to safety standards, labeling requirements, and testing procedures. For instance, in the United States (US), the FDA regulates hair care products and enforces regulations related to safety, labeling, and product claims. In the case of drug products, the US FDA has established a comprehensive regulatory framework for drug development, approval, and post-marketing surveillance. The European Union (EU) has stringent regulations for cosmetic products sold within its borders. European Medicines Agency (EMA), the regulatory body for drugs within the EU, regulates the hair care products under the category of drugs. Bureau of Indian Standards (BIS) in India has laid down specifications for various products under the cosmetic hair care products category, and the Central Drugs Standards Control Organization (CDSCO) regulates both drug and cosmetic products in India.

13.4 Key Agencies Around the World

For the purpose of this chapter, three key markets have been considered—the US, the EU, and India. Table 13.1 captures the key elements with regards to the regulatory overview for each of these geographies.

Table 13.1 Regulatory overview for the US, EU and India

Geography	United States	European Union	India
Regulatory Authority	US FDA (U.S. Food and Drug Administration)	EMA (European Medicines Agency)	CDSCO (Central Drugs Standards Control Organization)
Applicable Law/ Guidelines for "Cosmetic"	The US Federal Food, Drug and Cosmetic Act (Office of the Commissioner 2018)	EC Regulation 1223/2009 (Legislation.gov.uk 2009)	The Drugs and Cosmetics Act, 1940 (CDSCO n.d.-a); The Cosmetics Rules, 2020 (Cosmetics Rules 2020) and Indian Standards 4707 (Part 2) (IS 4704 (Part2) 2017)
Applicable Law/ Guidelines for "Drug"	The US Federal Food, Drug and Cosmetic Act	Directive 2001/83/EC; EC Regulation 726/2004	The Drugs and Cosmetics Act, 1940; The Drugs Rules, 1945
Definition of "Cosmetic"	Products that are applied to the human body in various ways (like rubbing, pouring, sprinkling, or spraying) for purposes such as cleansing, beautifying, making someone more attractive, or changing their appearance (Office of the Commissioner 2018)	Cosmetics are substances or mixtures meant to be used on the outside of the body (like skin, hair, nails, lips, and external genital organs) or on the teeth and mouth lining. Their main purposes are to clean, perfume, change appearance, protect, maintain, or correct body odors (Legislation.gov.uk 2009)	Cosmetic is any product meant to be applied to the human body in various ways (like rubbing, pouring, sprinkling, or spraying) for purposes such as cleansing, beautifying, making someone more attractive, or changing their appearance. This also includes any item intended to be used as a part of a cosmetic product (CDSCO n.d.-a; Cosmetics Rules 2020)

(continued)

Table 13.1 (continued)

Geography	United States	European Union	India
Definition of "Drug"	A drug is any substance used to diagnose, cure, reduce, treat, or prevent diseases. It also includes substances (other than food) that can change how the body works or its structure, whether in humans or animals.	A drug is any substance or a combination of substances that is claimed to have the ability to treat or prevent diseases in humans; or a drug is any substance or combination of substances used in humans to restore, correct, or change how the body works. This can be done through pharmacological (drug-related), immunological (immune system-related), or metabolic (body's chemical processes) actions. It also includes substances used for medical diagnosis.	Medicines for humans or animals, used either inside or outside the body, including those for diagnosing, treating, reducing, or preventing diseases, and even those used to repel insects like mosquitoes. Substances (excluding food) that can change how the body works or are used to kill pests or insects that cause diseases, as specified by the government. Ingredients used to make medicines, including empty capsules Devices used for diagnosing, treating, reducing, or preventing diseases in humans or animals, as specified by the government.

13.5 United States

The US is a major country in North America. With around 334 million people (US Census Bureau 2021), the US is the third most populated country in the world (Wikipedia 2024a) and has a GDP of $26.949 trillion (IMF 2023a), making it the second largest economy globally (Wikipedia 2024b).

In the US, the FDA regulates all cosmetic products, including hair care products. The FDA is a federal agency responsible for ensuring the safety, effectiveness, and security of human and veterinary drugs, biological products, and medical devices.

The US Federal Food, Drug, and Cosmetic Act (FD&C Act) ensures the safety of food, drugs, medical devices, and cosmetics in the US. According to this act, cosmetics are defined as products intended to be applied to the human body for cleansing, beautifying, promoting attractiveness, or altering appearance. This includes a wide range of items like skin moisturizers, perfumes, lipsticks, nail polishes, makeup, shampoos, hair colors, and deodorants. It also covers any substance used as a component of a cosmetic product.

In the US, the FDA does not "approve" cosmetic products under the FD&C Act. However, the act prohibits the sale of misbranded and adulterated cosmetics. Manufacturers do not need FDA approval for cosmetic products and ingredients, except for color additives, before selling them.

The FD&C Act and the Fair Packaging and Labeling Act (FPL Act) are the two key laws for cosmetics in the US. The FPL Act requires that all consumer commodities have a label with:

1. The product's identity.
2. The name and location of the manufacturer, packer, or distributor.
3. The net quantity of contents.

Consumer commodities include any food, drug, device, or cosmetic intended for retail sale for personal use. Labels can be any written, printed, or graphic material attached to the product or its packaging (Center for Food Safety and Applied Nutrition 2022c; Congress.gov 1965).

Under the FD&C Act and the FPL Act, the FDA regulates cosmetic products in the US through the Center for Food Safety and Applied Nutrition, which also regulates food. Cosmetic products do not need pre-market approval by the FDA unless they have a "structure or function claim" that classifies them as drugs (or cosmeceuticals). However, color additives must be approved by the FDA before being used in cosmetic products sold in the US (Ross 2006).

In the US, cosmetic manufacturers can use any ingredient in their products, except for banned or restricted substances and color additives, as long as:

1. The ingredients and the final product are safe when used as directed or under normal conditions.
2. The product is correctly labeled.
3. The ingredient does not make the product adulterated or falsely labeled according to FDA regulations.

Manufacturers and distributors are legally required to ensure their products are safe. However, the FDA does not mandate specific tests to prove the safety of each product or ingredient, nor do manufacturers have to share their safety data with the FDA.

The cosmetic industry is primarily responsible for ensuring the safety of its products. However, the FDA steps in when necessary to protect public health. The FDA recommends that manufacturers conduct all necessary testing to ensure product and ingredient safety.

Manufacturers can establish safety using existing toxicological data for individual ingredients and similar formulations, along with any additional tests as needed. Regulations also limit or ban certain ingredients and require cautionary statements on labels for specific products.

The FD&C Act does not mandate animal testing for cosmetic safety (Center for Food Safety and Applied Nutrition 2022d). Manufacturers may choose to use animal testing to ensure the safety of new products, especially if other methods are insufficient. The FDA supports legal, regulatory, and policy frameworks for animal testing and encourages humane and minimal use of animals, ensuring the most valuable scientific data is obtained.

The FDA has the authority to inspect cosmetic facilities and imported cosmetics (Center for Food Safety and Applied Nutrition 2022a) to ensure they comply with laws and regulations, are safe, and correctly labeled. These inspections can happen at any time without prior notice and are influenced by factors like the nature of the products, complaints, the company's compliance history, and available resources.

Although cosmetics do not need pre-market approval, the FDA collects samples for testing during inspections and in response to complaints about adverse effects. The FDA also conducts research on cosmetic products and ingredients to address any safety concerns.

The FDA also does not have direct authority to order recalls of cosmetic products (Center for Food Safety and Applied Nutrition 2022b). In case it is found that a certain manufacturer is selling cosmetic products that are adulterated or misbranded, the manufacturer may be sued by the FDA or asked by the FDA to recall the products from the market. Marketing of adulterated or misbranded cosmetics in the US is prohibited by the FD&C Act. The act defines the terms "adulteration" and "misbranding" as follows.

A cosmetic product is considered "Adulterated" under FD&C Act, Section 601, for four main reasons:

1. It is injurious to consumers under conditions of prescribed use because either the cosmetic product or its container consists of a potentially harmful substance.
2. It contains dirty, filthy, or decomposed material.
3. It contains non-permitted or non-certified color additives.
4. It is manufactured or stored in insanitary conditions because of which it has become injurious to consumers or contains decomposing or filthy substances.
5. This provision does not apply to coal-tar hair dyes because they carry appropriate disclaimers and details of testing.

A cosmetic product is deemed as "Misbranded" under FD&C Act, Section 602, when:

1. The labelling is false or misleading.
2. It fails to state explicitly any information that is required by or under authority of this act.
3. It contains misleading container presentation or fill.

Under the FD&C Act, a product may be deemed misbranded if it does not disclose essential information. This includes, for instance, any instructions for safe usage and necessary warning labels that ensure the product can be used safely.

Furthermore, the FPL Act mandates that the FDA requires cosmetics to include a list of ingredients (as per Title 21, Code of Federal Regulations (CFR), Section 701.3). If cosmetics do not adhere to the FPL Act, they are considered misbranded (refer to FPL Act, Section 1456). This is not applicable to cosmetics meant exclusively for professional, institutional, or promotional use, such as free samples or hotel toiletries. The FDA is authorized to take regulatory actions against any cosmetics found to violate these statutes, as well as individuals or entities involved in distribution (Ferreira et al. 2022).

Cosmetic manufacturers are legally not obligated to register their facilities or submit their product formulations with the FDA, nor do they need a registration number to import cosmetics into the US. However, they are invited to join the FDA's Voluntary Cosmetic Registration Program (VCRP) (Center for Food Safety and Applied Nutrition 2023).

The VCRP is a reporting tool available for manufacturers, packers, and distributors of cosmetics in the US. This tool helps FDA in overseeing the cosmetic industry and ensuring good-quality cosmetics. Using the VCRP tool, the stakeholders can submit details about their products and register their manufacturing or packaging locations in the VCRP database.

In US market, a pre-market approval is mandatory for a hair care product categorized as a drug. The FDA's Center for Drug Evaluation and Research (CDER) evaluates new drugs, both prescription and OTC drugs, before they enter the market. CDER's evaluation ensures that dishonesty is avoided by providing access to correct information for doctors and patients. CDER ensures that products are genuine and effective, and their health benefits outweigh their risks. CDER is responsible for the evaluation of both prescription and OTC drugs (Clark et al. 2015; Center for Drug Evaluation and Research 2016).

The FDA has a structured process to approve drugs (Center for Drug Evaluation and Research 2017).

1. Preclinical research: This involves laboratory and animal testing to gather initial data on the drug's safety and biological activity.
2. Investigational new drug (IND) application: Based on pre-clinical findings, this application includes details on animal pharmacology, toxicology, drug composition, stability, and intended clinical uses.
3. Clinical trials: If the IND is approved, the drug undergoes several phases of clinical trials to test its safety and effectiveness in humans.
4. Preclinical research: This involves laboratory and animal testing to gather initial data on the drug's safety and biological activity.
5. IND application: Based on pre-clinical findings, this application includes details on animal pharmacology, toxicology, drug composition, stability, and intended clinical uses.
6. Post-market safety monitoring: Even after approval, the drug is monitored for any long-term or rare side effects.

Not every drug that starts this process makes it to the market, as it must demonstrate clear clinical benefits to progress through each stage.

To obtain FDA approval for marketing a new drug in the US, a drug sponsor submits a New Drug Application (NDA). This application includes comprehensive data from animal and human studies, analyses, and details on the drug's pharmacokinetics and manufacturing process. Upon receiving an NDA, the FDA has 60 days to decide if it qualifies for review. An FDA review team then evaluates the sponsor's evidence on the drug's safety and effectiveness. The FDA also reviews the drug's professional labeling to ensure it effectively communicates essential information to healthcare providers and consumers. Additionally, the FDA inspects the manufacturing facilities. If the review is satisfactory and no issues are found, the FDA will approve the application. If there are concerns, the FDA will send a response letter outlining the discrepancies to the sponsor.

The post-marketing phase is crucial in the lifecycle of any FDA-approved drug. Since clinical trials cannot predict every possible effect, continuous monitoring of drug safety after it hits the market is essential. The FDA's post-marketing safety system is vital for detecting severe, unexpected adverse reactions and taking necessary actions. Once a drug is approved by the FDA, the post-marketing surveillance phase begins, during which the drug's sponsor must regularly provide safety updates to the FDA.

The FDA has a system called MedWatch for people, including doctors and consumers, to report any bad reactions to drugs. If new serious risks are found, they are added to the drug's label, and the public is informed. Sometimes, the drug's use might need to be limited, and in rare cases, it might be taken off the market completely (Center for Drug Evaluation and Research 2022).

13.6 The European Union

The EU is a group of 27 countries in Europe that work together politically and economically. It has about 449 million people (Eurostat n.d.), which is around 6% of the world's population (European Demographic Datasheet 2022). In 2023, the EU had a GDP of $25.399 trillion (IMF 2022).

The EMA oversees the evaluation and supervision of pharmaceutical products in the EU, but it does not regulate cosmetics. Instead, the European Commission (EC) sets the rules for cosmetic products, including hair care items, to be sold in the European market (EU Cosmetics n.d.; European Directorate for the Quality of Medicines and HealthCare n.d.).

In the EU, Regulation EC 1223/2009 (Cosmetic Design Europe 2023; EUR-LeX 2009) governs the manufacture, labeling, and supply of cosmetics and personal care products. This regulation applies to everyone, whether individuals or companies, who imports or produces cosmetic products. To sell their products in the EU, they must comply with these rules.

The first step in meeting compliance requirements is to check if the product you want to make or import fits the definition of a "cosmetic product" according to the regulation (Cosmetics Europe—The Personal Care Association n.d.-a). According to EC 1223/2009, a "cosmetic product" is any substance or mixture meant to be used on the external parts of the human body (like skin, hair, nails, lips, and external genital organs) or on teeth and the inside of the mouth. The purpose is mainly to clean, perfume, change appearance, protect, keep in good condition, or correct body odors.

Hair care products are covered under the same rules as other cosmetics in the EU. To sell a cosmetic product in the European market, you need the following (European Commission n.d.-a).

1. A "responsible person" to ensure compliance.
2. A "product information file" (PIF).
3. Notification of the product on the "Cosmetic Products Notification Portal" (CPNP).
4. Correct labeling of the product.

Article 4 of the regulation mandates that manufacturers must appoint a "responsible person" to ensure their cosmetic products comply with Regulation EC 1223/2009. This person handles compliance before and after the product is on the market and resolves any issues.

Article 11 says the responsible person must keep a PIF for each cosmetic product sold in the EU. These files must be kept for 10 years after the last batch is sold and be available for inspection. The PIF should include a product description, safety assessments, proof of the product's effectiveness, and any animal testing data.

Article 3 maintains that all cosmetics must be safe for human health when used as intended or in a reasonably expected way. Article 8 of the regulation ensures manufacturers must follow good manufacturing practices (GMPs) to ensure product safety, which means following established standards.

Article 10, on the other hand, says that before selling a cosmetic product, the responsible person must ensure its safety by conducting a thorough safety assessment and creating a detailed safety report. This report should consider how the product will be used and the absorption of each ingredient. The safety report must be updated with any new information after the product is on the market.

Before selling a cosmetic product, the responsible person must submit the following information to the EC via the CPNP):

1. The category and name(s) of the cosmetic product.
2. The name and address of the responsible person where the PIF is accessible.
3. The country of origin, if the product is imported.
4. The EU member state where the product will be sold.
5. Contact details of a person to reach in case of an emergency.
6. Information on any nanomaterials in the product, including their chemical name and expected exposure conditions.
7. The name and Chemical Abstract Service (CAS) number or European Community (EC) number of any substances classified as carcinogenic, mutagenic, or toxic for reproduction (CMR).
8. The product's formulation to allow for proper medical treatment if needed.

Before selling a cosmetic product, the responsible person must inform the EC about the original labeling and, if possible, provide a photo of the packaging. The labeling must follow the guidelines in Article 19 of EC 1223/2009 and be clear and permanent.

The label must include:

1. Name and address of the responsible person and the country of origin (for imported products).
2. Weight or volume of the product at the time of packaging.
3. Expiration date.
4. Precautions for use.
5. Batch number.

6. Function of the product unless it is obvious.
7. List of ingredients in descending order by weight; ingredients less than 1% can be listed in any order after the main ingredients.

The function, amount, precautions, and warnings must be translated into the language of the destination country. Some countries require a full translation of all label elements. The EU has restrictions on certain substances in cosmetic products under Article 14 of the regulation. These include:

1. Prohibited substances listed in Annex II.
2. Restricted substances not used according to Annex III.
3. Colorants not listed in Annex IV or not used according to its conditions.
4. Preservatives not listed in Annex V or not used according to its conditions.
5. UV filters not listed in Annex VI or not used according to its conditions.

Hair coloring products must comply with Annex IV if included. They cannot contain unlisted hair colorants or those not used according to the Annex. In Europe, cosmetic products and their ingredients cannot be tested on animals, as per Article 18. Products or ingredients tested on animals cannot be sold in the European market (Cosmetics Europe—The Personal Care Association n.d.-b; European Commission n.d.-b).

The EC and the EMA do not directly handle the market placement of cosmetic hair care products. However, the EMA plays a key role when these products are classified as drugs. The regulatory framework for medicines in Europe involves about 50 regulatory authorities from 30 countries (Wikipedia 2024c) in the European Economic Area (EEA), which includes the 27 EU Member States, Iceland, Liechtenstein, and Norway. This network, along with the EC and EMA, makes the European regulatory system unique (Wikipedia 2024d). The EU used to have 28 members until the United Kingdom (UK) in 2020. Now, the Medicines and Healthcare products Regulatory Agency (MHRA) is the sole regulatory body for the UK (Wikipedia 2023).

The EMA works with the 27 EU Member States to share knowledge in evaluating new medicines, ensuring their safety, and managing public health crises. They exchange regulatory information to avoid redundant efforts, share responsibilities, and effectively regulate medicines across the EU.

To protect public health and ensure high-quality, safe, and effective medicines in Europe, every medicine must be authorized before being sold in the EU. There are different ways to get this authorization, including centralized, decentralized, and mutual recognition processes.

The centralized route allows a medicine to be distributed across the EU after a single assessment and authorization that is recognized EU-wide. This lets pharmaceutical companies submit one application through the EMA and then distribute the medicine throughout the European Economic Area, making it accessible to patients and healthcare professionals with a single marketing authorization (European Medicines Agency n.d.-a).

Before applying, applicants should check if their product qualifies for the centralized evaluation process. If approved, they must notify the EMA about their intent to submit a Marketing Authorization Application (MAA). The Committee for Medicinal Products for Human Use (CHMP) and the Pharmacovigilance Risk Assessment Committee (PRAC) assign reviewers to examine the scientific data. Applicants attend pre-submission meetings with the EMA to ensure their submissions meet the legal and regulatory standards, speeding up the validation process. They must also confirm or update the EMA on their submission date. Applications are submitted in the electronic common technical document (eCTD) format through the EMA's e-Submission gateway. The EMA checks if the application includes all necessary elements for a scientific review and may request additional information if needed.

The CHMP reviews applications submitted via the centralized procedure, incorporating feedback from the PRAC on the risk management plan and from the Committee for Advanced Therapies (CAT) on advanced-therapy medicinal products. After a typical evaluation period of 210 days, the CHMP provides a scientific assessment, which the EMA forwards to the EC for the final marketing authorization.

The EC makes the final decision on granting centralized marketing authorizations based on the EMA's recommendations. Once approved, the authorization is recognized across all EU Member States and EEA countries, including Iceland, Liechtenstein, and Norway.

For companies seeking authorization in multiple Member States, there are two options: the decentralized procedure and the mutual recognition procedure. The decentralized procedure allows for concurrent authorization requests in several EU Member States for medicines not yet approved in any EU country and not covered by the centralized procedure. The mutual recognition procedure allows a medicine authorized in one EU Member State to be recognized in others, using shared scientific evaluations.

After the EMA's assessment, a European Public Assessment Report (EPAR) is published for approved or denied medicines (European Medicines Agency n.d.-b). Similarly, a Public Assessment Report is available for medicines authorized by an individual Member State. EPARs are translated into all 24 official EU languages for accessibility.

13.7 India

India, officially the Republic of India, is a country in South Asia. It is the seventh largest country by area, the most populous country as of June 2023 (Wikipedia 2024e), and the world's largest democracy. In 2023, India had a GDP of $13.119 trillion, making it the third largest economy in the world (IMF 2023b).

The demand for organic hair care products is growing in India (India Hair Care and Styling Products Market Insights n.d.). Environmental factors like pollution and UV exposure cause hair damage, leading to issues like dryness, dandruff, and premature greying (Statista n.d.). The frequent use of chemical products, heat styling, and hair dyes also contributes to hair problems. Because of this, more people are choosing organic hair care products that promise to reduce damage and promote healthy hair.

Consumers are becoming more aware of these products and are willing to spend more on items that offer quick and effective results. Manufacturers are responding by creating products that meet these preferences. Hair oils are especially popular for their ability to nourish the scalp, strengthen hair, provide relaxation, and reduce damage. Companies are developing hair oils with herbal and natural ingredients to meet this demand.

In India, most hair care products are considered cosmetics. There are also specialized products for treating specific hair issues, as well as herbal and natural "AYUSH" products, which are popular in the market. All these products, whether cosmetic, drug, or traditional remedy, are regulated by the Drugs and Cosmetics Act 1940, the Drugs Rules 1945, and the Cosmetics Rules 2020. Other rules, like the New Drugs and Clinical Trials Rules 2019, Legal Metrology (Packaged Commodities) Rules 2011, and Plastic Waste Management Rules 2016, are also important during the product's lifecycle in India.

The Drug Controller General of India (DCGI) regulates the pharmaceutical and cosmetic industry through the CDSCO (ChemLinked n.d.; CDSCO n.d.-b). Unlike the US and EU, India requires cosmetic products to be approved before they can be sold. The responsibility of approving cosmetic products to be manufactured and sold in India is with the State Licensing Authorities (Drug Controllers for the respective States) while the Central Licensing Authority (DCGI) reserves the power to grant licenses to import cosmetic product into India.

In India, a cosmetic is any product meant to be applied to the human body for cleaning, beautifying, making more attractive, or changing the appearance. This includes any component used in cosmetics. A "New Cosmetic" is a product with an ingredient that has never been used anywhere in the world or is not recognized for use in cosmetics in any national or international literature (CDSCO n.d.-a; Cosmetics Rules 2020).

The Bureau of Indian Standards (BIS) is India's National Standards Body under the Ministry of Consumer Affairs. One of its main roles is to create, recognize, and promote Indian Standards. BIS has created over 20,000 standards to help improve the quality of products and services in various sectors.

For cosmetic products, including hair care items, BIS has set specific standards that must be followed according to the Ninth Schedule of the Cosmetics Rules 2020. These standards include:

1. Soap-based Shampoo (IS: 7669).
2. Surfactant-based Shampoo (IS: 7884).
3. Hair Creams (IS: 7679).
4. Liquid Oxidation Hair Dyes (IS: 8481).
5. Aftershave Lotion (IS: 9255).
6. Pomades and Brilliantines (IS: 9339).
7. Chemical Depilatories (IS: 9636).

8. Shaving Creams (IS: 9740).
9. Powder Hair Dye (IS: 10350).
10. Henna Powder (IS: 11142).
11. Coldwax Hair Remover (IS: 15152).
12. Emulsion-type Oxidation Hair Dyes (IS: 15205).
13. Baby Hair Shampoo (IS: 17117).

These standards ensure that hair care products in India meet quality and safety requirements.

The raw materials used in hair care cosmetic products must comply with IS 4704 (Part 2) and only include dyes, colors, and pigments specified in IS 4707 (Part 1) and listed in the tenth schedule of the Cosmetics Rules 2020 (IS 4704 (Part2) 2017).

13.7.1 Importing Hair Care Products

To import a hair care product into India, it must be registered using Form COS-1 (or COS-3 for a "New Cosmetic") with the CLA. The CLA reviews the application and grants a registration certificate and import license.

13.7.2 Manufacturing Hair Care Products

To manufacture a hair care product for sale in India, an application must be submitted to the State Licensing Authority (SLA) using Form COS-5 (or COS-6 for a loan license) as per Rule 23. The application must include a self-declaration of compliance with GMPs in Form COS-7, along with other required information and documents. The SLA reviews the application and typically grants the license within 45 days using Form COS-8 (or COS-9 for a loan license), allowing the product to be manufactured and sold in India indefinitely. Within 30 days of issuing the license, the SLA will inspect the manufacturing site to verify the information provided in Form COS-7. If any discrepancies are found, the license may be withheld or cancelled.

Labeling Requirements (Rule 34 of the Cosmetics Rules 2020):

1. Name of the cosmetic.
2. Name and address of the manufacturer.
3. Expiry date or "use before"/"use by" date.
4. Distinctive batch number.
5. Manufacturing license number.
6. Country of origin and import registration number (for imported products).
7. Declaration of net contents (as per LMPC Rules).
8. Directions for safe use, warnings, and cautions.
9. List of ingredients.

Hair care products must also meet the labelling requirements specified in the relevant Indian standards.

Voluntary Recall: Manufacturers can withdraw products from the market if they believe the product may pose a health risk, after informing the SLA or the CLA.

Hair Care Products as Drugs: Hair care products classified as drugs must follow the regulatory pathway for drugs in India.

A "drug" in India, under the D&C Act 1940, is defined to include (Central Drugs Standard Control Organization 1945):

1. **Medicines** for humans or animals, used either inside or outside the body, including those for diagnosing, treating, or preventing diseases, and even those used to repel insects like mosquitoes.
2. **Substances** (excluding food) that can change how the body works or are used to kill pests or insects that cause diseases, as specified by the government.
3. **Ingredients** used to make medicines, including empty capsules.
4. **Devices** used for diagnosing, treating, or preventing diseases in humans or animals, as specified by the government.

13.7.3 Importing Hair Care Drug Products

1. To import hair care drug products into India, you need a license called Form-10.
2. To get this license, you must apply using Form-8 and include a copy of your Registration Certificate (Form-41), which you get by submitting Form-40.
3. The license will follow the rules set in the Drug Rules 1945.

13.7.4 Manufacturing Hair Care Drug Products

1. To manufacture hair care drug products in India, you need a license called Form-25.
2. You apply for this license using Form-24.
3. Your application must meet the requirements of Rule 71, and your manufacturing site must pass an inspection.
4. The license is valid for 5 years and can be renewed.

13.7.5 Labeling Requirements

Hair care drug products must follow the labeling rules in Rule 96 of the Drug Rules 1945. If a product is labeled as a cosmetic but contains a therapeutic substance, it is considered a drug and must follow the Drug Rules 1945, even if it also meets cosmetic regulations.

List of the labeling requirements for hair care drug products:

1. Name of the drug or its combination and brand name.
2. Net contents statement.

3. Active ingredients content.
4. Manufacturer's name and address.
5. Batch or lot number.
6. Manufacturing license number.
7. Import license number, and the importer's name and address (for imported products).
8. Manufacture and expiry dates (expiry not more than 60 months from manufacture).
9. Statement: "FOR EXTERNAL USE ONLY".

If the product contains a drug listed under Schedule H, the label must have the "Rx" symbol on the top-left corner. It should also include the following warning in black text inside a red box: "SCHEDULE H PRESCRIPTION DRUG—CAUTION—Not to be sold by retail without the prescription of a Registered Medical Practitioner". New Drugs and Clinical Trial Rules, 2019, provide a detailed definition of what constitutes a "new drug" in India.

A "new drug" in India is defined under the New Drugs and Clinical Trials (NDCT) Rules, 2019 as:

1. New drug: A drug (including active ingredients or plant-based drugs) not significantly used in India and not yet approved as safe and effective by the CLA.
2. Modified claims: A drug already approved for certain uses but proposed to be marketed with new or changed claims, such as different uses, ways of taking it, or dosages.
3. Fixed-dose combination: A combination of two or more drugs, approved separately, but proposed to be combined for the first time or with changed ratios.
4. Modified release: A drug with a new way of releasing its active ingredients over time or a novel delivery system.
5. Advanced products: Includes vaccines, genetically engineered products, living modified organisms, monoclonal antibodies, stem cell products, gene therapies, or transplants from other species intended for use as drugs.

A new chemical entity in India is any substance that has not been approved as a drug by any country's drug regulatory authority and is being developed as a new drug for the first time, proving its safety and effectiveness. If hair care drug products are made with new drug substances or new drugs are developed for hair care, they must follow the rules and regulations under the New Drugs and Clinical Trial Rules, 2019.

13.8 Conclusion

As highlighted in the preceding sections, the hair care industry operates on an international scale and is subject to regulation by various authorities worldwide. The regulatory landscapes differ greatly across regions, influencing the industry's competitive edge and financial success. Key markets like the US, EU, and India have stringent regulatory standards, which have been the focus of this discussion. The key differences with respect to these markets have been captured in Table 13.2.

Table 13.2 Hair Care Products Regulation - Key differences between US, EU and India

	United States	European Union	India
Hair care cosmetic products			
Agency involvement	FDA does not "approve" cosmetics. Provides resources to enable listing of product in the market	EMA not involved. EC Regulations play a vital role and provide guidance on how to list the product in the market	CDSCO/DCGI is directly involved in approving a cosmetic product. Governed through the Drugs & Cosmetics Act, the Cosmetics Rules and BIS specifications
Pre-market approval for cosmetic product	Not required	Not required	Required through State Licensing Authorities (SLAs)
Labelling of cosmetic products	As per FD&C Act and FPL Act	As per EC Regulation 1223/2009	As per Cometics Rules, 2020 and specific Indian Standards (Ninth Schedule) when applicable (Additionally, LMPC Rules and PWM Rules)
Post-marketing reporting system	Voluntary, through VCRP	EC may enforce recalls in case of non-compliance	Voluntary, addressed in Cosmetics Rules, 2020
Hair care drug products			
Pre-market approval for drug product	Required	Required	Required
Marketing authorization process	Single process	Multiple ways—centralized, decentralized, mutual recognition	Single process
Mode of submission of application	Electronic and physical	Electronic	Electronic
eCTD implemented?	Yes	Yes	Yes
Labelling of drug products	As per 21 CFR Part 201	As per Directive 2001/83/EC	As per Rule 96 of DR 1945
Post-marketing reporting system	Online ADR reporting through MedWatch	Online ADR reporting through EudraVigilence	Online ADR reporting through Vigiflow/Vigibase
Periodic safety reports	Through (Periodic Adverse Drug Experience Report (PADER)	Through Periodic Benefit Risk Evaluation Report (PBRER)	Through Periodic Safety Update Report (PSUR)

These regulatory variances can affect the safety and availability of hair care products in these areas. It is crucial for the hair care sector to adhere to each nation's regulatory demands to maintain product safety and quality, thereby preventing negative health outcomes for consumers. Although the pharmaceutical sector presents a distinct challenge for hair care, with heightened safety concerns leading to stricter regulations, it also serves as a platform for high-quality products to demonstrate their proven safety and effectiveness. Looking forward, as the global landscape evolves, the hair care industry must stay agile, continuously updating practices to align with the latest regulatory changes and consumer health considerations, ensuring that it not only meets but exceeds safety standards, fostering trust and loyalty among consumers.

References

CDSCO (n.d.-a) Drugs and Cosmetics Act 1940. https://cdsco.gov.in/opencms/opencms/en/Acts-and-rules/

CDSCO (n.d.-b) About cosmetics. https://cdsco.gov.in/opencms/opencms/en/Cosmetics/cosmetics/

Center for Drug Evaluation and Research (2016) FDA drug approval process infographic (horizontal). U.S. Food and Drug Administration. https://www.fda.gov/drugs/information-consumers-and-patients-drugs/fda-drug-approval-process-infographic-horizontal

Center for Drug Evaluation and Research (2017) The FDA's drug review process: ensuring drugs are safe and effective. U.S. Food and Drug Administration. https://www.fda.gov/drugs/information-consumers-and-patients-drugs/fdas-drug-review-process-ensuring-drugs-are-safe-and-effective

Center for Drug Evaluation and Research (2022) In: https://www.fda.gov/drugs/development-approval-process-drugs/how-drugs-are-developed-and-approved (ed) How drugs are developed and approved. U.S. Food and Drug Administration

Center for Food Safety and Applied Nutrition (2022a) Inspection of cosmetics. U.S. Food and Drug Administration. https://www.fda.gov/cosmetics/cosmetics-compliance-enforcement/inspection-cosmetics

Center for Food Safety and Applied Nutrition (2022b) FDA recall policy for cosmetics. U.S. Food and Drug Administration. https://www.fda.gov/cosmetics/cosmetics-recalls-alerts/fda-recall-policy-cosmetics

Center for Food Safety and Applied Nutrition (2022c) In: https://www.fda.gov/cosmetics/cosmetics-labeling-regulations/summary-cosmetics-labeling-requirements (ed) Summary of cosmetics labeling requirements. U.S. Food and Drug Administration

Center for Food Safety and Applied Nutrition (2022d) Animal testing and cosmetics. U.S. Food and Drug Administration. https://www.fda.gov/cosmetics/product-testing-cosmetics/animal-testing-cosmetics#:~:text=The%20FD%26C%20Act%20does%20not%20specifically%20require%20the,effective%20for%20substantiating%20the%20safety%20of%20their%20products

Center for Food Safety and Applied Nutrition (2023) Voluntary cosmetic registration program. U.S. Food and Drug Administration. https://www.fda.gov/cosmetics/voluntary-cosmetic-registration-program

Central Drugs Standard Control Organization (1945) Drugs rules. https://cdsco.gov.in/opencms/opencms/en/Acts-and-rules/Drugs-Rules/

ChemLinked (n.d.) India Cosmetic Regulation | ChemLinked. https://cosmetic.chemlinked.com/cosmepedia/india-cosmetic-regulation

Clark GW, Pope SM, Jaboori KA (2015) Diagnosis and treatment of seborrheic dermatitis. AAFP. https://www.aafp.org/pubs/afp/issues/2015/0201/p185.html

Congress.gov (1965) Text—S.985—89th Congress (1965–1966): An Act to regulate interstate and foreign commerce by preventing the use of unfair or deceptive methods of packaging or labeling of certain consumer commodities distributed in such commerce, and for other purposes. Congress.gov | Library of Congress. https://www.congress.gov/bill/89th-congress/senate-bill/985/text?r=1&s=2

Cosmetic Design Europe (2023) New changes to European cosmetics legislation. cosmeticsdesign-europe.com. https://www.cosmeticsdesign-europe.com/Article/2023/10/23/New-changes-to-European-cosmetics-legislation

Cosmetics Europe, Renner G, Jacques M (2021) An overview of existing cosmetics frameworks: common principles and compatibility. https://share.ansi.org/Shared%20Documents/Standards%20Activities/International%20Standardization/Standards%20Alliance/Standards%20Alliance%20Phase%202%20(SA2)/Activities/ARSO/ARSO%20PCPC%20Cosmetics%20Webinar%20Series/Webinar%202/Webinar2_Cosmetics_Europe_Final_Presentation.pdf

Cosmetics Europe—The Personal Care Association (n.d.-a) Understanding the cosmetics regulation. https://cosmeticseurope.eu/cosmetics-industry/understanding-cosmetics-regulation/

Cosmetics Europe—The Personal Care Association (n.d.-b) Alternatives to animal testing—the only way forward for the EU. https://cosmeticseurope.eu/statement-aat-only-way-forward-eu

Cosmetics Rules (2020). https://cdsco.gov.in/opencms/opencms/en/Acts-and-rules/Cosmetics-Rules/

Drugs.com (n.d.) List of 21 alopecia medications compared. https://www.drugs.com/condition/alopecia.html

EU Cosmetics (n.d.) International Trade Administration | Trade.gov. https://www.trade.gov/eu-cosmetics

EUR-LeX (2009) Regulation (EC) No 1223/2009 of the European Parliament. https://eur-lex.europa.eu/; https://eur-lex.europa.eu/legal-content/EN/TXT/?uri=CELEX%3A02009R1223-20230816

European Commission (n.d.-a) CPNP—Cosmetic products notification portal. https://webgate.ec.europa.eu/cpnp/

European Commission (n.d.-b) Ban on animal testing. Internal Market, Industry, Entrepreneurship and SMEs. https://single-market-economy.ec.europa.eu/sectors/cosmetics/ban-animal-testing_en

European Demographic Datasheet (2022). https://www.populationeurope.org/en/

European Directorate for the Quality of Medicines and HealthCare (n.d.) European legal framework for cosmetic products—European Directorate for the Quality of Medicines and HealthCare—EDQM. https://www.edqm.eu/en/european-legal-framework-for-cosmetic-products

European Medicines Agency (n.d.-a) Marketing authorisation. https://www.ema.europa.eu/en/human-regulatory-overview/marketing-authorisation

European Medicines Agency (n.d.-b) European Public Assessment Report. https://www.ema.europa.eu/en/glossary/european-public-assessment-report

Eurostat (n.d.) Regional demographic statistics. ec.europa.eu. https://ec.europa.eu/eurostat/databrowser/view/tps00001/default/table?lang=en

Ferreira M, Matos AR, Couras A, Marto J, Ribeiro HM (2022) Overview of cosmetic regulatory frameworks around the world. Cosmetics 9(4):72. https://doi.org/10.3390/cosmetics9040072

Fortune Business Insights (2030) Hair Care Market Size, Share | Global Industry Trends https://www.fortunebusinessinsights.com/hair-care-market-102555

IMF (2022) Report for selected countries and subjects. https://www.imf.org/en/Publications/WEO/weo-database/2022/October/weo-report?a=1&c=998,&s=NGDPD,PPPGDP,PPPPC,&sy=2021&ey=2022&ssm=0&scsm=0&scc=0&ssd=1&ssc=0&sic=0&sort=country&ds=.&br=1

IMF (2023a) Report for selected countries and subjects. https://www.imf.org/en/Publications/WEO/weo-database/2023/October/weo-report?c=111,&s=NGDPD,PPPGDP,NGDPDPC,PPPPC,&sy=2020&ey=2028&ssm=0&scsm=1&scc=0&ssd=1&ssc=0&sic=0&sort=country&ds=.&br=1

IMF (2023b) Report for selected countries and subjects. https://www.imf.org/en/Publications/WEO/weo-database/2023/October/weo-report?c=534,&s=NGDP_RPCH,NGDPD,PPPGDP,NGDPDPC,PPPPC,&sy=2021&ey=2028&ssm=0&scsm=1&scc=0&ssd=1&ssc=0&sic=0&sort=country&ds=.&br=1

India Hair Care and Styling Products Market Insights (n.d.). https://www.mordorintelligence.com/industry-reports/india-hair-care-and-styling-products-market-industry

IS 4704 (Part2) (2017). https://www.services.bis.gov.in/php/BIS_2.0/StandardsFormulationV2/doc_details_outside.php?ID=MTc2NDQ%3D

Legislation.gov.uk (2009) European Regulation (EC) No 1223/2009 of the European Parliament and of the Council of 30 November 2009 on cosmetic products. https://www.legislation.gov.uk/eur/2009/1223/contents

Muinos L (2022) Hair loss medications: where to get, effectiveness, and side effects. Healthline. https://www.healthline.com/health/hair-loss-medication#medications

Office of the Commissioner (2018) Federal food, drug, and cosmetic act (FD&C act). U.S. Food and Drug Administration. https://www.fda.gov/regulatory-information/laws-enforced-fda/federal-food-drug-and-cosmetic-act-fdc-act

Ross GL (2006) A perspective on the safety of cosmetic products: a position paper of the American Council on Science and Health. Int J Toxicol 25(4):269–277. https://doi.org/10.1080/10915810600746049

Statista (n.d.) Hair care—India | Statista market forecast. https://www.statista.com/outlook/cmo/beauty-personal-care/personal-care/hair-care/india

Stylecraze (2023) 10 Best medicines for hair growth with good results. https://www.stylecraze.com/articles/effective-medicines-for-hair-growth/

TechsciResearch Pvt Ltd. (2029) India Hair Care Market by Size, Share and Forecast | TechSCI Research. TechsciResearch Pvt Ltd. https://www.techsciresearch.com/report/india-hair-care-market/4896.html

US Census Bureau (2021). 2020 Census apportionment results delivered to the President. Census.gov. https://www.census.gov/newsroom/press-releases/2021/2020-census-apportionment-results.html

Wikipedia (2023) Wikipedia contributors. Medicines and Healthcare products Regulatory Agency. https://en.wikipedia.org/wiki/Medicines_and_Healthcare_products_Regulatory_Agency

Wikipedia (2024a) Wikipedia contributors. List of countries and dependencies by population. https://en.wikipedia.org/wiki/List_of_countries_and_dependencies_by_population

Wikipedia (2024b) Wikipedia contributors. List of countries by GDP (PPP). https://en.wikipedia.org/wiki/List_of_countries_by_GDP_(PPP)

Wikipedia (2024c) Wikipedia contributors. European Union. https://en.wikipedia.org/wiki/European_Union

Wikipedia (2024d) Wikipedia contributors. European Medicines Agency. https://en.wikipedia.org/wiki/European_Medicines_Agency

Wikipedia (2024e) Wikipedia contributors. List of countries by population (United Nations). https://en.wikipedia.org/wiki/List_of_countries_by_population_(United_Nations)

Strategic Developments for Pre-clinical Safety/Efficacy Studies of Hair Care Products

14

Ankita Srivastava, Ankur Kumar Srivastava, and A. B. Pant

Abstract

Hair plays a substantial role in human biology and social interactions. In addition to providing a physical barrier, protecting the head from sunburn and temperature extremes, hair also reflects an individual's personality, culture, and overall health. With the diversity in hair types and concerns, the hair care industry has developed a wide range of products designed to cater to specific needs. With the development of novel formulations in hair care products for both cosmetic and medicinal needs, testing their safety and efficacy before human application is essential. Moreover, these hair care products must adhere to the regulatory standards set by health authorities such as the U.S. Food and Drug Administration (US FDA), European Cosmetic and Perfumery Association (COLIPA), Central Drugs Standard Control Organization (CDSCO), etc., which involve monitoring the stringent testing for safety and efficacy before being marketed to consumers. Hair product safety and efficacy testing utilizes pre-clinical and clinical models to assess whether the product delivers the promised benefits, such as improved hair growth rate, hair loss prevention, moisture retention, etc., through laboratory measurements and user feedback. These models range from two-dimensional (2D) monolayer cultures of primary cell types in hair follicles (HFs) such as dermal papilla, dermal fibroblast outer root sheath keratinocytes, or ex vivo

A. Srivastava · A. K. Srivastava
Systems Toxicology Group, CSIR-Indian Institute of Toxicology Research, Lucknow, Uttar Pradesh, India

A. B. Pant (✉)
Systems Toxicology Group, CSIR-Indian Institute of Toxicology Research, Lucknow, Uttar Pradesh, India

Academy of Scientific and Innovative Research (AcSIR), Ghaziabad, Uttar Pradesh, India
e-mail: abpant@iitr.res.in

A. Dwivedi et al. (eds.), *Hair Care Products*,
https://doi.org/10.1007/978-981-97-6740-3_14

organ cultures of isolated HFs to wild-type or transgenic animal models. This chapter discusses the available experimental models and progress made in evolving new models to observe the impact of tested hair products on different stages of hair cycles. Further, the critical issues associated with hair health, the rationale, advantages, and shortcomings of the developed experimental models for hair care products, and the resulting progress in their development have also been discussed.

Keywords

Hair care products · Hair cosmetics · Hair safety/toxicity · Regulatory requirements · Drugs and cosmetics act and rules · Experimental models for hair care

14.1 Introduction

Hair, a distinctive feature exclusive to mammals, plays a multi-faceted role in human physiology and social dynamics. As part of the integumentary system, each hair penetrates the dermis, anchoring within a hair follicle (HF). The human body houses approximately five million such follicles, which guard against environmental elements like cold and ultraviolet (UV) radiation and secrete sebum. While primarily serving as a protective barrier for the skin against mechanical damage, this filamentous biomaterial also aids in thermoregulation. Its sensory capabilities enhance the tactile perception of the skin. Beyond its physiological functions, hair significantly influences sexual and social interactions, as evidenced by the profound psychological effects observed in conditions like hirsutism and alopecia. Notably, studies have indicated a correlation between personality disorders and androgenetic alopecia (AGA), highlighting its impact beyond mere physical appearance. The psychological implications of hair, particularly in cases of irregular growth or structural anomalies, are substantial (Kim and Dao 2020). This increased concern about appearance among consumers has given the hair care product industry an unexpected hike, claiming to resolve their problem efficiently in no time. Before being released in the market for human application, these products must pass the inevitable testing for their safety and efficacy. Before 2014, the testing involved both in vitro and in vivo models; however, in 2014, the European Union (EU) banned animal testing and the marketing and import of cosmetic products, including hair care products tested on animals. Later, this initiative was accepted by many other countries, including India. Alternatively, the European Centre for the Validation of Alternative Methods (ECVAM) proposed a few in vitro assays (reconstructed epithelial tissue EpiSkin™ and Epiderm™ and human reconstructed corneal epithelium EpiOcular™, etc.) as alternative models to test the safety and efficacy of cosmetic products. This chapter of the book entails detailed information on the most relevant cell-based in vitro models to predict the safety, toxicity, and efficacy of constituents in hair care products. The observed endpoints related to each model to predict the impact of tested hair products on different stages of hair cycles, the pros and limitations of each model, and the resulting progress in their development are also discussed.

14.2 Hair Anatomy, Physiology, Growth, and Development

14.2.1 Anatomy and Physiology

Hair constitutes a specialized derivative of the epidermis. The hair structure is characterized by fine, flexible tubes made up of dead, wholly keratinized epithelial cells on the outside. In contrast, inside the skin, it is connected to distinct, living HFs. These follicles are cylindrical epithelial structures that penetrate the dermis and sub-cutaneous fat, expanding at their base to form the hair bulb that encapsulates the mesenchymal-derived dermal papilla (DP). Hair encompasses two primary constructions: the externally visible hair shaft and the sub-epidermal HF. The follicle itself houses both an inner and outer root sheath (ORS). Above the epidermal layer, the hair shaft manifests as a slender, pliable cylinder composed of non-living, keratinized epithelial cells. In contrast, beneath the epidermis, it integrates into the living follicle, broadening into the hair bulb at its base. The shaft is characterized by a cortex, encasing cuticle cells, and, in some instances, a central medulla in thicker hair variants. The cortex constitutes most of the hair fiber and is crucial in defining physical and mechanical attributes of the hair, including strength, texture, and color. It primarily consists of microfibrils, aggregates of microfibrils rooted in a matrix (Martel et al. 2017).

The follicle serves as the foundational structure for hair growth, with its histological arrangement including both the outside and inside of the root sheaths. The outside of the root sheath is acknowledged for its reservoir of multi-potent stem cells, capable of differentiating into various cell forms like melanocytes and keratinocytes. The bulge area between the arrector pili muscle' insertion point and the sebaceous gland (SG) duct is the primary location of these stem cells. Three layers from inside to outside, the Henle layer, the Huxley layer, and the cuticle, provide anchorage for the hair shaft to the follicle. This anchoring is reinforced by keratins and trichohyalin produced by the inner root sheath (IRS) cells. The cuticle, comprising flattened, overlapping cells, sheaths the hair shaft from the base to the tip. It plays a significant role cosmetically by imparting a straight appearance to the hair. The hair bulb, situated at the follicle's base, actively produces hair. It extends into the skin's dermal layer, enveloping the DP, a mesenchyme-derived structure rich in nerve fibers and blood capillaries for nutrient delivery. The hair papilla is crucial for hair development, orchestrating essential cues that determine the hair shaft's thickness and hue via a sophisticated mix of growth elements, such as insulin-like growth factor (IGF), stem cell factor, keratinocyte growth factor, and bone morphogenetic protein. Divided into two regions by the Auber line, the hair bulb features cells beneath this demarcation that are still in the differentiation process. These developing cells form the follicle's growth center, or matrix, marked by their cell division activity and upward progression, during which they grow, lengthen vertically, and integrate into the hair shaft (Welle 2023). The innervation of HFs is analogous to the dermal nerve network, encompassing sensory afferents and autonomic sympathetic fibers. Stimulation of these follicles augments tactile perception. The autonomic fibers predominantly modulate the arrector pili muscles, whose contraction results

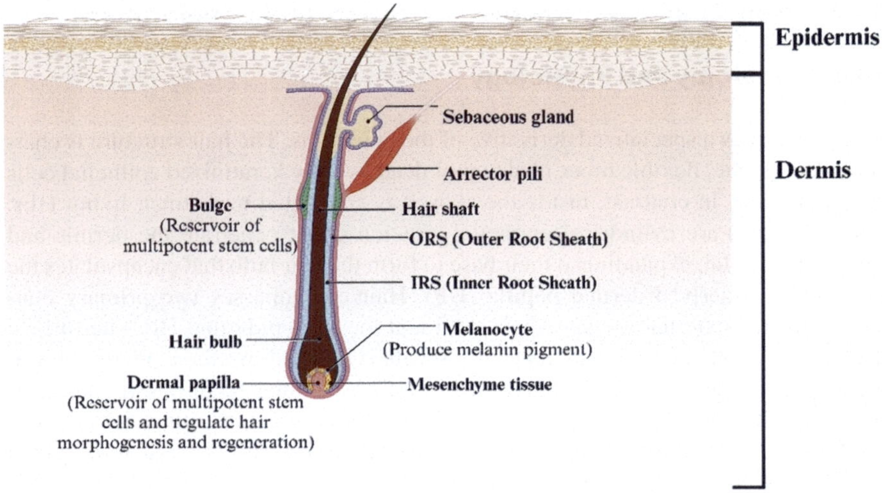

Fig. 14.1 Structure of hair

in the hair assuming an erect position. This reaction, primarily a remnant of evolutionary adaptation in our fur-bearing ancestors, was beneficial for thermal insulation and intimidation tactics in ancestral environments. Blood supply to the follicles stems from minor arterioles rooted in the sub-cutaneous adipose tissue. These blood vessels facilitate follicular sustenance by supplying oxygen and nutrients, removing metabolic byproducts, and fostering growth (Zhang et al. 2021a) (Fig. 14.1).

14.2.2 Growth and Development

Hair growth is predisposed by a combination of factors, including vascular, hormonal, and neural signals, age, and dietary patterns. Typically, the human scalp loses about 100 to 200 HFs daily. In contrast, hair grows at an average pace of 0.35 mm/day, equating to roughly 1 cm/month or 15 cm annually (Hoover et al. 2023). At the time of birth, the human body is mainly covered with fine, short, vellus hairs along with terminal hairs found in areas like the scalp, eyelashes, and eyebrows. The onset of puberty triggers hormonal changes, specifically androgens, which transform specific vellus hairs in regions like the face (including beard growth), underarms, genital area, and torso into terminal hairs. These terminal hairs are characterized by their greater length, thickness, and coloration (Breehl and Caban 2018). The development of HFs follows a cyclic pattern, with alternating stages of swift growth and hair shaft elongation and times of inactivity and decline, influenced by signals that induce programmed cell death. This repetitive cycle is segmented into three distinct phases: the anagen phase, which is the growth period; the catagen phase, which serves as a transitional stage; and the telogen phase, which is a period of rest.

During the anagen phase, the HF produces a new shaft. The duration of this phase varies; on the scalp, it ranges from 2 to 6 years, while for eyebrows and eyelashes, it spans only a few months. The presence of the inferior segment of the HF marks this phase. Activation of the anagen phase involves the DP signaling the bulge's multi-potent epithelial stem cells. These activated stem cells allow the inferior segment of the follicle to extend downward, encircling the DP. Subsequently, the DP prompts the matrix cells within the bulb to multiply, differentiate, and ascend, creating a new hair strand. The catagen phase, the transitional or regressive stage, is the briefest, lasting mere weeks. In this phase, matrix cell division halts, and the inferior segment of the follicle starts to deteriorate, eventually disappearing. As the follicle regresses, the DP ascends towards the bulge. This process results in forming a club hair, identifiable by a hard, white node at its end. The final stage is the telogen phase or the resting period. During this time, effectively lifeless club hairs are retained. On the scalp, they typically remain for about 100 days before shedding, paving the way for the anagen phase to restart with the development of new hair (Geyfman et al. 2015).

14.3 Issue of Concern

14.3.1 Hair Loss/Alopecia

Hair loss, also known as alopecia, refers to the loss of hair from the scalp or body. This condition can range from mild to severe and can affect both genders. Alopecia is the medical term for hair loss. It can occur in a localized area or affect the entire scalp or body. Hair loss occurs when the hair growth cycle is disrupted. Key hair-related diseases are broadly categorized into non-scarring and scarring types (Rahangdale and Wankhade 2023).

14.3.2 Types of Alopecia

There are two types of alopecia, cicatricial alopecia (CA) and non-cicatricial alopecia (non-CA). Alopecia without scarring, also known as non-CA, is a more prevalent form observed in clinical settings. This condition may arise due to various factors such as AGA, alopecia areata (AA), telogen effluvium (TE), or anagen effluvium. In contrast, scarring alopecia, which is less common, leads to irreversible hair loss and dermal scarring, exemplified by conditions like lichen planopilaris or frontal fibrosing alopecia. These are some of the most common types of hair loss, but others are less frequent or more specific types (Alessandrini et al. 2021a).

Non-cicatricial Hair Diseases: Non-scarring alopecia, a reversible form of hair loss, is distinguished by a disruption in the normal hair growth cycle while maintaining the integrity of HFs. This category of alopecia can be further classified into localized and diffuse types. Localized non-scarring alopecia encompasses conditions like AA, tinea capitis, and trichotillomania. In contrast, diffuse non-scarring alopecia covers broader conditions such as alopecia totalis or universalis, and variations like telogen or anagen effluvium. These classifications are detailed in Fig. 14.2. Among these types, androgenic alopecia is identified as the most prevalent (Cardoso et al. 2021).

Androgen-Related Alopecia: AGA, prevalent in 80% of men and 50% of women, is primarily influenced by genetic and age-related factors that disrupt the androgen mechanism responsible for transforming vellus hair into terminal hair. This disruption leads to follicle miniaturization and hair regression. Characterized by continuous hair loss, this condition involves mast cells and lymphocytes surrounding the miniaturized follicles, which are stem cell-rich (Gentile and Garcovich 2019; Alessandrini et al. 2021a, b). Genetic predisposition plays a crucial role, as evidenced by a heritability rate of 0.81 in twin studies, indicating significant familial aggregation. Notably, variations in the androgen receptor (AR) gene and 5α-reductase gene have been implicated. Additionally, associations between triple-repeat polymorphisms, single-nucleotide polymorphisms, and the AR gene on the X chromosome have been observed. The ectodysplasin A2 receptor (EDAR2) gene also shows polymorphic variations linked to AGA, although its precise function remains unclear. Alterations in the genes involved in the Wnt (Wingless/Integrated) pathway, which plays a significant role in controlling DP cells (DPCs) and the metabolism of androgens, have been linked to AGA. Elevated levels of prostaglandin synthase (PGDS) and its derivative, prostaglandin D2 (PGD2), could potentially

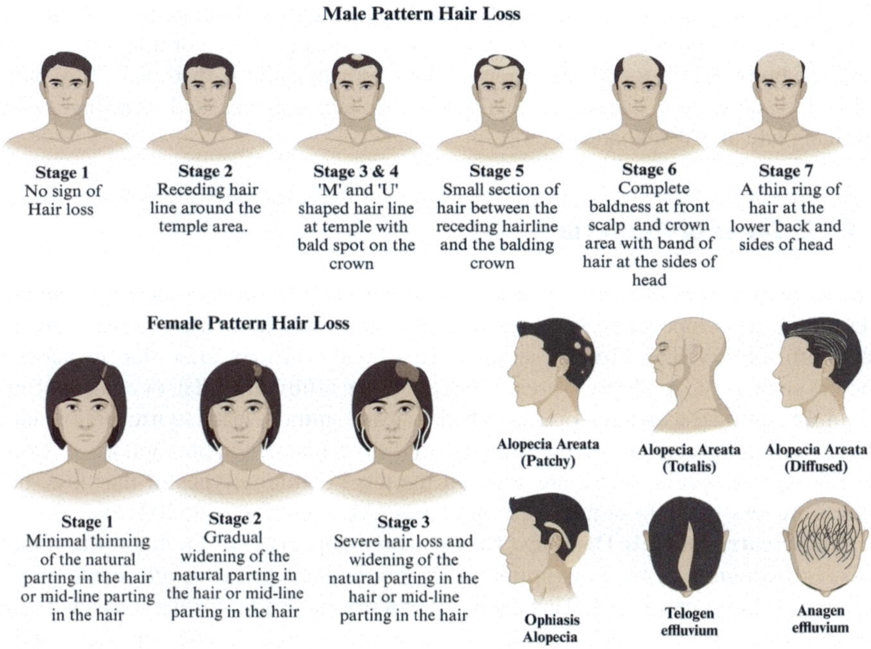

Fig. 14.2 Types of hair-related problems in humans

impede hair growth by triggering early onset of the catagen phase (Martinez-Jacobo et al. 2018).

Alopecia Areata: AA manifests in various forms, each characterized by distinct patterns and extents of hair loss. This condition typically initiates with sporadic, patchy hair loss. The principal forms of AA include:

Alopecia Areata (Patchy)*:* This type is marked by one or more circular or oval patches of hair loss, approximately the size of a coin. These patches can appear on the scalp or other hair-bearing regions of the body. Patchy AA might evolve into more extensive forms, namely alopecia totalis (complete scalp hair loss) or alopecia universalis (total body hair loss). However, it often remains confined to discrete patches. Persistent patchy AA is identified by ongoing patchy hair loss on the scalp without progression to more severe forms like totalis or universalis.

Alopecia Areata Totalis: Alopecia areata totalis represents a more extreme variant of AA, characterized by complete loss of hair on the scalp. This condition, typically involving hair loss from various body parts, is understood to be an autoimmune disorder. In this scenario, the body's immune system erroneously targets the HFs, resulting in hair loss. Although the precise triggers of this immune reaction remain unclear, it is speculated to be influenced by a combination of genetic predispositions and environmental elements.

Diffuse Alopecia Areata: Characterized by rapid, widespread thinning of hair across the scalp, diffuse AA presents a diagnostic challenge as it closely resembles other hair loss disorders, such as TE or AGA (commonly known as male or female pattern baldness).

Ophiasis Alopecia: A distinct form of AA, ophiasis alopecia features hair loss in a band-like pattern along the sides and rear of the scalp (Lintzeri et al. 2022).

Telogen Effluvium: TE represents a form of alopecia characterized by non-scarring hair shedding, triggered by various physiological stressors including childbirth, significant weight fluctuation, mental strain, extended pharmacotherapy, and medical conditions like thyroid dysregulation. It is also associated with dietary changes and extended periods of fasting. While typically reversible, this condition significantly impacts the psychological well-being of individuals. The underlying pathophysiological mechanisms of TE remain elusive. The complexity and multiplicity of potential etiological factors contribute to the challenges in accurately diagnosing and effectively managing this disorder (Yin et al. 2021).

Anagen Effluvium: It is a specific type of hair loss primarily caused by exposure to certain chemicals that are often used in chemotherapy for cancer treatment. This condition differs from more common forms of hair loss in that it rapidly affects hairs in the active growth phase, known as the anagen phase. Chemotherapy drugs are designed to target rapidly dividing cells, a characteristic of cancer cells. However, these drugs also affect other rapidly dividing cells in the body, including HFs. As a result, hair fibers, which are in the midst of growth, become weakened and break off at or near the scalp. This leads to widespread hair loss, often occurring within days to weeks of starting treatment. Unlike other hair loss types, anagen effluvium is typically reversible, with hair regrowth often resuming after chemotherapy is completed, although the new hair might initially have a different texture or color (Starace et al. 2022).

Cicatricial Hair Diseases: Scarring hair loss conditions, known as CAs, represent a group of diseases marked by permanent loss of hair due to the destruction of HFs followed by fibrotic scarring. These conditions are divided into two main categories: primary CA (PCA) and secondary CA (SCA) (Fanti et al. 2018). PCA includes various inflammatory diseases, each characterized by distinct clinical and histopathological features. The causes of these diseases are mostly unknown, but they specifically attack and destroy HFs, contributing to about 7% of all hair loss cases. PCA is further classified into three types: lymphocytic, neutrophilic, and mixed (i.e., lymphocytic + neutrophilic) (Civaş et al. 2018). Some clinical conditions of PCA include chronic cutaneous lupus erythematosus, lichen planopilaris, classic Brocq pseudopelade, folliculitis decalvans, and dissection folliculitis (Villablanca et al. 2017). In contrast, SCA arises from conditions that primarily impact the dermis, leading to indirect follicular damage. These include a range of factors such as genodermatoses, developmental anomalies leading to permanent alopecia, physical and chemical traumas, infections, inflammatory dermatoses, drug reactions, and neoplastic processes. This classification system and its components are detailed in Table 14.1.

14.3.3 Dandruff

Dandruff is a common scalp condition that is usually more of a nuisance than a serious health concern. It is characterized by flaking and sometimes mild itchiness. The exact cause of dandruff is not fully understood, but it is believed to be associated with a variety of factors, including:

14.3.4 Microbial Factors

Fungal: *M. furfur* is identified as a primary contributor to dandruff. This fungus influences dandruff through one or both of the mechanisms detailed below:

M. furfur activates a scalp enzyme known as lipase, which oxidizes sebum triglycerides into unsaturated and saturated fatty acids. The fungus utilizes saturated fatty acids for its growth, while unsaturated ones, like oleic and arachidonic acids, cause skin irritation and enhance inflammatory reactions. This process leads to scalp irritation, inflammation, and fungal proliferation, resulting in the formation of dry, flaky dandruff in the hair and on clothing. Additionally, *M. furfur* impacts the natural process of shedding dead skin cells. Normally, certain scalp enzymes dissolve connections between dead cells, allowing them to slough off individually. However, *M. furfur* interferes with this function, preventing the separation of these connections. Consequently, coenocytes clump together and shed in visible clusters, forming the white flakes commonly associated with dandruff (Vest and Krauland 2022).

Table 14.1 Classification of cicatricial alopecia

Primary cicatricial alopecia (Working classification of PCA by North American Hair Research Society)	Secondary cicatricial alopecia								Source	Medical conditions/Causative agents
Lymphocytic	Chronic cutaneous lupus erythematosus	Lichen planopilaris	Frontal fibrosing alopecia	Graham little syndrome	Classic pseudopelade (Brocq)	Central centrifugal cicatricial alopecia	Alopeciamucinosa	Keratosis follicularis spinulosa decalvans	Physical or chemical trauma	• Burns, toxic/corrosive substances (e.g., acid or alkali burns) • Ischemia/pressure • Traction/trichotillomania (end stage)
									Ionizing radiation	Radiation
									Infections	• Bacterial infections • Viral infections • Fungal infections (tinea capitis in particular • Deep trichophytosis caused by zoophilic pathogens)

(continued)

Table 14.1 (continued)

Primary cicatricial alopecia (Working classification of PCA by North American Hair Research Society)	Secondary cicatricial alopecia			Source	Medical conditions/Causative agents
Neutrophilic	Folliculitis decalvans	Dissecting cellulitis/folliculitis		Malignant and benign tumors	• Primary tumors • Metastases • Lymphoproliferative diseases • Epidermal and organoid nevi
				Genodermatoses	• Aplasia cutis congenita • Ectodermal dysplasia • Ichthyosis, epidermolysis bullosa • Darier's disease • Incontinentia pigmenti • Hyalinosis cutis et mucosae
				Granulomatous diseases	• Sarcoidosis • Necrobiosis lipoidica
Mixed	Folliculitis (acne) keloidalis	Folliculitis (acne) necrotica	Erosive pustular dermatosis	Autoimmune diseases	• Graft-versus-host disease, scleroderma (*en coup de sabre*) • Lichen sclerosus blistering dermatoses (cicatricial pemphigoid)
Non-specific	Idiopathic scarring alopecia with inconclusive clinical and histopathological findings	End stage of various inflammatory scarring alopecia		Deposition dermatoses	• Amyloidosis • Mucinosis
				Inflammatory diseases	• Psoriasis • Pityriasis amiantacea

Bacterial: The imbalance between the two primary bacterial colonies present on the scalp, specifically *Propionibacterium acnes* and *Staphylococcus epidermidis*, is also hypothesized to be a contributing factor in the development of dandruff. Studies suggest that when the natural balance of these bacteria is disrupted, it could lead to scalp conditions such as dandruff. Normally, these bacterial populations coexist in a harmonious balance, each playing a role in maintaining the health of the scalp's microbiome. *P. acnes* is typically involved in the breakdown of fatty acids, while *S. epidermidis* is known for its role in skin barrier protection and immunity. However, when the equilibrium between these two bacterial groups is disturbed, it may result in scalp inflammation and flaking, commonly recognized as symptoms of dandruff. This disequilibrium can be influenced by various factors, including changes in sebum composition, hormonal fluctuations, stress, and external environmental conditions, all of which can alter the scalp's microbiota and potentially exacerbate dandruff (Clavaud et al. 2013 Xu et al. 2016).

14.3.5 Non-microbial Factors

Seborrheic Dermatitis: Seborrheic dermatitis (SD) and dandruff represent different stages of a common condition affecting the seborrheic areas of the body. While dandruff is limited to itchy, flaking skin on the scalp without visible inflammation, SD can impact both the scalp and other areas, and is characterized by itching, flaking or scaling skin, and inflammation. The development of SD and dandruff involves a mix of factors, including sebaceous secretion, fungal colonization on the skin surface, individual susceptibility, and their interactions (Borda and Wikramanayake 2015).

Dry Skin: Dry skin typically produces smaller, less greasy flakes compared to other dandruff causes. Additionally, symptoms of dry skin, like flaking, often appear on other body areas, including the arms and legs.

Sensitivity to Hair Care Products: Contact dermatitis is a skin reaction that occurs when the skin comes into contact with certain substances, leading to symptoms such as redness, itching, and scaling. This condition can be triggered by various irritants or allergens, including chemicals found in hair care products and hair dyes. It manifests as an inflammatory response in the affected area of the scalp.

Other Skin Conditions: Individuals suffering from skin disorders such as psoriasis or eczema are at a higher risk of experiencing dandruff. These conditions, known for causing inflammation and irritation of the skin, can exacerbate the likelihood and severity of dandruff, a common scalp issue caused by various skin disorders. In most cases, dandruff can be controlled with over-the-counter dandruff shampoos and good scalp hygiene. However, if your dandruff is severe, persistent, or accompanied by other symptoms like intense itching or signs of infection (like redness, swelling, or tenderness), it is advisable to see a healthcare provider. They can determine whether you have a more serious condition that needs specific treatment. For most people, dandruff is a chronic condition that can be controlled but not cured. The goal of treatment is to reduce the visible flakes and scalp irritation. It is also worth noting that dandruff is not contagious and does not reflect poor personal hygiene.

14.3.6 Other Hair Problems

Hair health reflects the internal biological processes and external environmental interactions. Common hair issues like hair breakage, split ends, dryness and frizziness, and oily or greasy hair can be indicative of both physiological conditions and external factors.

Hair Breakage and Split Ends: Hair breakage and split ends are primarily caused by physical and chemical stress. Physical stress includes harsh brushing, tight hairstyles, and excessive heat styling, while chemical stress involves treatments like coloring, perming, or straightening. These factors weaken the hair shaft, particularly the cuticle layer, leading to breakage and split ends.

Dryness and Frizziness: Dryness and frizziness often result from a lack of moisture in the hair. Environmental factors like low humidity, sun exposure, and wind, as well as heat styling, can strip hair of its natural oils. Certain hair types, particularly curly or wavy hair, are more prone to dryness and frizziness due to their structure.

Oily/Greasy Hair: Oily or greasy hair is a common concern for many people and can be influenced by a variety of factors. Understanding the causes and exploring potential solutions can be helpful in managing this condition. There are various factors, such as sebaceous gland activity, genetic predisposition, hormonal changes, poor hair care habits, diet and lifestyle, which may cause the oily and greasy hair condition. The SGs in the scalp produce sebum, a natural oil that moisturizes and protects hair. Overactive glands can lead to an excess production of sebum, resulting in oily hair. Genetic predisposition can play a significant role in how oily your scalp gets. Some people are simply more prone to having an oily scalp due to their genetic makeup. Hormonal fluctuations, such as those experienced during puberty, pregnancy, or menopause, can increase sebum production. Over-washing or under-washing your hair, using harsh hair products, or frequently touching your hair can contribute to oil buildup and creates oily hair (Lanjewar et al. 2020).

Head Lice Infestation and Its Impact on Hair Health: Head lice infestation, a common issue particularly prevalent among school-aged children, poses questions regarding its impact on hair health. Head lice infestation does not directly cause hair damage, its secondary effects can impact hair health. Contrary to popular belief, head lice do not cause hair loss or significant hair damage. Their primary impact is on the scalp, where they may induce itching due to their bites. Persistent scratching due to itchiness can lead to scalp inflammation and secondary bacterial infections. These conditions, rather than the lice themselves, might indirectly affect hair health. Apart from this, in the attempts to eradicate lice, individuals may resort to aggressive combing or the use of harsh chemical treatments, which can cause mechanical and chemical damage to hair (Trüeb et al. 2023).

14.4 Hair Care Products

Hair serves as a critical indicator of overall health and contributes significantly to physical attractiveness. With the diversity in hair types and concerns, the hair care industry has developed a wide range of products, each designed to cater to specific needs. The landscape of hair care and treatment has undergone significant transformation, driven by ongoing research and development in creating novel formulations to address both cosmetic and medicinal needs. Traditionally, hair and scalp care has centered on the use of shampoos for effective yet gentle cleansing. Over time, the role of shampoo has evolved beyond merely cleansing; it now plays a crucial part in maintaining hair health and aesthetics, enhancing shine, and facilitating manageability. Along with the cleanliness and maintenance, hair dyeing and bleaching, for aesthetic purposes, have been supplemented by advanced treatments. These modern approaches primarily target hair loss prevention, enhancement of HF health, and treatment of hair-related infestations.

14.4.1 Classification of Hair Care Products

The classification of the hair care products is based on various factors like their functionality, chemical composition, the type of hair intended, and their use as a therapeutic module for baldness. Hair, at its molecular level, consists of several components that determine how different hair cosmetic products perform. Presently, the primary categories of hair cosmetic formulations encompass agents for hair cleansing (notably shampoos), hair conditioning (utilizing conditioners), hair styling (including waving and straightening products), anti-hair loss treatments, and products for hair dyeing, coloring, and bleaching. These formulations are designed to serve various hair care objectives, including cleansing and maintenance, stimulation of hair growth, and alteration of hair color.

Shampoos: Formulated to cleanse the hair and scalp, shampoos often contain ingredients that interact with the exocuticle to remove dirt, oil, and product residues. Shampoo formulations typically consist of several key components. These include cleansing agents, which effectively remove dirt and oil; foaming agents, which help in creating a rich lather; conditioning elements, which improve hair manageability and softness; thickening agents, which enhance the texture of the shampoo; opacifiers, which contribute to the shampoo's visual appeal; sequestering agents, which bind metal ions and improve the shampoo's effectiveness; pH adjusters to ensure the product is gentle on hair and scalp; preservatives, which extend the shelf life of the shampoo; and various additives that may be included for specific benefits like fragrance or color (Draelos 2010; Gubitosa et al. 2019).

Conditioners: Conditioners work in reducing hair frizz, enhancing manageability, and imparting gloss. Conditioners are particularly beneficial for mitigating the impact of chemical or mechanical stress on hair, such as that caused by permanent dyeing, bleaching, or extensive brushing. The hair shaft is characteristically negatively charged, while conditioning agents carry a positive charge. The action of conditioners involves neutralizing the hair fiber's negative charge through the addition of positive charges and by smoothing the cuticle. This process decreases the fiber's hydrophilicity, leading to less moisture absorption (Dias 2015). Conditioner formulas often consist of positively charged surfactants such as cetyltrimethyl ammonium chloride and stearamidopropyl dimethylamine, along with others like behentrimonium and propyltrimonium. These agents, combined with polymers like polyvinylpyrrolidone and hydrolyzed proteins, help to even out irregularities in the hair shaft, leading to a smoother hair surface. This smoothing effect not only imparts a glossy sheen but also minimizes static, electrically charged nature. Additionally, conditioners contain moisturizing agents derived from either natural sources like olive, jojoba, or grape seed oil, or synthetic ones such as silicones, including dimethicone and cyclomethicone. Dimethicone is especially common in hair care products for its ability to shield hair from wear and tear. Other ingredients in conditioners comprise fatty alcohols like cetyl and stearyl alcohol, various waxes including carnauba and paraffin, thickeners such as guar gum, salts (e.g., sodium chloride), and emulsifying agents like ethoxylated fatty alcohols including ceteareth-20 and polysorbate-80 (Alessandrini et al. 2021a, b).

14.4.2 Hair Color Products

Hair Dyes: Hair dyes, popular among all genders, are used for color changes or to counteract graying. These dyes are categorized into temporary, semi-temporary, and permanent types, differentiated by their ability to penetrate the hair cortex, a process facilitated by an alkaline pH. Common alkalizing agents in these products include ammonia or ethanolamine. Temporary dyes, often in shampoo, lotion, or foam forms, alter hair color briefly, usually until the next wash. Their primary uses include adding highlights, neutralizing yellow tones in white hair, or concealing minor graying. They bind to cationic polymers to enhance hair affinity and are mixed in surfactant-based solutions. Semi-permanent dyes, made of low-molecular weight nitroaromatic amines or aromatic dyes, do not bond to hair proteins and are water-soluble, fading after 6–10 washes. These dyes, free from ammonia and ethanolamine, may contain hydrogen peroxide or resorcinol. Permanent hair dyes chemically alter hair color through oxidation, allowing color molecules to infiltrate the hair. These dyes, lasting 4–6 weeks, can darken or lighten hair and cover significant graying. The coloring process involves molecule oxidation within the hair shaft, simulating natural melanin. Commonly, these dyes contain derivatives of p-phenylenediamine for darker shades or p-aminophenol for lighter tones. Additional ingredients in dyes include hydrogen peroxide, which activates color reactions and lightens natural hair color, and ammonia, an alkalizer. Chelating agents stabilize peroxide, and solvents and surfactants enhance dye application and cleanliness. However, hair dyes can complicate certain dermatoscopic diagnoses of hair diseases. For example, permanent dyes can mimic symptoms of AA, while semi-permanent dyes can resemble skin conditions. Prolonged scalp exposure to dyes can lead to contact dermatitis and, in severe cases, facial edema (Rust and Schlatter 2022).

Bleaching Agents: These products facilitate irreversible hair decolorization, devoid of any pigment addition, through the utilization of chemicals such as hydrogen peroxide, ammonia, and persulfate salts. This multi-stage process involves gradual color transformation over time. Specifically, hydrogen peroxide acts as a bleaching agent, eliminating the natural hair color. This oxidizing agent, when in an alkaline medium, infiltrates the hair cortex following the lifting of the cuticle scales. It then alters the melanin chemically, resulting in the dissolution of melanin granules and the creation of microscopic voids within the cortex. Consequently, frequent application of these bleaching treatments may lead to increased hair porosity and the perpetual elevation of cuticle scales, culminating in hair damage known as weathering.

14.5 Hair Styles

14.5.1 Permanent Waves

Applying cosmetic products such as lotions or sprays on the surface can temporarily alter hair styling, but these changes are usually undone by the first wash. To achieve a lasting transformation in hair shape, it is necessary to alter its internal chemical composition through a specific process. A permanent wave procedure effectively curls hair by chemically breaking and reconfiguring the hair shaft's robust disulfide bonds. An alkaline-based waving solution, often ammonium thioglycolate, is applied next. This solution lifts the cuticle scales, penetrates the cortex, and breaks the disulfide bonds that maintain the hair's natural shape. Given its intensity, hair waving is one of the riskiest cosmetic hair treatments and requires vigilant monitoring.

14.5.2 Relaxers

The process of hair relaxation involves a series of chemical reactions that alter the structure of hair. Initially, this involves the breaking down of disulfide bonds within the hair fibers. These disulfide bonds are responsible for maintaining the hair's natural curl or wave pattern. Once these bonds are broken, the hair can be reshaped. The next phase in hair relaxation is the reformation of hair structure. During this stage, the hair is physically manipulated to take on a new shape, typically transitioning from a curly or coiled state to a straighter one. Following the physical reshaping, the hair undergoes a chemical process where the broken disulfide bonds are reformed, but in a new alignment that supports the straightened structure. This process of hair relaxation is not always a one-time event. For hair that is tightly curled, multiple treatments or extended exposure to the relaxing chemicals may be necessary to achieve the desired level of straightness. However, this repeated or prolonged chemical processing can lead to what is known as hair weathering. Hair weathering encompasses various types of damage, including dryness, brittleness, and breakage, which can compromise the overall health and appearance of the hair (Paula et al. 2022).

14.5.3 Hair Straightening/Perming

Hair transformation techniques like perming and straightening significantly modify the hair's physical structure. These methods work by disrupting the disulfide linkages within keratin filaments, impacting the hair's structural integrity and its cystine levels. The process reshapes these bonds to achieve the desired hair form. Typically, these treatments require renewal every 3–4 months. During this process, the essential hydrophobic layer, known as 18-methyleicosanoic acid (18-MEA), is permanently lost and cannot be replaced. Consequently, the hair becomes more absorbent to water, swells more readily, and its tensile strength diminishes. Chemical straightening, also known as lanthionization, is a favored technique for people with frizzy or difficult-to-manage hair, resulting in the conversion of approximately 35% of cysteine bonds into lanthionine. The initial stage involves hair shaft swelling and cuticle lifting, achieved using strong alkaline agents such as 1.5–3.5% sodium/potassium hydroxide (lye-based), or alternatives like guanidine/lithium hydroxide (no-lye). Lye-based straighteners, known for causing significant irritation, necessitate the application of petrolatum to the scalp and hairline for protection. This step is not required for no-lye treatments, except at the hairline. These straighteners have a pH range between 12.5 and 14, making them highly caustic to the skin, scalp, and eyes, thus requiring professional handling in a salon setting. An emerging alternative involves the use of thioglycolic acid salts as reducing agents instead of ammonia. These have a milder pH of 9–9.5, are less caustic, and yield effective curling/straightening outcomes. For home use, products containing ammonium bisulfite + sulfite are available. These have a lower alkaline pH (below 9), making them safer but offering only mild and temporary straightening effects (Tated et al. 2021).

14.6 Strategies in Regenerative Medicine for Promoting Hair Growth and Restoration

Alopecia is linked to a range of psychological, medical, and economic impacts on individuals and society. This condition stems from various causes and exhibits a complex genetic background, making traditional medical treatments somewhat limited. In response, regenerative medicine is investigating new therapeutic approaches for alopecia. Increasing research supports the potential of treatments such as mesenchymal stem cell (MSC) implantation, therapies based on MSC-derived secretions, and treatments using platelet-rich plasma (PRP) derived from blood. Promising regenerative medicine-based treatments for alopecia currently include stem cell and PRP therapies. The therapeutic benefits of stem cells were primarily due to their ability to move to and differentiate within damaged tissues (a process known as homing). PRP works by releasing cytokines and various growth factors, including platelet-derived growth factor (PDGF), transforming growth factor-beta (TGF-β), vascular endothelial growth factor (VEGF), epidermal growth factor (EGF), IGF, and fibroblast growth factor (FGF). These play significant roles in tissue regeneration, including hair. Presently, stem cell therapies predominantly employ stem cells derived from adult tissues, specifically mesenchymal stromal cells (MSCs), along with stem cell-derived chemokines or growth factors (secretome), and regenerative biomaterials like lipogem therapy. This approach also leverages the existing stem cell niche (microenvironment) to initiate a significant healing process. Stem cell-based treatments for alopecia may be either autologous or allogeneic (Vasanthan et al. 2020).

14.6.1 Stem Cell-Based Therapies

Stem cell-based therapies have emerged as a promising avenue for treating alopecia, a condition characterized by hair loss. These therapies leverage the unique regenerative abilities of stem cells to potentially restore hair growth. It has been reported that stem cells derived from the HF bulge can be stimulated to differentiate into HF cells, thereby promoting hair regrowth. Furthermore, a study highlighted the role of adipose-derived stem cells (ADSCs) in enhancing the proliferation and differentiation of DPCs, which are crucial for HF development and growth. These findings suggest that stem cell therapies could target the underlying mechanisms of hair loss, offering a novel approach to treat alopecia (Park and Choi 2022).

Treatment of Alopecia using Adipose-Derived Mesenchymal Stem Cells (AD-MSCs): Therapeutic strategies in regenerative medicine are advancing, with stem cell-based treatments, biomaterials, and growth factor methodologies expanding their scope beyond dermatological applications. ADSCs are increasingly utilized in adult stem cell therapies, owing to their plentiful supply and ease of procurement. For volumetric deficits, adipose tissue is particularly effective, as it adapts to the defect's shape, aiding in both volume restoration and the regeneration of the recipient tissue. When focusing on regeneration rather than just volume restoration, alternatives like the stromal vascular fraction (SVF), abundant in stem cells, purified MSCs, or their cell-free by-products are often chosen. The success of cell-free treatments is linked to the natural tissue repair capabilities of stem cells, primarily through their secretions. These secretions include mitogenic factors, cytokines, chemokines, and components of the extracellular matrix (ECM). The ability of the secretome to target multiple aspects simultaneously has demonstrated encouraging outcomes in both pre-clinical and clinical studies. These studies address the complex pathology of various conditions such as atopic dermatitis (AD), vitiligo, psoriasis, acne, lichen sclerosus (LS), chronic wounds, and alopecia (Bellei et al. 2022).

Treatment of Alopecia using Hair Follicle Stem Cells (HFSCs)**:** Hair follicle stem cells (HFSCs), primarily located in the bulge area of HFs, consist of both epithelial and melanocytic stem cells. Typically, in a quiescent state, these cells possess the inherent capability to migrate, proliferate, and differentiate, contributing to the maintenance of physiological balance. The inactive HFSCs in the bulge region give rise to "early progenitor cells" or "primed stem cells," which subsequently generate rapidly dividing progenitor cells in the hair matrix. These progenitor cells undergo further differentiation and ascend toward the surface, forming the IRS and hair shaft. Concurrently, as matrix cells continue differentiating, the bulge-located stem cells execute self-renewal during the anagen phase, eventually returning to a dormant state within the bulge. Some of these dormant HFSCs migrate from the bulge, initiating the formation of a new bulge and a novel set of primed stem cells, thereby initiating a new growth cycle. This cyclical process is observed throughout the anagen, catagen, and telogen phases. Beyond their role in HF equilibrium, HFSCs are also instrumental in generating the interfollicular epidermis, SGs, and facilitating the wound healing processes (Gentile et al. 2020).

Treatment of Alopecia using Bone Marrow-Derived Mesenchymal Stem Cells (BM-MSCs): Bone marrow is traditionally recognized as a primary source of MSCs. Research in various animal models has demonstrated that intradermal injection of BM-MSCs can catalyze the transition from the telogen to the anagen phase of hair growth, and activate genes linked to hair regeneration. BM-MSCs are predominantly osteogenic, whereas ADSCs exhibit angiogenic properties.(Xia et al. 2020). BM-MSCs and HFSCs have shown promise in alopecia treatment. HFs from the occipital region, resistant to AGA, are typically preferred for FSC therapies (Elmaadawi et al. 2018). In skin and appendage development, Wnt and Bone Morphogenetic Protein (BMP) signaling pathways are crucial. Cultured MSCs can differentiate into DP-like tissues (DPLTs), resembling human DPCs (hDPCs) in structure and function. Factors such as growth stimulating-, anti-inflammatory-, and angiogenic agents are instrumental in enhancing hair growth. Bone marrow aspirate concentrate (BMAC) contains progenitor cells, constituting 0.001% to 0.01% of its total, and is rich in growth factors like PDGF, TGF-β, and BMP-2 and -7, known for their anabolic and anti-inflammatory properties. MSCs also significantly produce IL-1Ra and IL-1 molecules, which inhibit IL-1 catabolism. It is found that the DPLTs were comparable to natural DPCs in morphology and protein composition, confirmed through microscopy and immunohistochemistry. These transplanted DPLTs effectively stimulated HF growth. Despite the potential of BMAC and bone marrow mononuclear cells (BMMCs) in hair restoration, there is a scarcity of human trials to ascertain their efficacy fully. Understanding the biology of BMAC is crucial for its clinical application. Future hair restoration strategies might focus on regenerative molecules. Known risks during bone marrow aspiration include bleeding, infection, and persistent pain. However, the autologous and accessible nature of BMAC positions it as a significant future player in hair restoration. In a similar fashion, various studies support the use of umbilical cord blood mesenchymal stem cells (hUCB-MSCs), hDPCs, and human induced pluripotent stem cells (hiPSCs) for the treatment of alopecia (Shimizu et al. 2022).

14.6.2 Cell-Free Therapies of Stem Cells

Cell-free therapies in hair regeneration represent a cutting-edge approach in the treatment of hair loss and baldness. Unlike traditional stem cell therapies, which involve the transplantation of live cells, cell-free techniques focus on the use of extracellular vesicles (EVs), such as exosomes, derived from stem cells. These vesicles are rich in growth factors, cytokines, and other signaling molecules that are pivotal in hair growth and follicle regeneration. By harnessing the regenerative properties of stem cells without the complexities and risks associated with live cell transplantation, cell-free therapies offer a safer, more controlled method of treatment. Research has shown promising results in stimulating hair growth, improving hair density, and revitalizing dormant HFs. This approach circumvents issues like immune rejection and potential tumor formation, making it a highly promising avenue in regenerative medicine and cosmetology. As research advances, cell-free therapies could revolutionize the way hair loss is treated, offering a minimally invasive, effective solution for millions struggling with hair loss (Salhab et al. 2022).

14.7 International and National Regulations for Hair Care Formulation/Product

Hair care formulations are designed to cleanse, modify, foster growth, and manage hair through topical application. These formulations are generally categorized into three main groups: hair grooming products, hair cleansing agents, and hair growth enhancers. Hair shampoos, falling under the hair cleansers category, are designed to remove dirt and oil from hair. Hair growth enhancers include oils, lotions, and creams, whereas hair grooming products encompass hair dyes, permanent waves, straighteners, relaxers, tonics, and dressings (Alessandrini and Piraccini 2016). In ancient times, noticeable makeup was often not considered "proper." Products like shampoo, lotions, creams, and makeup were typically made at home. These items were unregulated and, in some cases, considered to be potentially harmful. The earliest known hair care products date back to ancient civilizations, with the first recorded use being in ancient Egypt. Egyptians used hair oils made from plant-based ingredients like olive oil and almond oil, often infused with fragrant herbs to nourish and condition the hair. They also utilized other substances such as animal fats and honey for their hair care routines. These early formulations were rudimentary compared to modern products, but they laid the foundation for the development of more sophisticated hair care treatments in subsequent cultures and eras. During the Middle Ages in Europe, hair care methods were relatively basic, with the use of herbs and rudimentary soaps made from animal fats and lye for cleansing. The application of oils for hair treatment was less prevalent during this period. The hair care practices became more intricate, particularly within the aristocratic circles in the Renaissance era. There was a shift toward the use of sophisticated, often fragranced oils and lotions aimed at maintaining hair health and aesthetics. In India, shampoo has been used since 1500 AD. A mixture of reetha (soapberries), amla (gooseberry), hibiscus, shikakai (acacia), and other plants were used to cleanse and nourish hair.

The 18th and 19th centuries witnessed the rise of wigs and hair powders, predominantly among the European elite, leading to a diminished emphasis on natural hair care However, by the late nineteenth century, there was a resurgence of interest in natural hair, catalyzing the evolution of contemporary hair care products. In 1898, the German chemist Hans Schwarzkopf developed the first commercial shampoo. It was a water-soluble powder shampoo that was very successful. In 1927, he invented the liquid shampoo. Permanent waving technologies were introduced to hair salons in the early 1900s. A pivotal advancement was the invention of synthetic shampoo in the early twentieth century, initially available in powder form and later evolving into liquid solutions. This was followed by the development of conditioners, offering a more comprehensive approach to hair care.(Mahomoodally and Ramjuttun 2017) From the mid-twentieth century onwards, the hair care industry underwent rapid expansion, with the creation of diverse products catering to specific hair types and needs. This era introduced innovations such as hair sprays, permanent waves, and hair colorants. Advances in scientific research and technology led to the development of specialized products focused on hair repair, volumizing, moisturizing, and styling. In recent times, spanning the late 20th and early 21st centuries, there has been a growing trend toward natural and organic hair care products, reflecting heightened health and environmental consciousness. Concurrently, advancements in biochemistry and nanotechnology have facilitated the development of sophisticated hair care solutions, targeting specific hair issues at the molecular level. These innovations represent a significant leap in the field of trichology, offering targeted and effective hair care options (Kapoor 2005).

The global hair care market, as per a Fortune Business Insights study, was estimated at $91.60 billion in 2022, with projections to reach $147.49 billion by 2030, growing at a CAGR of 5.8% during the forecast period. The study highlights that hair care products are essential for maintaining hair health and hygiene for both men and women and safeguarding hair against damage. These products range from shampoos, conditioners, oils, serums, and more, designed for hair nourishment and available in various forms like liquids, gels, creams, and lotions. The market offers diverse products tailored for specific needs, including shampoos for dry or oily hair, conditioners for damaged scalp care, and specialized products for colored or curly hair (Haircare_Market_Report 2023; MMR_Market_Report 2023; _Market_Outlook_2024–2032 2023). The global surge in hair-related issues, particularly hair loss and premature greying, has amplified the importance of hair care cosmetics. Their use is essential not only to address these concerns but also to preserve the hair's natural characteristics. The escalating demand for hair care products has heightened regulatory and industrial vigilance to ensure the safety and efficacy of these products in the global market. To ensure the safeguarding of consumer health and to uphold environmental protection standards, stringent regulatory frameworks are implemented for hair care products on both international and national levels. These regulations primarily aim to mitigate potential risks posed by harmful substances, ensuring that such products are formulated within the bounds of established safety parameters. Additionally, these guidelines enforce the accuracy of product claims regarding efficacy, fostering a market environment characterized by transparency and reliability. This approach not only facilitates informed decision-making among consumers but also stimulates innovation within the industry, as manufacturers strive to develop novel, compliant products. Furthermore, the standardization of quality and safety benchmarks across different regions plays a crucial role in harmonizing the global market, thereby streamlining international trade in this sector. The fundamental objective of regulatory compliance in the hair care product industry is to ensure the provision of products that are safe, effective, and of high quality for human use. This objective is encompassed within cosmetic legislations, which include key regulatory requirements such as accurate nomenclature and comprehensive labeling. These legislations are instrumental in governing the presence and distribution of cosmetic/hair care products in the marketplace. Moreover, the process of conducting safety assessments for these products is influenced by the varying standards and guidelines set forth by regulatory bodies worldwide. This variability highlights the need for meticulous adherence to diverse regulatory frameworks to maintain the overarching goal of consumer safety and product integrity.

14.7.1 Regulatory Frameworks for Hair Care Products

The rising demand for hair care products underscores the crucial importance of addressing their safety and toxicity concerns. Given the potential risks associated with unregulated use, it is imperative that these products are overseen by competent authorities with diligence and precision. This necessitates thoroughly assessing hair care products to ensure their safety and efficacy. The current regulatory framework for hair care products varies by country but typically involves oversight by government agencies responsible for ensuring product safety and efficacy (Fig. 14.2). The regulatory frameworks of various countries are as follows.

United States of America (USA): In the United States, the regulatory framework governing cosmetics is primarily outlined by the Federal Food, Drug, and Cosmetic (FD&C) Act, alongside the Fair Packaging and Labeling Act. These regulations are systematically compiled within Title 21 of the Code of Federal Regulations (21 CFR, parts 700–740). The FD&C Act specifically mandates stringent guidelines against the distribution of adulterated or inaccurately labeled cosmetics. Coal-tar hair dyes are subject to special labeling requirements, including statutory warning notices. However, those offering guidance for patch testing are exempt from the adulteration clause despite potential skin irritancy or other harmful effects.

Additionally, certain cosmetic products like hair treatments, makeup items, and skin cleansing or moisturizing creams, which are not designated for retail sale, are not bound by these labeling requirements, provided they are not sold directly to end-users. The Tariff Act of 1930 further stipulates that all imported goods must display the country of origin on their labeling (Cosmetics 2018). In 2009, the Certification, Accreditation, and Compliance Committee of the United States Department of Agriculture (USDA) set forth specific guidelines addressing the issue of inaccurately labeled organic personal care products (PCPs). The committee emphasized that the USDA holds exclusive authority over certifying any product's organic status. This issue mainly arises in products like shampoos and conditioners, which may claim to be made of 100% organic ingredients yet fail to display the official organic certification or the USDA Organic Seal. This lack of consistent verification and labeling has led to consumer uncertainty regarding the authenticity of organic claims in PCPs. The USDA committee proposed revisions to clarify the term "organic" usage and suggested limiting the use of the USDA Organic Seal. However, the National Organic Standards Board (a Federal Advisory Board of United States) has not formally adopted these suggestions and do not currently represent official USDA policy.

Notably, while the US Food and Drug Administration (US FDA) does not pre-approve cosmetics or their components, except for color additives, before market entry, the law imposes a mandatory safety evaluation for each cosmetic product. This underscores the prioritization of consumer safety. Cosmetic companies adhere to these regulations and frequently implement safety measures that exceed the legal requirements, demonstrating a commitment to consumer well-being and product integrity. These companies are responsible for ensuring that their products do not contain prohibited substances, underlining their accountability in maintaining high safety standards. The Modernization of Cosmetics Regulation Act of 2022 (MoCRA) represents a significant update to the regulatory framework for cosmetics, including hair care products, in the United States. The enactment of the Modernization of Cosmetics Regulation Act (MoCRA) of 2022 on December 29, 2022, as a component of the Consolidated Appropriations Act, 2023, marks a significant shift in the regulatory landscape for PCPs under the US FDA. This legislation is especially relevant to dermatologists who frequently encounter and manage adverse reactions to PCPs, including emollients, shampoos, and cleansers, which are often integral to dermatological treatment plans (US FDA 2005).

Keratin-based hair treatment products, designed for smoothing permed hair and straightening curly hair, have recently garnered increased attention. Analytical studies have revealed that numerous keratin hair care products incorporate free formaldehyde, either in a dissolved state or within a chemical compound capable of converting into gaseous formaldehyde. These products may be marketed with or without a "formaldehyde-free" label. Formaldehyde is recognized as a respiratory irritant and a classified human carcinogen. Comprehensive investigations conducted by the Occupational Safety and Health Administration (OSHA), the National Institute for Occupational Safety and Health (NIOSH), and Health Canada have determined that products labeled as "formaldehyde-free" may still contain detectable levels of formaldehyde. The US FDA oversees the regulation of cosmetic products containing formaldehyde. Additionally, legislative members have urged the FDA to implement enforcement actions against the distribution and use of these keratin hair treatment products due to potential health risks (Cosmetics and U.S. Law 2022).

The European Union (EU): The EU standardized its regulations for hair care products in the early 1970s. The Cosmetics Directive was established in 1976, reevaluated in 2009 to enable further harmonization, and replaced by the EU-wide Cosmetics Products Regulation on July, 12, 2013. The new regulation, EU Regulation (EC) No 1223/2009, replaced the Cosmetics Directive (76/768/EEC), and most of its provisions became applicable from July 11, 2013. Adopted on November 30, 2009, the EU introduced the updated EU Cosmetic Products Regulation, formally known as EU Regulation (EC) No 1223/2009. This regulation, which largely occurred on July 11, 2013, supersedes the previous Cosmetics Directive (76/768/EEC). It encompasses standardized notifications and provides more explicit guidelines regarding the composition and structure of product safety evaluations and Product Information Files (PIFs). Additionally, it enhances the collaborative market surveillance initiatives among the competent authorities of the EU member states (EU Cosmetics Regulations and Registration 2023). The directive mandates that all products released in the EU market must be safe for human use and not cause harm. The EU, similar to the United States, holds manufacturers responsible for proving the safety of their products. Designated authorities enforce national regulations in each member country. These regulations cover a wide range of products, from basic hygiene items like soap, deodorant, shampoo, and toothpaste to luxury beauty products, ensuring consumer safety and market integrity. The EU framework encompasses rules for market entry, regulatory alignment, and international trade relations. The European Commission collaborates with stakeholders in the cosmetics industry at both EU and international levels. Furthermore, the EU legislation stipulates that all products on the market must be registered with the Cosmetic Products Notification Portal (CPNP) (European-Commission 2024). Authorities in countries like France and Germany have been instructed to take action against certain products, explicitly issuing warnings against using hair smoothing products containing high levels of formaldehyde. Both France and Ireland have implemented steps to withdraw these products from their markets (EWG 2011). Additionally, the Scientific Committee on Consumer Safety (SCCS) has issued its guidelines regarding hair care products. According to these guidelines, as per the Regulation (EC) No. 1223/2009, these products are not recommended for eyelashes due to varying risk levels depending on the application area, differing between scalp hair and eyelashes. Moreover, the SCCP has advised the Commission to consider further restrictions on using substances in hair dyes. The SCCP has also suggested conducting a comprehensive safety evaluation of hair dye substances to assess their potential genotoxic or carcinogenic effects (EU-Report 2014).

Japan: In Japan, the Ministry of Health, Labor, and Welfare (MHLW) oversees the regulation of cosmetic products under the Pharmaceutical and Medical Devices Law (PMDL, formerly known as the Pharmaceutical Affairs Law). According to this law, beauty products are categorized into quasi-drugs and cosmetics, each with distinct regulatory requirements. Cosmetics, designed for enhancing appearance and maintaining healthy skin and hair, encompass a range of products, including hair dyes, shampoos, and treatments (Hu 2021; Law 2015). Japan has established a list of prohibited ingredients, covering specific UV filters and preservatives. The responsibility for ensuring product safety now falls on the manufacturers, allowing them to use any ingredient deemed safe in their formulations. A license and pre-market approval from the MHLW is mandatory for product manufacturing. Since 2001, Japanese cosmetic companies must register their product brands before commencing manufacturing or importation. The costly pre-market registration process has been supplanted by a more cost-effective post-market surveillance system, ensuring consumer safety without the high expenses previously associated with the process (Law 2015).

China: In recent decades, the Chinese market has experienced a significant surge in demand for hair care products. This surge has positioned China as a global leader in terms of market share, offering substantial opportunities for business and import activities (China_Report 2023) In China, the cosmetic industry, including hair care products, is regulated under a comprehensive set of laws and regulations. This regulatory framework consists of various key documents, such as the Hygiene Supervision Regulations (first issued in 1990 and later updated in 2005), the Hygienic Standard for Cosmetics (2007), and the Administration of Hygienic Licenses (amended in 2010). Additionally, the framework includes the Standard Chinese Names of International Cosmetics Ingredients Inventory (2010), the Guideline for Risk Evaluation of Substances (2010), the Guidelines for the Registration and Evaluation of New Cosmetic Ingredient (2011), and the Cosmetics Technical Requirement Standard (2011). Manufacturers of finished cosmetic products must acquire a hygiene license record-keeping certificate from the China Food and Drug Administration (CFDA). This requirement is obligatory for general cosmetics, such as hair care items. However, products classified as special cosmetics, like hair dyes, hair growth enhancers, and perm products, may obtain an optional hygienic license (Guo 2016).

Canada: In Canada, hair care products are generally classified as cosmetics. They are regulated by Health Canada's Cosmetics Program, which operates under the key guidelines set forth in the Food and Drugs Act and the Cosmetic Regulations and governed by Canadian government. The oversight and regulation of cosmetics, including the registration of Cosmetic Notification Form (CNF), fall within the jurisdiction of Health Canada. This federal body plays a crucial role in upholding and normalizing the health standards for Canadians (Canadian's_Cosmetic_Regulations 2024). This program delineates specific requirements for labeling, producing, distributing, and selling cosmetic products in Canada. Similar to practices in the United States, Canadian manufacturers are primarily responsible for ensuring that their products adhere to all relevant laws and regulations. Additionally, they must prove that their products are safe for consumer use. Product Safety Officers manage the enforcement and organization of these regulations. Notably, in 2011, Health Canada issued a significant advisory regarding keratin hair treatments. This advisory specifically targeted hair smoothing products containing formaldehyde exceeding 0.2%, recommending their prohibition. Cosmetic products distributed in Canada must comply with the standards established in the Food and Drugs Act and the Cosmetic Regulations. Additionally, these products must adhere to the Consumer Packaging and Labelling Act stipulations and its relevant regulations. Furthermore, the ingredients used in cosmetics and personal care items might also fall under the regulations of the Canadian Environmental Protection Act (Canadian Cosmetic Products Regulations 2021).

Australia: In Australia, the government conducts thorough evaluations of cosmetic products and ingredients, whether they are produced domestically or imported, to guarantee their safety for users, employees, and the environment, as outlined in 2021 (DHAC 2021). This scrutiny is carried out under the Australian Industrial

Chemicals Introduction Scheme (AICIS), which oversees the regulation of cosmetic ingredients, including those labeled as "natural," categorizing them as industrial chemicals as per the Industrial Chemicals Act, as stated in the "Australia Cosmetic Regulation" document of 2021 (Australia Cosmetic Regulation 2023). The Therapeutic Goods Administration (TGA) reviews cosmetic products claiming therapeutic benefits. For queries about ingredients in personal care items that do not fall under therapeutic goods, AICIS is the appropriate contact point. Inquiries about the labeling or safety of cosmetic products should be directed to the Australian Competition and Consumer Commission (ACCC). While Australia does not have dedicated regulations for hair care products, the "Infection prevention and control guidelines for hair, beauty, tattooing and skin penetration industries," established in 2019, offers guidance to professionals in these sectors. This helps them adhere to the standards set in the Public Health and Wellbeing Act 2008 and the corresponding Regulations of 2019, ensuring compliance and maintaining high health and safety standards.

Brazil: In Brazil, the oversight of cosmetics is managed by the Brazilian Health Regulatory Agency (Anvisa), as outlined in Resolution RDC7/2015 and its amendment, Resolution RDC 237/2018. These documents provide the framework for the authorization of personal hygiene products, cosmetics, and fragrances in the Brazilian market. The oversight of hair care product regulations in Brazil is conducted by the Brazilian Health Regulatory Agency (ANVISA), (Ferreira 2021). Personal care items, cosmetics, and perfumes are categorized into two classes—Class I and Class II. The registration process for these products in Brazil is determined based on their classification. ANVISA mandates that PCPs, cosmetics, and perfumes adhere to the specific provisions of Class I and Class II. This includes compliance with the lists of approved preservatives, colorants, and restricted substances. These guidelines ensure that PCPs, cosmetics, and fragrances adhere to the established conditions and limitations. Furthermore, ANVISA plays a crucial role in upholding the standards set by the Brazilian Health Surveillance Agency. This involves the sanitary regulation of the production and consumption of various products and services, including related environments, processes, materials, and technologies. Additionally, ANVISA's responsibilities extend to coordinating, overseeing, and managing activities linked to the registration, information, inspection, setting of norms and standards, and enforcement of regulations pertaining to the sanitary surveillance of PCPs, cosmetics, perfumes, and sanitizers.

India: Hair care products are classified as cosmetics under the Drugs and Cosmetics Act of 1940. The oversight and regulation of cosmetics fall under the purview of the 1940 Drugs and Cosmetics Act and its subsequent 1945 Rules. The chief governing authority in this domain is the Central Drugs Standard Control Organization (CDSCO). Over the years, India has progressively updated its cosmetic regulations to align with international norms. These regulations include mandatory licensing for the production and distribution of cosmetic products. The CDSCO maintains a detailed list of ingredients that are allowed or forbidden in cosmetics and sets specific standards for product labeling (Kavya 2022). Despite these developments, India's regulatory landscape continues to evolve, focusing on adopting global practices such as prohibiting animal testing for cosmetic products. The 1940 Drugs and Cosmetics Act currently serves as the primary legislative framework in this sector. The Drugs and Cosmetics Rules of 1945 outline the procedures to be followed in manufacturing cosmetic products in India. Under Schedule M-II, cosmetics are categorized into 11 distinct classes, encompassing a range of hair care products like shampoos, oils, creams, lotions, dyes, etc. Manufacturers must obtain a license from a designated Licensing Authority to produce any of these items. Moreover, a government notification [(G.S.R. 426) (E)] issued on May 19, 2010, mandates the registration of all imported cosmetics (Welankiwar 2013). The CDSCO further regulates this. The Bureau of Indian Standards (BIS) plays a crucial role in ensuring the safe use of ingredients in cosmetics, setting guidelines that manufacturers must follow. As of April 1, 2013, it became obligatory for all imported cosmetics to be registered with the Drug Controller General of India (DCGI). However, domestic cosmetics in India are exempt from this registration. Nonetheless, products such as shampoos, hair oils, and dyes do require a pre-manufacturing loan license. Additionally, the manufacturing and importing products containing certain harmful substances like hexachlorophene, arsenic, lead, and mercury are stringently controlled (India Cosmetic Regulation 2020). Along with the countries mentioned above, different countries have their regulatory guidelines for cosmetics and hair care products, overseen by the respective regulatory bodies. These guidelines vary from country to country and are tailored to meet each region's specific legal, cultural, and scientific needs. The regulatory guidelines followed in several other countries are depicted in Fig. 14.3.

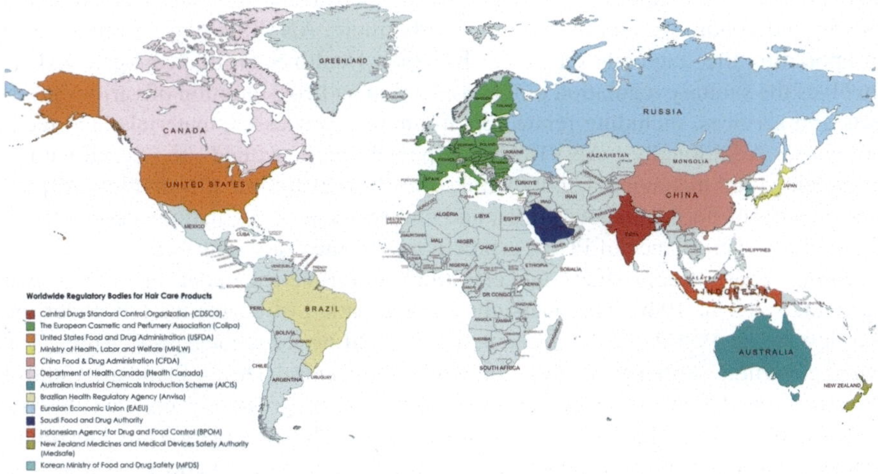

Figure 2: Current operational regulatory bodies worldwide for the hair care products testing, approval and marketing .

Fig. 14.3 Regulatory Guidelines and Acts/Regulations followed by various other countries

14.8 Evaluation of Safety and Efficacy of Hair Care Products

Alopecia, a condition characterized by hair loss, impacts individuals of both genders across various age groups. Currently, the US FDA has permitted only two medications, (FN) and minoxidil, for the treatment of alopecia (Lee et al. 2018). However, despite the partial efficacy and potential side effects of these two, there have been significant efforts in hair research for creating novel hair growth agents. The limitations of standard hair care products, including limited and temporary effectiveness, variable efficiency, and side effects, highlight the critical need for new and efficacious hair growth modulators. Parallelly, efficient pre-clinical screening models for reliable data related to safe use are also needed to expedite research in this field. As the majority of countries worldwide have banned animal testing of cosmetic products and promote cruelty-free testing, the European Centre for the Validation of Alternative Methods (ECVAM) has listed the validated cell-based in vitro models for predicting the safety and toxicity of cosmetic ingredients, including hair products (Almeida et al. 2017) Researchers have also developed various pre-clinical models to study HF functions and the action mechanisms of hair growth formulations, as well as ex vivo cultures of human and mouse HFs (Fig. 14.4). Investigations into the hair growth-promoting effects of chemicals often involve primary DPC culture sourced from follicles from human scalp and rodent vibrissae follicles, and macaque skin (Madaan et al. 2018). However, the short lifecycle of primary cultures and the need for more cell quantities have led to the development of immortalized hDPC lines, achieved through the introduction of the human telomerase reverse transcriptase (hTERT) gene (Kwack et al. 2018). These immortalized DPC lines are now extensively utilized in hair growth studies and, along with primary DPCs, are commercially available, significantly advancing hair growth research.

14.8.1 Pre-clinical Testing

The chemical products that are being tested through various in vitro studies for their efficacy in hair growth promotion must pass the skin sensation test prior to being implemented for human use.

14.8.2 In Vivo Models

Rodent models have always remained an irreplaceable alternative to understanding the effects of chemical products prior to applying them for human use. Employing rodent models for cosmetic product testing offers numerous benefits in product development. However, most of the time, animals must bear unnecessary pain, allergic reactions, and even death during the testing. As cosmetic products, including hair products, come under luxury products and are not an essential commodity, in 2013, the EU banned animal testing and marketing of animal-tested cosmetics in

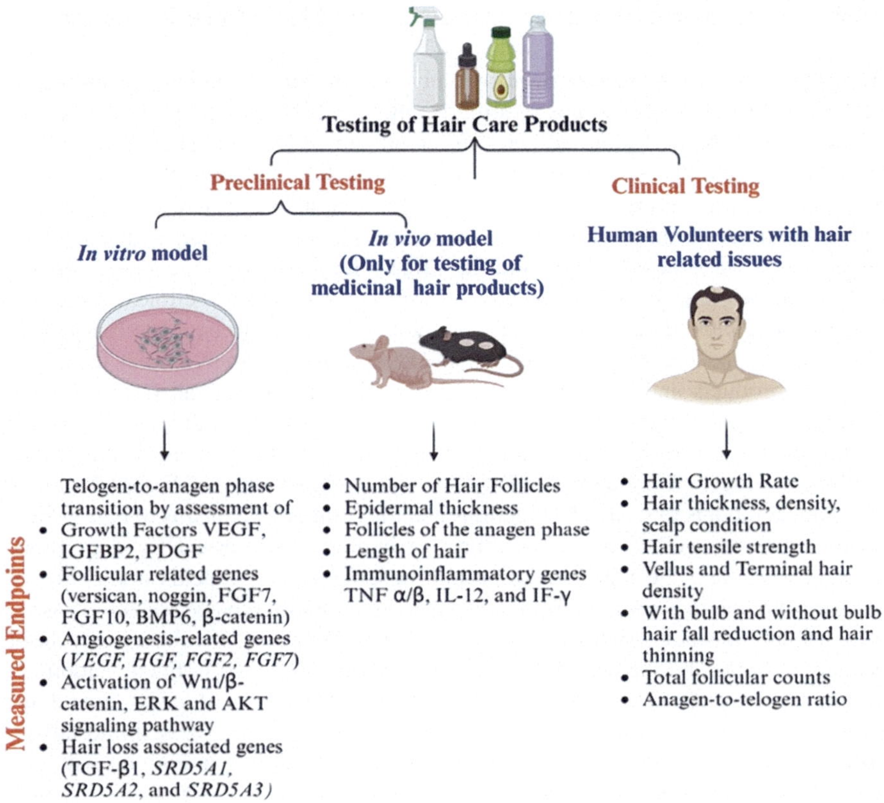

Fig. 14.4 Pictorial representation of pre-clinical and clinical models used to test the safety and efficacy of hair care products. The figure shows the molecular and physical markers used to analyze the overall impact of hair care products on hair growth physiology and appearance

the EU under the seventh Amendment to the EU Cosmetics Directive. Further, the United States, Brazil, and India (2014), Australia (2019), New Zealand (2015) and South Korea (2018), Taiwan (2019), Mexico (2021), and Guatemala (2017) have taken initiatives to completely ban or limit/regulate animal testing for cosmetic products (Silva and Tamburic 2022). In India, under the Prevention of Cruelty to Animals Act of 1960, the prohibition of testing of cosmetics on animals and import of animal-tested products was introduced in 2014. On the contrary, the hair product ingredients labeled as "industrial chemicals legislation" in EU reports are tested in an in vivo model under the Registration, Evaluation, Authorization and Restriction of Chemicals (REACH) regulation. Under REACH testing, in vivo testing can be employed for those hair product ingredients that are listed in the REACH dossier, and in vivo-based chemical testing is required to prove as "safe" for further human application (Knight et al. 2021). Alternatively, to support cruelty-free or animal-free testing, ECVAM has proposed to use validated cell-based in vitro models for predicting the safety and toxicity of cosmetic/hair product ingredients. For instance, as an alternative for topical toxicity, the in vitro human epidermis (HE) model (EpiSkin™, EpiDerm™, and SkinEthic™ Reconstructed Human Epidermis (RHE)) and human corneal epithelial (EpiOcular™) model have been used against in vivo rabbit skin and eye Draize test (Almeida et al. 2017). The details for *the* in vitro model used for testing hair care products are discussed below.

A shorter hair growth cycle in rodents compared to humans allows researchers rapid observation of the various stages of hair cycle; hence, for research purposes and to study the therapeutic mechanism of hair product ingredients in primary hair disease conditions, various models have been developed (Orasan et al. 2016). Rodents, primarily genetically modified strains, allow researchers to study the impact of specific genes on hair growth and response to hair drugs, providing valuable insights into the genetic factors involved. Rodent models enable researchers to conduct detailed histological analysis of HFs, allowing for a deeper understanding of how products affect hair at the cellular level. Currently, mouse models developed from genetic manipulations (transgenic) and hormonal disturbance (androgen-dependent) are widely exploited to study hair biochemistry (Orasan et al. 2016). For screening the newly developed hair drugs, androgen-induced mouse models are the best choices. In 2010, Crabtree and coworkers developed a transgenic AGA mice mode with keratin-5 promoter-driven human AR overexpression in the skin. Under the elevated levels of 5α—dihydrotestosterone (DHT), this model hindered hair regeneration, similar to the human AGA scalp (Crabtree et al. 2010). In K5-hAR transgenic mice, human AR (hAR) is expressed in the basal epidermis and follicle ORS but not in the hair follicle DP (HFDP) cells of these mice (Naito et al. 2008). Recently, a DHT-induced wild-type C57BL6 mouse model of AGA has been developed that imitated human AGA, with stuck hair regrowth, early recession, and HF miniaturization (Fu et al. 2021). For the evaluation of anagen induction capabilities of potential hair growth-promoting drugs, in vivo models involving mice with pigmented hair (wild type C57BL/6 and C3H strains) are frequently employed. The benefits of utilizing these mouse models for assessing anagen initiation include synchronized and short hair cycles and exclusive localization of melanocytes in HFs. The regular hair cycle of the C57BL/6 mouse bears is for the period of 7 weeks to 16 weeks, which represents a lengthier telogen phase compared to other rodent models. Thus, it is suitable to study the effect of hair products on the conversion from telogen to anagen follicle (Ohn et al. 2019). The C3H/HeJ inbred mouse strain and the Dundee experimental bald rat (DEBR) strain naturally progress toward adult-onset AA, an autoimmune disorder that attacks actively growing HFs. These models may also provide a better insight into the role of genetics and molecular mechanisms in AA disease, such as psycho emotional stressors and associated neuropeptides, immunopathologies, and autophagy (Gilhar et al. 2020; Gund and Christiano 2023). Though the rat DEBR model acquires spontaneous AA with higher frequency than other mouse models, their large size does not make them cost-effective for drug screening. The complete reliability of these models is still questioned due to inbred backgrounds and highly controlled pathogen-free environments (Sun et al. 2008). The golden Syrian hamster flank organ has been used to study the androgen-dependent responses of SGs and hair, quantitative assessment of androgenic and anti-androgenic drugs on hair development, processes accountable for the contrasting consequences of androgen on hair growth, and development of therapy against hirsutism (Lucky et al. 1986; Ohn et al. 2019). Wistar-Bratislava albino rats are used to analyze the hormonal effects on the hair cycle. A study using Wistar-Bratislava rats and mice, demonstrated varying levels of hair regrowth efficacy for test hair products in animal models, suggesting potential recommendations for their use in humans (Orasan et al. 2016). Other animal models used for learning hair dynamics included the stump-tailed macaque spontaneous baldness model (McElwee et al. 1999), the androgen-dependent mouse model (Truong et al. 2017), the human balding scalp grafted nude mouse model (Sintov et al. 2000), and the androchronogenetic alopecia mouse model (Park et al. 2003).

14.8.3 In Vitro Models for Test Product Sensitization

Human Epidermis (HE) Model for Dermal Sensitization Tests: The skin irritation test (SIT) is used to envisage the acute skin irritation for chemicals/substances. Previously, the in vivo Draize rabbit SIT (OECD TG 404) was used, but in 2010, the EU Commission approved a validated in vitro SkinEthic RHE model for replacing the in vivo equivalent (EC test method regulation as method B.46 and OECD TG 439). The method uses the SkinEthic reconstructed human epidermis model and involves the topical application of a chemical for 42 minutes followed by 42 hours incubation and then determination of cell viability via the MTT (3-(4, 5-dimethylthiazolyl-2)-2, 5-diphenyltetrazolium bromide) assay. MTT assay declares a chemical as an irritant if ≤50% of cells are viable or a non-irritant if >50% of cells are live (OECD 2021). Using RHE, a bottom-up and top-down testing strategy, 86 substances were assessed for their capacity for the European Union System of Classification of Chemicals implemented in the EU Classification, Labelling and Packaging (EU CLP) Regulation and UN GHS (United Nations Globally Harmonized System of Classification and Labelling of Chemicals) (Alepee et al. 2015). Another study conducted in Turkey has shown the suitability of the model for testing 60 widely used cosmetic products (shampoo and skin care products) (Kose et al. 2018). KeraSkin™-VM is a new, three-dimensional (3D) human epidermal corresponding model that exhibits native HE-like histology, such as the fully differentiated and multi-layered basal layer, stratum spinosum, granular layer, and stratum corneum (Jung et al. 2014). Though EU SIT does not approve KeraSkin™-VM a study has shown its equivalent sensitivity predictivity ability in accordance with OECD TG 439 PS estimated with 20 reference chemicals (Han et al. 2020).

Human-Derived Corneal Epidermal Model for Ocular Irritation Tests: This test predicts the ocular irritancy possibility of chemicals, substances, and mixtures. EU has approved the EpiOcular eye irritation test utilizing a 3D corneal tissue model consisting of normal human-derived epidermal keratinocytes (NHEKs) that are cultured at an air–liquid interface (OECD TG 492). The test substance is measured with a cytotoxicity MTT assay, with the outcome being used as a prediction for eye irritancy. A chemical is categorized as an irritant (GHS Category 1 or 2) if the tissue viability is ≤60% and as a non-irritant (GHS unclassified) if the viability is >60%. The successful validation and further applicability of this model for testing were confirmed by the reproducible data obtained after single exposure of 20 cosmetic test chemicals to EpiOcular OCL-200 tissues and produced 99.7% agreement in prediction (NC/I) after 298 independent trials in seven independent laboratories (Pfannenbecker et al. 2013).

14.8.4 In Vitro Models for Product Efficacy on Hair

The major in vitro cell model for testing proposed ingredients in test products includes DPCs, dermal fibroblast (DF), follicular epidermal keratinocytes, and ORS cells. These cells, cultured in either a two-dimensional (2D) monolayer or a 3D culture, are currently being exploited for animal cruelty-free testing of hair care products (Fig. 14.5).

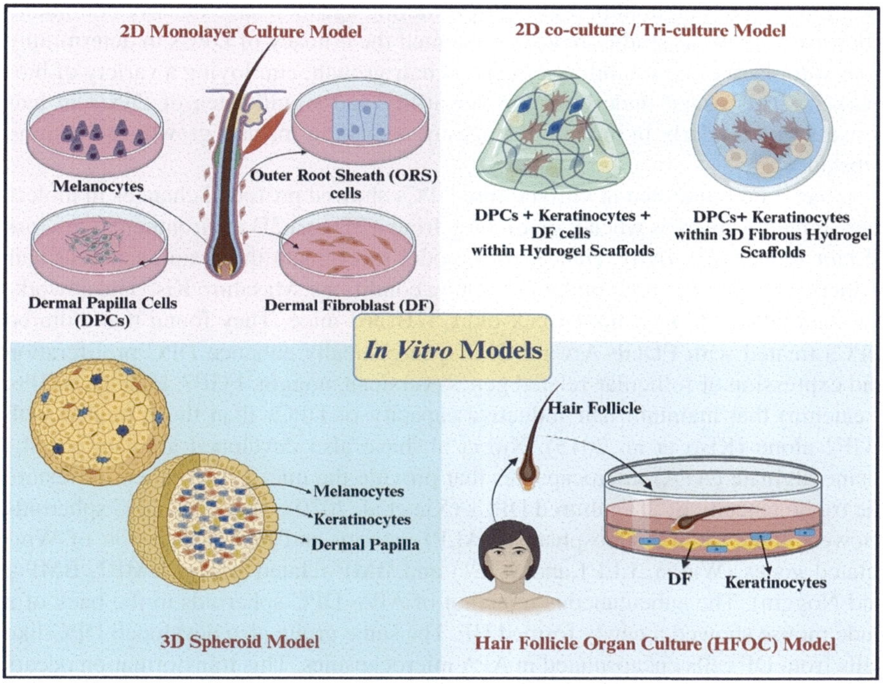

Fig. 14.5 The overview of in vitro models used to test hair care products for their safety and efficacy. The models include 2D monolayer culture of different types of cells present in the bulb of the HF and shaft, 2D co-culture/triculture of these different types of cells within the artificial scaffold, 3D spheroid made up of major types of cells in the HF, and ex vivo culture of whole HF isolated from the scalp of human

Hair Follicle Dermal Papilla Cells/Immortalized Rat Vibrissa Dermal Papilla Cells: Hair follicle DP cells (HFDPCs) are positioned at the HF base. DPCs are the main mesenchymal element of hair essential for hair morphogenesis and regeneration. These cells are the storehouse of multi-potent stem cells and require nutrients and growth factors. They primarily govern the three phases of hair growth: growth phase (anagen), regression/transition phase (catagen), and rest phase (telogen). During the telogen-to-anagen phase transition, HFDPCs regulate hair growth by balancing HF cycling through growth factor VEGF, IGFBP2, PDGF, and VEGF and modulation of the Wnt/β-catenin pathway (Singh et al. 2022). DPCs serve as an excellent model to investigate the mechanical action of compounds on the functioning of human follicles. Due to their essential role in hair growth, DPCs, as in vitro screening models, have been used to assess the effect of hair growth modulating agents at the cellular and molecular levels. The controlling effects of hair growth-promoting ingredients in hair care products on DPCs could be observed by measuring the changes in growth markers in comparison with untreated DPCs, or in the stress-induced model where cell proliferation, apoptosis, and other mechanistic features may be evaluated for hair growth assessment (Madaan et al. 2018). Hence, DPCs function as an effective preliminary screening tool, enabling the identification of potential hair growth-promoting agents from extensive compound libraries. Numerous studies have documented the efficacy of DPCs in determining both stimulatory and inhibitory effects on hair growth, employing a variety of biomarkers. This usage underlines the versatility and significance of DPCs in hair research, particularly in the initial stages of screening for hair growth-modulating substances.

It has been found that in vitro culture DPCs showed profound changes in molecular signature changes when transitioning from a 3D to a 2D environment, with loss of hair inductive activity. Hence, as a model to evaluate the maintenance of hair inductive activity of DPCs under 2D culture conditions, Masahiro Kiso and coworkers used vibrissal DP from 4-week-old C57BL/6J mice. They found that cultured DPCs treated with PDGF-AA + FGF2 synergistically enhance DPC proliferation and expression of follicular-related genes (versican, noggin, FGF7, FGF10, BMP6, β-catenin) that maintain hair inductive capacity of DPCs than those treated with FGF2 alone (Kiso et al. 2015). Xie et al. have also developed alginate-poly-L-lysine-alginate (APA) microcapsules that provide the microenvironment to restore the trichogenecity of 2D cultured DPCs (Xie et al. 2020). The APA–DPC spheroids showed high Alkaline Phosphatase (ALP) activity and high expression of Wnt-related genes (Wnt5a, LEF1,and WIF1) and BMP-related genes (BMP2, BMP4, and Noggin). The subcutaneous injection of APA–DPC spheroids to the back of a nude mouse showed a newly formed HF. The same group also developed DPC-like cells from DF cells encapsulated in APA microcapsules. This transformation occurs through activation of Wnt signaling. These APA microcapsules inserted subcutaneously into the hypodermis of nude mice showed HF-like complexes formation. This novel de novo folliculogenesis strategy could provide a massive quantity of safe and functional DPCs and overcome the DPC shortage in hair product testing and regenerative drug-based strategies (Xie et al. 2020).

For AGA medication, FN stands as the sole oral therapeutic agent sanctioned by the US Federal Drug Administration (US FDA). It primarily impedes the transformation of testosterone (T), an androgen, into DHT for selective inhibition of the type II 5α-reductase enzyme. A daily dose of 1 mg reduces HF maturation by diminishing the concentration of DHT up to 64% in scalp. It has been found that FN (50 and 100 μM) significantly increased stem cell-like aggregation behavior in DPCs. The activation of the Wnt/β-catenin signaling pathway (β-catenin and pAKT) stimulated DPCs, leading to HF formation and prolongation of the anagen phase. It has been found that FN induced the expression of stem cell transcription factor proteins (integrin-β1, Nanog, and Sox-2), substantiating its role in stem cell dynamics and HF development (Rattanachitthawat et al. 2019). Similarly, minoxidil or 2,4-diamino-6-piperidino-pyrimidine-3-oxide is another US FDA-approved drug for treating AGA. It possesses growth-promoting properties by vasodilation and increasing blood flow to the DP. Moreover, minoxidil promotes angiogenesis via direct inhibition of hypoxia-inducible factor (HIF), prolyl hydroxylase-2 (PHD-2), and associated induction of VEGF in human keratinocytes and DPCs (Yum et al. 2017). It has been proven that minoxidil also induces DPC proliferation by increasing the levels of phosphorylated Extracellular signal-Regulated Kinase (pERK), phosphorylated AKT/Protein Kinase B (pAKT), and B-cell lymphoma family proteins like Bcl2 and prolonging the anagen phase (Han et al. 2004; Yoo et al. 2007).

Various plant-based extracts have been studied for their growth-promoting property (Park and Lee 2021). DPCs are integral to HF function, supporting keratinocyte proliferation, facilitating the development of new HFs, and maintaining hair growth. Their role underscores the complex interplay of cellular and molecular mechanisms involved in hair biology. *Morus alba* root extract (MARE) has chlorogenic acid and umbelliferon, which exert their antioxidant activity and promote hair growth. It has been used as an ingredient in hair products, shampoos, conditioners, and hair masks for hair repair and as a hair growth promotor (Hyun et al. 2021). A study has shown the therapeutic effect of 10%–20% of MARE (MARE) with 80% viability of cultured HFDPC cells. MARE promoted telogen-to-anagen transition and hair growth by regulating and secretion of development- and angiogenesis-related genes (VEGF, HGF, FGF2, FGF7) and activating β-catenin in HFDPCs. Moreover, conditioned media (CM) of MARE treated HFDPCs, when supplemented to human umbilical vein endothelial cells and normal human epithelial keratinocytes, led to upregulation in the expression of involucrin, a maker of keratinocyte differentiation and promoting the anagen state (Hyun et al. 2021). The therapeutic properties of extracts of camellia flowers and oil from camellia seeds have been traditionally used for hair and skin in Asian countries. It is used in products for hair care, such as oil, serum, and mask (Pilgrim™). In a recent report on the efficacy of extracts of camellia flowers and oil from camellia seeds in promoting hair health, which has been evaluated in both in vitro and in vivo systems, the in vitro finding showed that 1% *Camellia japonica* placenta extract (CJPE) increased the expression of Aquaporin (AQP) 3 (water- and glycerol-transporting protein) and HF cell growth marker genes [CTNNB1, KGF, and VEGF] in HFDPC. The clinical tests with 42 adult female subjects demonstrated that the product containing 0.5% CJPE increased the scalp moisture content and reduced sebum content, dead scalp keratin, and erythema. The RNA-seq-based transcriptomic study showed that CJPE in HFDPC increased the expression of HF growth-regulating genes, lipid storage, cholesterol efflux, hormones, AMPK signaling pathway, and axon guidance (Cho et al. 2021).

A study found that the *Oryza sativa* L. cv. KDML105 bran extract contains various bioactive compounds, including phenolic acids, flavonoids, and tocopherols, which have antioxidant and anti-inflammatory properties and are used in hair oil products (mamaearth®). The extract was also found to have metal chelating activity, which may contribute to its antioxidant effects. In HFDPCs, the quote was shown to upregulate the expression of hair growth genes (IGF-1, CTNNB1, and VEGF) and downregulate the hair loss-associated genes (TGF-β1, SRD5A1, SRD5A2, and SRD5A3). These findings suggest that the *O. sativa* L. cv. KDML105 bran extract may have potential applications in developing hair growth products or treatments (Khantham et al. 2023). *Houttuynia cordata*, also known as heartleaf, lizard tail, or fish mint, is a medicinal plant traditionally used as herbal medicine in Asia for alopecia treatment. A study on the hair growth-promoting impact of *H. cordata* extract in cultured human HFDPCs revealed that the extract stimulates the proliferation of DPCs by elevating the mitochondrial membrane potential and levels of NADH and ATP. The section also modulated the expression of genes related to cell cycle, anti-apoptosis, growth factors, and signal transduction, including CDK4, Bcl2, p16, p21, PDGF-aa, VEGF, ERK, and AKT phosphorylation. Moreover, the extract prolonged the anagen stage in cultured DPC spheroids. These findings suggest that the *H. cordata* extract has the potential to be a natural and effective agent for promoting hair growth and may offer a promising alternative to current treatments for alopecia (Kim et al. 2019). The seed extract of *Silybum marianum* has been traditionally used in European countries as herbal medicine (Himalayan Organics™) to treat liver ailments (You et al. 2023). A recent study has identified apigenin as the dominant component in the *S. marianum* flower extract (SMFE). The in vitro study using HFDPCs has demonstrated that 100 μg/mL of SMFE induced the expression of anagen promoter genes IGF-1 and FGF-7 and repressed the expression of catagen inducer gene TGF-β1 and hence promoted increased HFDPC proliferation and hair growth. Moreover, 100 μg/mL of SMFE also showed a 31% increase in the proliferation of senescent HFDPCs and a 49% decrease in levels of senescence-associated IL-6 production compared to non-treated senescent HFDPCs. The in vitro finding was validatory in clinical testing in 42 volunteers aged 8–54 years with AGA. SMFE was found to increase hair density by 55% in the "test product" application group compared to the "placebo product" application group (You et al. 2023). Another pre-clinical study evaluated of the effects of a novel complex of active herbal ingredients used in various hair products and its cosmetic formulations (shampoo and tonic) on hair loss and aging (Honda Igarashi et al. 2021). The formulation applied to cultured HFDPCs and cultured human scalp showed a reduction in oxidative stress and increased mitochondrial activity (94.29% increased cytochrome C oxidase I) in UV-exposed HFDPCs and prevented follicular loss (reduction in DHT levels) and aging (increased Collagen XVIIα1 (COL17A1) expression) in cultured human scalp (Honda Igarashi et al. 2021).

Similarly, due to proven antioxidant capacity, sinapinic acid (SA), a hydroxycinnamic acid, is also used as an ingredient of hair care products (e.g. Keracyn™). Research has shown that SA induces HFDPC proliferation up to 150% at the dose 100 μM for 72 h. This self-induced cell proliferation was attributed to Wnt/β-catenin signaling pathway-induced secretion of growth factors by IGF-1 and VEGF and expression of cell cycle progression gene cyclin D1 by HFDPCs (Hyun et al. 2021). A study investigated the effect of a non-crosslinked hyaluronic acid formulation, HYDRO DELUXE BIO, on human DP cells (HFDPCs) and their potential to promote hair growth. Its safety for use was assessed at concentrations of 40 and 80 mg/ml, demonstrating good biocompatibility and non-cytotoxicity. The formulation showed a defensive impact against UVB-induced cytotoxicity with cell viability >80% compared to untreated and UVB-irradiated control cells and significantly increased the expression of VEGF in HFDPCs, which is known to play a crucial role in angiogenesis hair cycle (Zerbinati et al. 2021).

CARI ONE® is an herbal remedial formulation with a combination of natural components, including *Phellinus linteus*, *Cocos nucifera* oil, *Polygonum multiflorum* root, *Sophora angustifolia* root, *Punica granatum*, bamboo vinegar, *Morus bombycis* Koidz., molasses, *Ficus carica*, and *Ginkgo biloba*. A study scientifically examined the potential hair growth-promoting properties of CARI ONE and its constituent ingredients. The Korean Food and Drug Administration (KFDA) approved it as a quasi-drug for topical application to the scalp. It showed hair growth-promoting properties in telogenic 6-week-old C57BL/6N mice by triggering the anagen phase in resting HFs via β-catenin and Trp-1 proteins expression in DPCs and melanin content in B16F10 cells (Park 2014). Furthermore, culture DPCs obtained from rat whiskers, when treated with 50 μg/mL of bovine lactoferrin (bLF), showed significantly increased proliferation via activation of genes associated with Erk, AKT, and Wnt signaling pathway. Moreover, application of bLF twice daily for a duration of 7–12 days to the depilated back of both young (2-month-old) and aged (1-year-old) C57BL/6 mice showed a significant positive impact on hair growth irrespective of the generation of mice (Huang et al. 2019).

Dermal Fibroblasts (DFs): Around the bulbs of HFs, different cell types, such as DFs, endothelial cells, adipocytes, and immune cells, significantly add to HF cycling and morphogenesis. Through secretion of growth agents or EVs, DFs produce an ECM that cooperates with hair follicular cells and excites them to start the hair growth cycle (Driskell et al. 2011; Tracy et al. 2016). As mentioned, the research group that tested the hair growth power of MARE in HFDPCs also studied the angiogenic capacity of DFs on human umbilical vein endothelial cells (HUVECs) and HFDPCs. It has been found that DFs treated with 20% of MARE promoted cell proliferation, angiogenesis (VEGF and FGF7), and anti-apoptotic signaling (Bcl2) in DFs. The HUVECs treated with CM of MARE-activated DFs in a hypoxic environment exhibited denser tubular formation, improved cell viability, and enhanced migration capacity compared to cells cultured under normoxic or hypoxic conditions without CM treatment. Under hypoxic conditions, HFDPCs exposed to CM from MARE-treated DFs showed significant upregulation of critical angiogenesis and growth-promoting genes, such as VEGF, FGF7, and HGF. Additionally, the expression of IGF-1, for sustaining hair growth, increased in the CM-treated group, while pro-apoptotic caspase-3 expression decreased. Furthermore, the study found that activation of ERK and AKT, crucial for HFDPC proliferation and survival, was enhanced in the CM-treated group. These findings suggest that using HFDPCs, the efficacy of the test product MARE showed promising hair growth by modulating the secretion of growth factors and signaling pathways in HFDPCs (Im et al. 2022).

Human Hair Follicular Epidermal Keratinocyte (HHFK) Cells/Human/ Immortalized Keratinocytes Cell Line (HaCaT): A healthy communication between specialized DP cells containing mesenchymal tissue and keratinocytes containing epithelial tissues is vital for the growth and development of HFs. The specificity of follicular keratinocytes is owing to CD34 expression, which is absent in dermal keratinocytes and shows high proliferative capacity. Secreted molecular players such as Wnt, sonic hedgehog, and bone morphogenetic proteins play a crucial role in signaling between DPCs and epithelial cells to maintain and regenerate hair growth (Soteriou et al. 2016). Additionally, various secretory factors, including exosomes from hair follicular epidermal keratinocyte (HFK) cells, also induce the expression of hair cycle-related gene Wnt gene family member 5A (WNT5A), FGF2, and VEGF in HFDPC when co-cultured with HFK cells or supplemented with HFK-conditioned medium (Tomochika et al. 2022). Keratinocytes, as a part of HFs and hair shafts, are crucial for investigating keratin expression patterns, cellular differentiation, and interactions with other follicular components such as DPCs. Hence, the human hair follicular epidermal keratinocyte (HHFK) and human epidermal keratinocytes (HaCaT) cell models are pivotal in in vitro systems for studying hair growth and its underlying mechanisms. Moreover, these HHFKs were also used to generate iPSCs that could further re-differentiate in vitro into keratinocytes. These iPSc-derived keratinocytes are the valued supply of keratinocytes for in vitro hair cloning or may be used to treat hair baldness or for grafting in burn patients (Lee et al. 2016).

Phyllotex™ is an herbal formulation comprising four plant extracts (*Euterpe oleracea*, *Olea europea*, *Tabebuia impetiginosa*, *Coffea arabica*), essential micronutrients (Zn, Mg, pantothenic acid, and vitamin D3), and p-coumaric acid, a natural antioxidant and anti-inflammatory agent (Godarzi et al. 2020). It is used as a treatment for AGA, including women AGA. It has shown that Phyllotex™ stimulates proliferation of both DPCs and HaCaT cells by activating pERK1/2. The formulation mitigates oxidative damage and TGF-β1 without affecting DHT production in hDPCs more than minoxidil (Serruya and Maor 2021). A study has shown the proliferation and growth promoting properties of minoxidil in HaCaT cells; however, experimentally, it has been demonstrated that its hair growth capacity increased in combination with Retinol, a known antioxidant. Minoxidil (0.01 mM) plus retinol (1.7 pM to 1.7 nM) significantly increased the growth of HaCaT compared with control and greater than minoxidil and retinol alone. This synergistic effect of both hair growth promoters is ascribed to activation of the ERK–AKT-dependent pathways and anti-apoptotic mechanisms. The treatment of minoxidil plus retinol significantly upregulated pErk (204%) and pAKT (198%) compared to minoxidil (133% and 146%) and retinol (149% and 180%) alone in HaCaT; whereas, the combination increased Bcl2 (177%) expression (Yoo et al. 2007), An herbal constituent ginsenoside F2 of *Panax ginseng* induced significant proliferation of hDPC and HaCaT by 30% in comparison to the FN-treated group. This proliferation action was mediated by an increase in hair growth-promoting signaling related to β-catenin, Lef-1, and Dickkopf-1 (DKK-1) (Shin et al. 2014).

Outer Root Sheath Cells: The induction of the anagen phase, a critical stage of hair growth, necessitates stimulating HFSCs. These activated HFSCs produce new progeny cells essential for constructing new, regenerative HFs (Li et al. 2021). These HFSCs are predominantly located in the bulges of the ORS. Hence, cells from the ORS can be pivotal in screening for anagen induction capabilities of potential therapeutic agents. Plucked HFs with visible bulbs and an intact ORS should be used for culturing of ORS cells. These cultured ORS keratinocytes have been utilized as another in vitro model to elucidate the hair growth ability of new chemical products (Wagner et al. 2018). This approach is typically followed by evaluating the effectiveness of compounds in both ex vivo and in vivo models, thus facilitating a complete understanding of their potential to promote hair growth. The exosomes from ORS cells also play a crucial role in maintaining the hair inductivity of DPCs and improving the outcome for treating hair loss (Nilforoushzadeh et al. 2020). Using ORS cells as a model, studies have shown that immune-inflammatory stress in ORS cells may be responsible for AA disease (Hao et al. 2023). The inflammatory stress in ORS cells from AA patient's scalp is mediated by Sirtuin1 (SIRT1), substance P, and neurokinin-1 receptor (NK-1R), which decrease the expression of growth factors in ORS cells (Kim et al. 2022).

2D Co-culture/Tri-culture Models: Understanding the crucial role of bidirectional interactions between epidermal and dermal elements in hair development, especially the relationship between DPCs and ORS cells, it becomes evident that a combined model of these interactions would more closely mimic the natural in vivo conditions for hair growth. As previously discussed, this integrated approach could offer a more realistic representation than individual cell research models that examine these populations in isolation. Hence, expressing greater in vivo-like character in molecular expression patterns and maintenance of hair inductive activity of DPCs for hair cycle development co-culture may be beneficial. A co-culture model involving dermal papilla cells (DPCs), HaCaT keratinocytes, and human dermal fibroblasts (HDFs) has been developed using polyethylene glycol diacrylate (PEGDA) hydrogels. This model is designed to study the role of keratinocytes in maintaining cellular compartmentalization between DPCs and neighboring HDFs, which is critical for the growth and function of hair follicles (HFs) (Tan et al. 2019). 3D co-cultures of DPCs and keratinocytes have been employed to investigate the reciprocal connections between these epithelial and mesenchymal cells for follicle formation and hair cycling (Botchkarev and Kishimoto 2003). A research group developed microstructured poly (ethylene glycol) diacrylate (PEGDA) hydrogels with human HFs using soft lithography resembling the physiological structure of HFs in which the dermal and epidermal cell populations were encapsulated in different compartments to mimic complex hair structure (Pan et al. 2013). Lim et al. developed a novel technique for arranging DPCs and normal human epidermal keratinocytes

(NHEKs) within 3D fibrous hydrogel scaffolds. This structured environment, where DPCs and NHEKs could be positioned in a manner that closely mimics their natural arrangement in the hair, could help understanding epithelial and mesenchymal interactions (Lim et al. 2013). A co-culture system with DPCs and ORS cells treated with DHT was developed to mimic androgen-induced alopecia and to understand the molecular mechanisms underlying androgen-induced alopecia (Zhang et al. 2021b). A tri-co-culture of epidermal stem cells, ADSCs, and skin-derived precursors (SKPs) was employed to show growth-promoting capacity of amphiregulin via the Phosphoinositide 3-Kinase (PI3K) and Mitogen-Activated Protein Kinases (MAPK) pathways (Lu et al. 2021). Additionally, a co-culture system involving HFs and HF-derived MSCs (HF-MSCs) was employed to investigate the healing potential of MSCs in AA. This study revealed that co-culturing with HF-MSCs led to reduced inflammation and increased hair shaft elongation, suggesting a beneficial role of MSCs in mitigating AA symptoms (Deng et al. 2021). A co-culture of DPCs and adult keratinocytes from nude skin, grown on a poly-(ethylene-co-vinyl alcohol) (EVAL) surface, has been developed to further study follicular microtissue formation. This model may provide insights into the complex interactions between different cell types and their interactions with the substrate, which are essential for understanding the patterning of heterotypic cells into HFs (Yen et al. 2010).

Three-Dimensional (3D) Spheroid/Organoid Model: 3D cell culture systems have gained significant attention for their ability to replicate the tissue microenvironment. Distinct from the traditional 2D cultures, cells in 3D systems exhibit morphologies and physiological characteristics more like those observed in vivo. This resemblance to natural cellular environments is attributed to the 3D culture's capability to maintain cell organization in three dimensions, which enhances intercellular communication and significantly affects signal transduction, gene expression, and overall cell behavior. In hair research, DPCs cultured in a 3D spheroid structure, mainly through the hanging drop method, can restore some aspects of their hair-inducing capacity that is diminished in 2D cultures. This enhancement is mainly due to forming a condensate-like structure that mimics the actual DPC niche. The induction of genes related to collagen regulation, cell adhesion, and immune system and hair growth-associated genes confers the fitness of 3D spheroid arrangement for studying hair biochemistry and effect of hair-promoting agents such as minoxidil and 3,4,5-Tri-O-caffeoylquinic acid (TCQA) (Bejaoui et al. 2022). Claire A. Higgins and coworkers have developed a culture system that could reprogram the microenvironment around the DPCs to produce de novo HFs. Upon implantation of these spheroids in the skin of mice, it produced unpigmented hair shafts without any SG and showed high expression of DP marker alkaline phosphatase (Higgins et al. 2013). Gerd Lindner et al. constructed complete HF organoids in vitro by creating appropriate culture conditions, cell density, and a mixture of different cell types required for epithelial–mesenchyme interaction. DPCs obtained from human scalp samples were cultured in defined culture conditions to produce microfollicles, which were further complemented with MSCs, ECM proteins to form ECM, keratinocytes to form ORS, and melanocytes. Under controlled in vitro culture conditions, the hair-like structure developed completely resembled the original hair. It could be used for high-throughput screening of test products/drug candidates to determine their impact on hair growth (Lindner et al. 2011). Through microarray, it has been shown that 75% of gene expression overlapped between developed in vitro HFs and human clinical biopsy samples after treatment with minoxidil, suggesting the suitability of model pre-clinical substance testing (Atac et al. 2020). To create size- and shape-controlled microenvironments for 3D DPC spheroids and microfollicles, tissue-engineered scaffolds were developed to support de novo synthesis of HFs, for instance, photopolymerizable PEGDA hydrogels (Pan et al. 2013), human placenta ECM (HPECM) hydrogel (Zhang et al. 2019), matrigel (Miao et al. 2014), gelatin–alginate hydrogel (Kang et al. 2023), etc. To evaluate the suitability and applicability of the developed 3D models, Hyun Woo Joo and coworkers developed HF-like DPC spheroids from DPC-derived immortalized cell line (SV40T-hTERT DPc) and primary DPC spheroids. Both spheroids grew HF-like structures similar to those in vivo when co-cultured with a keratinocyte layer. This HF-like structure showed significant extension when treated with minoxidil, tofacitinib, and ascorbic acid (Joo et al. 2022). The in vitro, 3D human folliculoid microspheres, comprising DPCs and ORS cells grown within an ECM, offer a promising model that closely resembles in vivo conditions. Within this system, the application of calcitriol and cyclosporin A, hair growth-modulating agents, has been observed to impact the cellular dynamics in folliculoid microspheres related to hair development (Havlickova et al. 2009). Similarly, a 3D spheroid of human DPCs, HFSCs, and keratinocytes encased in a silk gelatin hydrogel has been utilized to mimic and to study the expression of various genes in DP cells within the HFs in an in vitro condition (Gupta et al. 2018). These 3D models represent significant progress in hair biology research, providing more accurate and relevant systems for investigating the complex interactions and mechanisms underlying hair growth at the cellular and molecular levels.

Hair Follicle Organ Culture (HFOC): Philpott et al. first developed an ex vivo model of hair using live HFs isolated from human scalp, grown under serum-free conditions (Philpott 1990). Hair follicle organ culture (HFOC) can be sustained for at least 6 days under serum-free conditions ex vivo. However, a degeneration of hair matrix cells is observed beyond this period. To preserve and extend the natural growth activity of the HFs, mainly sustaining the long-term anagen phase morphology and continuous hair elongation, growth factors such as IGF-1, hepatocyte growth factor (HGF), along with essential amino acids and B-group vitamins, have also been utilized (Yoo et al. 2010). The organ culture of HFs has significantly contributed to translational hair research, offering critical insights into a broad spectrum of biomedical fields beyond dermatology. This includes enhanced understanding in areas such as human hair biology and pathology, metabolomics, cell cycle control, immunology, endocrinology, toxicology, pharmacology, etc. (Mistriotis and Andreadis 2013). The ability of HF organ culture to model these diverse biological processes underscores its invaluable role in biomedical research. A pivotal aspect of this research is the ex vivo culture of HFs for the production of pigmented hair shafts, which necessitates intricate interactions between proliferating hair matrix keratinocytes and melanocytes within the hair bulb. This process has been extensively studied in mature whole skin from both mice and humans, emphasizing understanding the impact of physical and biological factors on hair growth (Langan et al. 2015). It is important to note that the human HFOC correlates predominantly with HFs in the anagen phase, the active growth period of the hair cycle. Assessments in other hair cycle stages may not demonstrate the same correlation due to differing biological activities inherent to those phases. In this context, rat vibrissae follicle culture emerges as a valuable organ culture model. These follicles provide a unique system that enables the observation of cyclical activity across all hair cycle phases. The rat vibrissae follicle model allows for a more comprehensive understanding of the hair growth cycle, encompassing the anagen, catagen (transitional period), and telogen phases (resting phase) (Philpott and Kealey 2000). While HFOC provides an invaluable tool for deciphering the complex biology of HFs, including their growth cycles, cellular interactions, and response to hormonal and molecular factors, it has certain limitations. Precisely, the culture conditions in HFOC may not perfectly replicate the natural interactions and signaling pathways in the skin, which may impact the translational relevance of the findings. This knowledge is crucial for guiding future research and interpreting results obtained from HFOC studies. Studies are available to test the efficacy of potent ingredients in hair care products. Resveratrol (RSV) is a powerful antioxidant, anti-inflammatory, and anti-apoptotic phytoconstituent. In experiments conducted on shaved C57BL/6 mice, RSV was found to effectively stimulate the transition of the hair cycle from the telogen phase (resting phase) to the anagen phase, the active growth phase. Additionally, studies using HFOC have demonstrated that RSV can increase the length of the hair shaft in HFs and also delay the onset of the catagen phase, the transitional phase in the hair cycle (Zhang et al. 2021c).

14.8.5 Clinical Studies Including Human Volunteers

While pre-clinical studies can provide initial safety data, the response in humans can be different due to physiological and genetic variations. Human subjects bring variability in genetics, lifestyle, and health status, which can affect the product's performance. Therefore, while in vitro and in vivo models can provide valuable insights, the ultimate efficacy and safety of hair care products must still be confirmed in human trials for meaningful product recommendations and claims. The effectiveness of a hair growth product must be validated in the intended user population. This includes assessing the rate and extent of hair growth, improvement in hair density, and understanding long-term safety and sustained efficacy. Additionally, evidence from human clinical trials is often mandatory for regulatory approval. Regulatory agencies around the globe require robust data demonstrating a product's safety and efficacy in humans before it can be marketed.

To study the effectiveness of the test product on humans, the volunteers with the most common hair-related issues or targeted hair problems must be involved. Specific criteria such as non-medical history or chronic illness, no social addiction such as cigarettes, alcohol, drugs, etc., no previous hair treatments (hair transplant, laser therapy) and use of medicated or prescription shampoos, hair care products, involvement in other therapeutic trials, no pregnant or breastfeeding females, etc., should be considered. An open-label, single-arm, non-comparative, safety and efficacy clinical study for the test product, ThriveCo Hair Serum 2.0, with active ingredients like Redensyl, AnaGain, Capilia Longa, Procapil, and 5 kDa HA was conducted. After ethical approval from US DHHS and CDSCO, the study included 29 subjects (15 females and 14 males) with hair loss (alopecia) throughout 60 days of examination. The study observed a significant increase in hair growing rate, thickness, density, and scalp condition after analyzing with the CASLite-Nova software and other parameters, viz., anagen-to-telogen ratio, hair fall rescue, hair and scalp appearance, hair strength, by day 30 and day 60 of product application (Merja et al. 2023). The same group also tested an anti-grey hair serum product, ThriveCo® Hair Prime Serum, containing Arcedin™ on 32 human subjects for treatment of hair graying. The product confirmed significant enhancements in hair development, visual appearance, and strength (Merja et al. 2023). Another clinical study evaluated the efficacy and safety of a test hair serum product containing Saberry® (*Emblica officinalis* extract), Cococin™ (freeze-dried *C. nucifera* water solids), and PeptiSeLect® (γ-L-Glutamyl-L-selenomethionine), as active ingredients in the form of water-soluble powder along with peanut shell extract, Sandalore® (sandalwood odorant), Cosmoperine® (tetrahydropiperine). After 90 days of test product application on 42 subjects (age 25–45 years with grade 3–6 ° of hair loss), a noteworthy recovery in hair growth rate, vellus and terminal hair density, with bulb and without bulb hair fall rescue and hair thinning was observed (Majeed et al. 2020) A clinical study evaluating the effectiveness and admissibility of a topical formulation containing caffeine and Procapil for improving male pattern hair loss on 20 eligible males for 12 weeks showed increase in terminal/vellus hair density and anagen/telogen ratio (Samadi et al. 2023). A novel formulation Pielus Anti-hair loss® lotion and shampoo [active ingredients NANO R (nano encapsulated Redensyl®), NANO F (SymHair® Force 1631 from Tahiti blue microalgae), and Baicapil] was assessed for the safety and efficacy on 45 participants (35 females and 10 males, between 20–59 years old) with clinical diagnosis of TE and AGA. The formulation showed an increase in total follicular counts, thickness, and tensile strength and decreased hair fall (Anzai et al. 2020). The safety and efficacy of a new formulation containing 1% cetirizine applied twice daily was compared with minoxidil 5% solutions for 16 weeks in 40 men, aged 18–49 years with AGA. The increase in total and vellus hair density and hair in the anagen phase in the 1% cetirizine group was comparable with that of the minoxidil groups (Mostafa et al. 2021). The effect of a novel combination of topical minoxidil solution (5%) and topical spironolactone solution (5%) in treating AGA in both sexes was found to be efficacious (Ammar et al. 2022). The efficacy of topical methotrexate 1% gel versus minoxidil 5% gel in treating AA in 50 patients showed no significant differences in the clinical and dermoscopic examination after treatment for 12 weeks (Toma et al. 2022).

A comparative study was conducted to evaluate the safety and effectiveness of a water-based topical 5% cetosomal minoxidil solution versus alcohol-based topical 5% minoxidil solution in 80 Indian male patients with AGA. The results showed that both formulations had similar efficacy, but the 5% cetosomal minoxidil demonstrated improved safety. This suggests that it may be a better alternative for patients sensitive to alcohol-based formulations in the treatment of AGA (Sattur et al. 2023). A study also observed improved hair growth rate, volume, and thickness in adult women with self-observed hair thinning after oral supplementation of Nutrafol® Women's Capsules (Ablon and Kogan 2018). Kerascalp hair serum, containing herbal esculin, ximenynic acid, and lauric acid, showed significant enhancement in the Anagen: Telogen ratio and hair density, thickness, and strength, with hair fall rescue in human subjects with mild to moderate AGA (Kohli et al. 2023). It has also been observed that the proposed topical "Redensyl, Capixyl, and Procapil (RCP) combination exhibited better clinical recovery in terms of hair growth compared to topical 5% minoxidil in men with AGA (Karaca and Akpolat 2019).

14.9 Conclusion

In hair growth research, both in vivo animal models and ex vivo HF cultures are indispensable for evaluating hair development dynamics. They are crucial in the research and involvement of new hair growth ingredients in products. The DPCs also aid as a pivotal in vitro model for investigating hair growth effects at the cellular and molecular levels. DPCs function effectively as a screening model, enabling the identification and shortlisting of potential hair growth-promoting compounds from extensive chemical libraries. The value of in vivo models can effectively work in assessing both stimulatory and inhibitory effects on hair growth by assessing hair quality parameters and biomarkers. The development of experimental models for assessing the safety and efficacy of hair care products is evolving rapidly, and future directions in this field hold promising potential for more accurate and ethical testing. Future research is likely to expand the use of (3D) HF models, which offer a more physiologically relevant environment than traditional 2D cultures. The iPSCs have revolutionized regenerative medicine, including the generation of HFs. This process holds immense potential for treating alopecia and other hair loss disorders. The iPSCs are derived from adult cells reprogrammed back into a pluripotent state, capable of differentiating into various cell types. Utilizing gene-editing technologies like CRISPR/Cas9 to create specific HF models with genetic variations can help understand how different hair types respond to hair care products. This approach could be beneficial for developing products tailored to specific hair conditions or types. With advances in genomics and personalized medicine, there is a trend toward developing customized hair care solutions based on individual genetic, epigenetic, and proteomic profiles.

References

Ablon G, Kogan S (2018) A six-month, randomized, double-blind, placebo-controlled study evaluating the safety and efficacy of a nutraceutical supplement for promoting hair growth in women with self-perceived thinning hair. J Drugs Dermatol 17(5):558–565. https://doi.org/10.1111/jdv.17079

Alepee N, Grandidier MH, Tornier C, Cotovio J (2015) An integrated testing strategy for in vitro skin corrosion and irritation assessment using SkinEthic reconstructed human epidermis. Toxicol In Vitro 29(7):1779–1792. https://doi.org/10.1016/j.tiv.2015.07.012

Alessandrini A, Piraccini BM (2016) Essential of hair care cosmetics. Cosmetics 3(4):34. https://doi.org/10.3390/cosmetics3040034

Alessandrini A, Bruni F, Piraccini B, Starace M (2021a) Common causes of hair loss–clinical manifestations, trichoscopy and therapy. J Eur Acad Dermatol Venereol 35(3):629–640. https://doi.org/10.1111/jdv.17079

Alessandrini A, Starace M, Piraccini BM (2021b) Hair cosmetics and cosmeceuticals. Techniques in the Evaluation and Management of Hair Diseases, pp 137–149

Almeida A, Sarmento B, Rodrigues F (2017) Insights on in vitro models for safety and toxicity assessment of cosmetic ingredients. Int J Pharm 519(1–2):178–185. https://doi.org/10.1016/j.ijpharm.2017.01.024

Ammar AM, Elshahid AR, Abdel-Dayem HA, Mohamed AA, Elsaie ML (2022) Dermoscopic evaluation of the efficacy of combination of topical spironolactone 5% and minoxidil 5% solutions in the treatment of androgenetic alopecia: a cross sectional-comparative study. J Cosmet Dermatol 21(11):5790–5799. https://doi.org/10.1111/jocd.15328

Anzai A, Pereira AF, Malaquias KR, Guerra LO, Mercuri M (2020) Efficacy and safety of a new formulation kit (shampoo + lotion) containing anti-inflammatory and antioxidant agents to treat hair loss. Dermatol Ther 33(3):e13293. https://doi.org/10.1111/dth.13293

Atac B, Kiss FM, Lam T, Fauler B, Edler C, Hu P, Tao TP, Jadicke M, Rutschle I, Azar RP, Youngquist S, Mielke T, Marx U, Lauster R, Lindner G, DiColandrea T (2020) The microfollicle: a model of the human hair follicle for in vitro studies. In Vitro Cell Dev Biol Anim 56(10):847–858. https://doi.org/10.1007/s11626-020-00513-x

Australia Cosmetic Regulation (2023). https://cosmetic.chemlinked.com/cosmepedia/australia-cosmetic-regulation. Ascessed 5 Jan 2023

Bejaoui M, Oliva AK, Ke MS, Ferdousi F, Isoda H (2022) 3D spheroid human dermal papilla cell as an effective model for the screening of hair growth promoting compounds: examples of Minoxidil and 3,4,5-tri-O-caffeoylquinic acid (TCQA). Cells 11(13). https://doi.org/10.3390/cells11132093

Bellei B, Migliano E, Picardo M (2022) Therapeutic potential of adipose tissue-derivatives in modern dermatology. Exp Dermatol 31(12):1837–1852. https://doi.org/10.1111/exd.14532

Borda LJ, Wikramanayake TC (2015) Seborrheic dermatitis and dandruff: a comprehensive review. J Clin Investig Dermatol 3(2). https://doi.org/10.13188/2373-1044.1000019

Botchkarev VA, Kishimoto J (2003) Molecular control of epithelial-mesenchymal interactions during hair follicle cycling. J Investig Dermatol Symp Proc 8(1):46–55. https://doi.org/10.1046/j.1523-1747.2003.12171.x

Breehl L, Caban O (2018) Physiology, Puberty. [Updated 2023 Mar 27]. In: StatPearls [Internet]. StatPearls Publishing, Treasure Island (FL). Available from: https://www.ncbi.nlm.nih.gov/books/NBK534827. Accessed on: 22 Aug 2024

Canadian Cosmetic Products Regulations (2021). https://criticalcatalyst.com/canadian-cosmetic-products-regulations/. Accessed 5 Jan 2024

Canadian's_Cosmetic_Regulations (2024) Cosmetic Regulations Canada and registration CNF services. https://cosmeregcom/cosmetic-regulations-canada-cnf. Ascessed 5 Jan 2024

Cardoso CO, Tolentino S, Gratieri T, Cunha-Filho M, Lopez RFV, Gelfuso GM (2021) Topical treatment for scarring and non-scarring alopecia: an overview of the current evidence. Clin Cosmet Investig Dermatol 14:485–499. https://doi.org/10.2147/CCID.S284435

China_Report (2023) China hair care market size and share analysis—growth trends and forecasts (2023–2028). https://www.mordorintelligence.com/industry-reports/china-hair-care-market-industry. Ascessed 5 Jan 2024
Cho WK, Kim HI, Paek SH, Kim SY, Hyun Seo H, Song J, Lee OH, Min J, Lee SJ, Jo Y, Choi H, Lee JH, Moh SH (2021) Gene expression profile of human follicle dermal papilla cells in response to Camellia japonica phytoplacenta extract. FEBS Open Bio 11(3):633–651. https://doi.org/10.1002/2211-5463.13076
Civaş E, Aykan A, Aksoy B, Eski M (2018) Surgical correction of primary cicatricial alopecia: experience with nine patients. Turk Arch Dermatol Venereol 52(3):85–90. https://doi.org/10.4274/turkderm.20270
Clavaud C, Jourdain R, Bar-Hen A, Tichit M, Bouchier C, Pouradier F, El Rawadi C, Guillot J, Ménard-Szczebara F, Breton L (2013) Dandruff is associated with disequilibrium in the proportion of the major bacterial and fungal populations colonizing the scalp. PLoS One 8(3):e58203. https://doi.org/10.1371/journal.pone.0058203
Cosmetics and U.S. Law (2022). https://www.fda.gov/cosmetics/cosmetics-laws-regulations/cosmetics-us-law. Ascessed 5 Jan 2024
Cosmetics FAO (2018) How Cosmetics are not FDA-approved, but are FDA-regulated. https://www.fda.gov/cosmetics/cosmetics-laws-regulations/fda-authority-over-cosmetics-how-cosmetics-are-not-fdaapproved-are-fda-regulated. Ascessed 5 Jan 2024
Crabtree JS, Kilbourne EJ, Peano BJ, Chippari S, Kenney T, McNally C, Wang W, Harris HA, Winneker RC, Nagpal S, Thompson CC (2010) A mouse model of androgenetic alopecia. Endocrinology 151(5):2373–2380. https://doi.org/10.1210/en.2009-1474
Deng W, Zhang Y, Wang W, Song A, Mukama O, Huang J, Han X, Deng S, Lin Z, Habimana JDD, Huang R, Peng K, Ni B, Zhang S, Yan X, Li J, Wu LP, Li Z (2021) Hair follicle-derived mesenchymal stem cells decrease alopecia areata mouse hair loss and reduce inflammation around the hair follicle. Stem Cell Res Ther 12(1):548. https://doi.org/10.1186/s13287-021-02614-0
DHAC, A (2021) Cosmetics, therapeutic goods administration (TGA). https://www.tgagovau/cosmetics. Ascessed 5 Jan 2024
Dias MFRG (2015) Hair cosmetics: an overview. Int J Trichology 7(1):2. https://doi.org/10.4103/0974-7753.153450
Draelos ZD (2010) Essentials of hair care often neglected: hair cleansing. Int J Trichology 2(1):24–29. https://doi.org/10.4103/0974-7753.66909
Driskell RR, Clavel C, Rendl M, Watt FM (2011) Hair follicle dermal papilla cells at a glance. J Cell Sci 124:1179–1182. https://doi.org/10.1242/jcs.082446
Elmaadawi IH, Mohamed BM, Ibrahim ZAS, Abdou SM, El Attar YA, Youssef A, Shamloula MM, Taha A, Metwally HG, El Afandy MM (2018) Stem cell therapy as a novel therapeutic intervention for resistant cases of alopecia areata and androgenetic alopecia. J Dermatol Treat 29(5):431–440. https://doi.org/10.1080/09546634.2016.1227419
EU Cosmetics Regulations and Registration (2023). https://www.cirs-reach.com/Cosmetics_Registration/eu_cosmetics_directive_cosmetics_registration.html. Ascessed 5 Jan 2024
EU-Report (2014) Opinion on sensitivity to hair dyes—consumer self testing, 18 December 2007. (SCCP/1104/07). European Commission. https://ec.europa.eu/health/ph_risk/committees/04_sccp/docs/sccp_o_114.pdf
European-Commission (2024) EU's role in cosmetic. http://wwweceuropaeu/growth/sectors/cosmetics_en. Ascessed 5 Jan 2024
EWG (2011) Hair straighteners markers and saloon cover up dangers. https://wwwewgorg/hair--straighteners/our-report/faqs-7/. Ascessed 5 Jan 2024
Fanti PA, Baraldi C, Misciali C, Piraccini BM (2018) Giornale Italiano di Dermatologia e Venereologia. J Giornale Italiano di Dermatologia 153(2):230–242. https://doi.org/10.23736/S0392-0488.18.05889-3
Ferreira A (2021) Overview of the registration process of cosmetics in Brazil. https://globalregulatorypartnerscom/wp-content/uploads/Cosmetics-Registration-in-Brazil_Version-12021-versionpdf. Ascessed 5 Jan 2024

Fu D, Huang J, Li K, Chen Y, He Y, Sun Y, Guo Y, Du L, Qu Q, Miao Y, Hu Z (2021) Dihydrotestosterone-induced hair regrowth inhibition by activating androgen receptor in C57BL6 mice simulates androgenetic alopecia. Biomed Pharmacother 137:111247. https://doi.org/10.1016/j.biopha.2021.111247

Gentile P, Garcovich S (2019) Advances in regenerative stem cell therapy in androgenic alopecia and hair loss: Wnt pathway, growth-factor, and mesenchymal stem cell signaling impact analysis on cell growth and hair follicle development. Cells 8(5):466. https://doi.org/10.3390/cells8050466

Gentile P, Scioli MG, Cervelli V, Orlandi A, Garcovich S (2020) Autologous micrografts from scalp tissue: trichoscopic and long-term clinical evaluation in male and female androgenetic alopecia. Biomed Res Int 2020:7397162. https://doi.org/10.1155/2020/7397162

Geyfman M, Plikus MV, Treffeisen E, Andersen B, Paus R (2015) Resting no more: re-defining telogen, the maintenance stage of the hair growth cycle. Biol Rev Camb Philos Soc 90(4):1179–1196. https://doi.org/10.1111/brv.12151

Gilhar A, Laufer Britva R, Keren A, Paus R (2020) Mouse models of alopecia Areata: C3H/HeJ mice versus the humanized AA mouse model. J Investig Dermatol Symp Proc 20(1):S11–S15. https://doi.org/10.1016/j.jisp.2020.05.001

Godarzi SM, Gorji AV, Gholizadeh B, Mard SA, Mansouri E (2020) Antioxidant effect of p-coumaric acid on interleukin 1-β and tumor necrosis factor-α in rats with renal ischemic reperfusion. Nefrologia (Engl Ed) 40(3):311–319. https://doi.org/10.1016/j.nefro.2019.10.003

Gubitosa J, Rizzi V, Fini P, Cosma P (2019) Hair care cosmetics: from traditional shampoo to solid clay and herbal shampoo, a review. Conmetics 6(1):13. https://doi.org/10.3390/cosmetics6010013

Gund R, Christiano AM (2023) Impaired autophagy promotes hair loss in the C3H/HeJ mouse model of alopecia areata. Autophagy 19(1):296–305. https://doi.org/10.1080/15548627.2022.2074104

Guo MA (2016) CFDA registration of imported Cosmetics in China—Hygiene License and Record-keeping Certificate. https://www.cirs-reach.com/Cosmetics_Registration/China_cosmetics_regulations_registration.html. Ascessed 5 Jan 2024

Gupta AC, Chawla S, Hegde A, Singh D, Bandyopadhyay B, Lakshmanan CC, Kalsi G, Ghosh S (2018) Establishment of an in vitro organoid model of dermal papilla of human hair follicle. J Cell Physiol 233(11):9015–9030. https://doi.org/10.1002/jcp.26853

Haircare_Market_Report (2023) Hair care Market size, Share | Global Industry Trends. https://wwwfortunebusinessinsightscom/hair-care-market-102555. Ascessed 5 Jan 2024

Han JH, Kwon OS, Chung JH, Cho KH, Eun HC, Kim KH (2004) Effect of minoxidil on proliferation and apoptosis in dermal papilla cells of human hair follicle. J Dermatol Sci 34(2):91–98. https://doi.org/10.1016/j.jdermsci.2004.01.002

Han J, Kim S, Lee SH, Kim JS, Chang YJ, Jeong TC, Kang MJ, Kim TS, Yoon HS, Lee GY, Bae S, Lim KM (2020) Me-too validation study for in vitro skin irritation test with a reconstructed human epidermis model, KeraSkin for OECD test guideline 439. Regul Toxicol Pharmacol 117:104725. https://doi.org/10.1016/j.yrtph.2020.104725

Hao L, Nam KH, Lee GJ, Kim D, Shin JM, Lee Y, Kim CD, Kim SJ, Yun SK, Park BH, Park J (2023) SIRT1 downregulation provokes immune-inflammatory responses in hair follicle outer root sheath cells and may contribute to development of alopecia areata. J Dermatol Sci 111(1):2–9. https://doi.org/10.1016/j.jdermsci.2023.05.005

Havlickova B, Biro T, Mescalchin A, Tschirschmann M, Mollenkopf H, Bettermann A, Pertile P, Lauster R, Bodo E, Paus R (2009) A human folliculoid microsphere assay for exploring epithelial-mesenchymal interactions in the human hair follicle. J Invest Dermatol 129(4):972–983. https://doi.org/10.1038/jid.2008.315

Higgins CA, Chen JC, Cerise JE, Jahoda CA, Christiano AM (2013) Microenvironmental reprogramming by three-dimensional culture enables dermal papilla cells to induce de novo human hair-follicle growth. Proc Natl Acad Sci U S A 110(49):19679–19688

Honda Igarashi M, da Silva SG, Mercuri M, Zuardi F, Facchini G, da Silva GH, Lucia Tabarini Alves Pinheiro A, Eberlin S (2021) Novel complex of cosmetic ingredients with promising action in preventing hair loss and follicular aging through mechanism involving enrichment of WNT/signaling, mitochondrial activity, and stem cells maintenance. J Cosmet Dermatol 20(7):2179–2189. https://doi.org/10.1111/jocd.13815

Hoover E, Alhajj M, Flores J (2023) Physiology, hair, vol 1. StatPearls Publishing, Treasure Island, FL

Hu A (2021) Japan cosmetics regulation. https://cosmeticchemlinkedcom/cosmepedia/japan-cosmetic-regulation. Ascessed 17 Dec 2023

Huang HC, Lin H, Huang MC (2019) Lactoferrin promotes hair growth in mice and increases dermal papilla cell proliferation through Erk/Akt and Wnt signaling pathways. Arch Dermatol Res 311(5):411–420. https://doi.org/10.1007/s00403-019-01920-1

Hyun J, Im J, Kim SW, Kim HY, Seo I, Bhang SH (2021) Morus alba root extract induces the Anagen phase in the human hair follicle dermal papilla cells. Pharmaceutics 13(8). https://doi.org/10.3390/pharmaceutics13081155

Im J, Hyun J, Kim SW, Bhang SH (2022) Enhancing the Angiogenic and proliferative capacity of dermal fibroblasts with mulberry (Morus alba. L) Root extract. Tissue Eng Regen Med 19(1):49–57. https://doi.org/10.1007/s13770-021-00404-6

India Cosmetic Regulation (2020). https://cosmetic.chemlinked.com/cosmepedia/india-cosmetic-regulation. Ascessed 5 Jan 2024

Joo HW, Kim MK, Bak SS, Sung YK (2022) Bioengineering of hair follicle-like structure for validation of hair growth promoting compounds. Bioengineering (Basel) 9(11). https://doi.org/10.3390/bioengineering9110645

Jung KM, Lee SH, Jang WH, Jung HS, Heo Y, Park YH, Bae S, Lim KM, Seok SH (2014) KeraSkin-VM: a novel reconstructed human epidermis model for skin irritation tests. Toxicol In Vitro 28(5):742–750. https://doi.org/10.1016/j.tiv.2014.02.014

Kang D, Liu Z, Qian C, Huang J, Zhou Y, Mao X, Qu Q, Liu B, Wang J, Hu Z, Miao Y (2023) 3D bioprinting of a gelatin-alginate hydrogel for tissue-engineered hair follicle regeneration. Acta Biomater 165:19–30. https://doi.org/10.1016/j.actbio.2022.03.011

Kapoor V (2005) Herbal cosmetics for skin and hair care. Indian J Nat Prod Resour

Karaca N, Akpolat ND (2019) A comparative study between topical 5% minoxidil and topical "Redensyl, Capixyl, and Procapil" combination in men with androgenetic alopecia. J Cosmetol Trichol:150370987

Kavya HSC (2022) Regulatory provisions for Cosmetics in India. Res J Top Cos Sci 13(1):14–20. https://doi.org/10.52711/2321-5844.2022.00003

Khantham C, Ruksiriwanich W, Sringarm K, Prom-u-thai C, Jamjod S, Arjin C, Muangsanguan A, Rachtanapun P, Jantanasakulwong K, Phimolsiripol Y, Barba FJ, Sommano SR, Chutoprapat R, Boonpisuttinant K (2023) Effects of bioactive composition in Oryza sativa L. cv. KDML105 bran extract on gene expression related to hair cycle in human hair follicle dermal papilla cells. Agronomy 13(2). https://doi.org/10.3390/agronomy13020295

Kim JY, Dao H (2020) Physiology, integument

Kim J, Shin JY, Choi YH, Jang M, Nam YJ, Lee SY, Jeon J, Jin MH, Lee S (2019) Hair growth promoting effect of Hottuynia cordata extract in cultured human hair follicle dermal papilla cells. Biol Pharm Bull 42(10):1665–1673. https://doi.org/10.1248/bpb.b19-00254

Kim C, Shin JM, Kim D, Park S, Hong D, Jung KE, Kim CD, Seo YJ, Lee Y (2022) Role of substance P in regulating micro-milieu of inflammation in alopecia Areata. Ann Dermatol 34(4):270–277. https://doi.org/10.5021/ad.21.161

Kiso M, Hamazaki TS, Itoh M, Kikuchi S, Nakagawa H, Okochi H (2015) Synergistic effect of PDGF and FGF2 for cell proliferation and hair inductive activity in murine vibrissal dermal papilla in vitro. J Dermatol Sci 79(2):110–118. https://doi.org/10.1016/j.jdermsci.2015.04.007

Knight J, Rovida C, Kreiling R, Zhu C, Knudsen M, Hartung T (2021) Continuing animal tests on cosmetic ingredients for REACH in the EU. ALTEX 38(4):653–668. https://doi.org/10.14573/altex.2104221

Kohli M, Ganjoo A, Sharma A, Patil CY, Sethi S, Patel B (2023) Prospective efficacy and safety study of an innovative Kerascalp hair growth serum in mild-to-moderate alopecia in India: regrowth study. Cureus 15(5):e38742. https://doi.org/10.7759/cureus.38742

Kose O, Erkekoglu P, Sabuncuoglu S, Kocer-Gumusel B (2018) Evaluation of skin irritation potentials of different cosmetic products in Turkish market by reconstructed human epidermis model. Regul Toxicol Pharmacol 98:268–273. https://doi.org/10.1016/j.yrtph.2018.08.010

Kwack MH, Yang JM, Won GH, Kim MK, Kim JC, Sung YK (2018) Establishment and characterization of five immortalized human scalp dermal papilla cell lines. Biochem Biophys Res Commun 496(2):346–351. https://doi.org/10.1016/j.bbrc.2018.01.058

Langan EA, Philpott MP, Kloepper JE, Paus R (2015) Human hair follicle organ culture: theory, application and perspectives. Exp Dermatol 24(12):903–911. https://doi.org/10.1111/exd.12836

Lanjewar A, Maurya S, Sharma D, Gaur A (2020) Review on hair problem and its solution. J Drug Delivery Therapeut 10(3-s):322–329. https://doi.org/10.22270/jddt.v10i3-s.4066

Law K (2015) Regulatory requirements for exporting Cosmetics to Japan. https://wwwscribdcom/document/234432932/Regulation-of-Cosmetics-in-Japan-pdf. Ascessed 5 Jan 2024

Lee H, Kim NH, Yang H, Bae SK, Heo Y, Choudhary I, Kwon YC, Byun JK, Yim HJ, Noh BS, Heo JD, Kim E, Kang C (2016) The hair growth-promoting effect of Rumex japonicus Houtt. Extract. Evid Based Complement Alternat Med 2016:1873746. https://doi.org/10.1155/2016/1873746

Lee SW, Juhasz M, Mobasher P, Ekelem C, Mesinkovska NA (2018) A systematic review of topical finasteride in the treatment of androgenetic alopecia in men and women. J Drugs Dermatol 17(4):457–463

Li H, Masieri FF, Schneider M, Bartella A, Gaus S, Hahnel S, Zimmerer R, Sack U, Maksimovic-Ivanic D, Mijatovic S, Simon JC, Lethaus B, Savkovic V (2021) The middle part of the plucked hair follicle outer root sheath is identified as an area rich in lineage-specific stem cell markers. Biomol Ther 11(2). https://doi.org/10.3390/biom11020154

Lim TC, Leong MF, Lu H, Du C, Gao S, Wan AC, Ying JY (2013) Follicular dermal papilla structures by organization of epithelial and mesenchymal cells in interfacial polyelectrolyte complex fibers. Biomaterials 34(29):7064–7072. https://doi.org/10.1016/j.biomaterials.2013.05.068

Lindner G, Horland R, Wagner I, Atac B, Lauster R (2011) De novo formation and ultra-structural characterization of a fiber-producing human hair follicle equivalent in vitro. J Biotechnol 152(3):108–112. https://doi.org/10.1016/j.jbiotec.2011.01.019

Lintzeri DA, Constantinou A, Hillmann K, Ghoreschi K, Vogt A, Blume-Peytavi U (2022) Alopecia areata–current understanding and management. J Dtsch Dermatol Ges 20(1):59–90. https://doi.org/10.1111/ddg.14689

Lu Q, Gao Y, Fan Z, Xiao X, Chen Y, Si Y, Kong D, Wang S, Liao M, Chen X (2021) Amphiregulin promotes hair regeneration of skin-derived precursors via the PI3K and MAPK pathways. Cell Prolif 54(9):e13106. https://doi.org/10.1111/cpr.13106

Lucky AW, McGuire J, Nydorf E, Halpert G, Nuck BA (1986) Hair follicle response of the golden Syrian hamster flank organ to continuous testosterone stimulation using silastic capsules. J Invest Dermatol 86(1):83–86. https://doi.org/10.1111/1523-1747.ep12283883

Madaan A, Verma R, Singh AT, Jaggi M (2018) Review of hair follicle dermal papilla cells as in vitro screening model for hair growth. Int J Cosmet Sci 40(5):429–450. https://doi.org/10.1111/ics.12489

Mahomoodally MF, Ramjuttun P (2017) Phytocosmetics from the African herbal pharmacopeia. Int J Phytocosmet Natural Ingredi 4(1):4–4. https://doi.org/10.15171/ijpni.2017.04

Majeed M, Majeed S, Nagabhushanam K, Mundkur L, Neupane P, Shah K (2020) Clinical study to evaluate the efficacy and safety of a hair serum product in healthy adult male and female volunteers with hair fall. Clin Cosmet Investig Dermatol 13:691–700. https://doi.org/10.2147/CCID.S271013

Market_Outlook_2024-2032, HC (2023) Hair care market report by product type (shampoo, hair color, conditioner, hair styling products, and others), distribution channel (supermarkets and hypermarkets, specialty stores, convenience stores, online retailers, and others), and region 2024–2032. https://wwwimarcgroupcom/hair-care-market. Ascessed 5 Jan 2024

Martel JL, Miao JH, Badri T (2017) Anatomy, hair follicle, 1st edn. StatPearls Publishing. https://www.ncbi.nlm.nih.gov/books/NBK470321/

Martinez-Jacobo L, Villarreal-Villarreal CD, Ortiz-Lopez R, Ocampo-Candiani J, Rojas-Martinez A (2018) Genetic and molecular aspects of androgenetic alopecia. Indian J Dermatol Venereol Leprol 84(3):263–268. https://doi.org/10.4103/ijdvl.IJDVL_262_17

McElwee KJ, Boggess D, Miller J, King LE Jr, Sundberg JP (1999) Spontaneous alopecia areata-like hair loss in one congenic and seven inbred laboratory mouse strains. J Investig Dermatol Symp Proc 4(3):202–206. https://doi.org/10.1038/sj.jidsp.5640211

Merja A, Patel N, Patel M, Patnaik S, Ahmed A, Maulekhi S (2023) Safety and efficacy of hair growth promoting product in adult human subject having hair fall complaints (alopecia). J Cosmet Dermatol 23:938–948. https://doi.org/10.1111/jocd.16084

Miao Y, Sun YB, Liu BC, Jiang JD, Hu ZQ (2014) Controllable production of transplantable adult human high-passage dermal papilla spheroids using 3D matrigel culture. Tissue Eng Part A 20(17–18):2329–2338. https://doi.org/10.1089/ten.TEA.2013.0547

Mistriotis P, Andreadis ST (2013) Hair follicle: a novel source of multipotent stem cells for tissue engineering and regenerative medicine. Tissue Eng Part B Rev 19(4):265–278. https://doi.org/10.1089/ten.TEB.2012.0422

MMR_Market_Report (2023) Hair care market: consumer demand and fashion and beauty. https://www.maximizemarketresearch.com/market-report/global-hair-care-market/22240/. Ascessed 5 Jan 2024

Mostafa DM, Samadi A, Niknam S, Nasrollahi SA, Guishard A, Firooz A (2021) Efficacy of cetirizine 1% versus minoxidil 5% topical solution in the treatment of male alopecia: a randomized, single-blind controlled study. J Pharm Pharm Sci 24:191–199. https://doi.org/10.18433/jpps31456

Naito A, Sato T, Matsumoto T, Takeyama K, Yoshino T, Kato S, Ohdera M (2008) Dihydrotestosterone inhibits murine hair growth via the androgen receptor. Br J Dermatol 159(2):300–305. https://doi.org/10.1111/j.1365-2133.2008.08671.x

Nilforoushzadeh MA, Aghdami N, Taghiabadi E (2020) Human hair outer root sheath cells and platelet-lysis exosomes promote hair inductivity of dermal papilla cell. Tissue Eng Regen Med 17(4):525–536. https://doi.org/10.1007/s13770-020-00266-4

OECD (2021) In vitro skin irritation: reconstructed human epidermis test method, vol 439. OECD

Ohn J, Kim KH, Kwon O (2019) Evaluating hair growth promoting effects of candidate substance: a review of research methods. J Dermatol Sci 93(3):144–149. https://doi.org/10.1016/j.jdermsci.2019.02.004

Orasan MS, Roman II, Coneac A, Muresan A, Orasan RI (2016) Hair loss and regeneration performed on animal models. Clujul Med 89(3):327–334. https://doi.org/10.15386/cjmed-583

Pan J, Yung Chan S, Common JE, Amini S, Miserez A, Birgitte Lane E, Kang L (2013) Fabrication of a 3D hair follicle-like hydrogel by soft lithography. J Biomed Mater Res A 101(11):3159–3169. https://doi.org/10.1002/jbm.a.34628

Park HJ (2014) CARI ONE induces anagen phase of telogenic hair follicles through regulation of beta-catenin, stimulation of dermal papilla cell proliferation, and melanogenesis. J Diet Suppl 11(4):320–333. https://doi.org/10.3109/19390211.2014.887600

Park B-S, Choi H-I (2022) Stem cell-based therapies for hair loss: what is the evidence from a clinical perspective? In: Francisco Jimenez CH (ed) Hair follicle regeneration, vol 72. Springer, pp 219–236

Park S, Lee J (2021) Modulation of hair growth promoting effect by natural products. Pharmaceutics 13(12):2163. https://doi.org/10.3390/pharmaceutics13122163

Park WS, Lee CH, Lee BG, Chang IS (2003) The extract of Thujae occidentalis semen inhibited 5alpha-reductase and androchronogenetic alopecia of B6CBAF1/j hybrid mouse. J Dermatol Sci 31(2):91–98. https://doi.org/10.1016/s0923-1811(02)00146-9

Paula JNHD, Basflio FMA, Mulinari-Brenner FA (2022) Effects of chemical straighteners on the hair shaft and scalp. An Bras Dermatol 97(2):193–203. https://doi.org/10.1016/j.abd.2021.02.010

Pfannenbecker U, Bessou-Touya S, Faller C, Harbell J, Jacob T, Raabe H, Tailhardat M, Alepee N, De Smedt A, De Wever B, Jones P, Kaluzhny Y, Le Varlet B, McNamee P, Marrec-Fairley M, Van Goethem F (2013) Cosmetics Europe multi-laboratory pre-validation of the EpiOcular reconstituted human tissue test method for the prediction of eye irritation. Toxicol In Vitro 27(2):619–626. https://doi.org/10.1016/j.tiv.2012.11.007

Philpott, MP, Martin RG, Terence K (1990) "Human hair growth in vitro." Journal of cell science 97(3):463–471

Philpott MP, Kealey T (2000) Cyclical changes in rat vibrissa follicles maintained in vitro. J Invest Dermatol 115(6):1152–1155. https://doi.org/10.1046/j.1523-1747.2000.00183.x

Rahangdale PC, Wankhade AM (2023) A review on-types and treatment of alopecia. Asian J Pharm Res 13(2):123–128. https://doi.org/10.52711/2231-5691.2023.00025

Rattanachitthawat N, Pinkhien T, Opanasopit P, Ngawhirunpat T, Chanvorachote P (2019) Finasteride enhances stem cell signals of human dermal papilla cells. In Vivo 33(4):1209–1220. https://doi.org/10.21873/invivo.11592

Rust RC, Schlatter H (2022) Hair dyes. In: Draelos Z (ed) Cosmetic dermatology: products procedures, vol 2. John Wiley & Sons, pp 239–249

Salhab O, Khayat L, Alaaeddine N (2022) Stem cell secretome as a mechanism for restoring hair loss due to stress, particularly alopecia areata: narrative review. J Biomed Sci 29(1):1–11. https://doi.org/10.1186/s12929-022-00863-6

Samadi A, Movaffaghi M, Kazemi F, Yazdanparast T, Ahmad Nasrollahi S, Firooz A (2023) Tolerability and efficacy assessment of an oral collagen supplement for the improvement of biophysical and ultrasonographic parameters of skin in middle eastern consumers. J Cosmet Dermatol 22(8):2252–2258. https://doi.org/10.1111/jocd.15700

Sattur S, Talathi A, Shetty G, Arsiwala S, Pereira R, Dhoot D (2023) Comparative clinical study evaluating the efficacy and safety of topical 5% cetosomal minoxidil and topical 5% alcohol-based minoxidil solutions for the treatment of androgenetic alopecia in Indian men. Cureus 15(10):e46568. https://doi.org/10.7759/cureus.46568

Serruya R, Maor Y (2021) Hair growth-promotion effects at the cellular level and antioxidant activity of the plant-based extract phyllotex. Heliyon 7(9):e07888. https://doi.org/10.1016/j.heliyon.2021.e07888

Shimizu Y, Ntege EH, Sunami H, Inoue Y (2022) Regenerative medicine strategies for hair growth and regeneration: a narrative review of literature. Regen Ther 21:527–539. https://doi.org/10.1016/j.reth.2022.10.005

Shin HS, Park SY, Hwang ES, Lee DG, Song HG, Mavlonov GT, Yi TH (2014) The inductive effect of ginsenoside F2 on hair growth by altering the WNT signal pathway in telogen mouse skin. Eur J Pharmacol 730:82–89. https://doi.org/10.1016/j.ejphar.2014.02.024

Silva RJ, Tamburic S (2022) A state-of-the-art review on the alternatives to animal testing for the safety assessment of cosmetics. Cosmetics 9(5):90. https://doi.org/10.3390/cosmetics9050090

Singh P, Bhat SS, Singh N, Venkanna BU, Mohamed R, Rao RP (2022) Cell-based model Systems for Validation of various efficacy-based claims for cosmetic ingredients. Cosmetics 9(5). https://doi.org/10.3390/cosmetics9050107

Sintov A, Serafimovich S, Gilhar A (2000) New topical antiandrogenic formulations can stimulate hair growth in human bald scalp grafted onto mice. Int J Pharm 194(1):125–134. https://doi.org/10.1016/s0378-5173(99)00359-2

Soteriou D, Kostic L, Sedov E, Yosefzon Y, Steller H, Fuchs Y (2016) Isolating hair follicle stem cells and epidermal keratinocytes from dorsal mouse skin. J Vis Exp 2016(110):53931. https://doi.org/10.3791/53931

Starace M, Orlando G, Bruni F, Alessandrini A, Piraccini B (2022) Anagen effluvium and the role of trichoscopy. Clin Exp Dermatol 47(3):589–591. https://doi.org/10.1111/ced.14982

Sun J, Silva KA, McElwee KJ, King LE Jr, Sundberg JP (2008) The C3H/HeJ mouse and DEBR rat models for alopecia areata: review of preclinical drug screening approaches and results. Exp Dermatol 17(10):793–805. https://doi.org/10.1111/j.1600-0625.2008.00773.x

Tan JJ, Common JE, Wu C, Ho PC, Kang LJCP (2019) Keratinocytes maintain compartmentalization between dermal papilla and fibroblasts in 3D heterotypic tri-cultures. Cell Prolif 52(5):e12668

Tated SN, Mahajan U, Walokar A, Khetade R (2021) A review on marketed hair products. Int J Appl Res 7:50–57

Toma DM, Atallah RB, Eldahshan RM (2022) Comparative study between topical methotrexate 1% gel and minoxidil 5% gel in the treatment of localized alopecia areata. Dermatol Ther 35(9):e15696. https://doi.org/10.1111/dth.15696

Tomochika M, Sumie R, Miyano R, Mochizuki Y, Inoue H, Kajikawa A (2022) Possible cellular communication with human follicle dermal papilla cells via secretions from human hair follicle keratinocytes. J St Marianna Univ 13(2):87–93. https://doi.org/10.17264/stmarieng.13.87

Tracy LE, Minasian RA, Caterson EJ (2016) Extracellular matrix and dermal fibroblast function in the healing wound. Adv Wound Care (New Rochelle) 5(3):119–136. https://doi.org/10.1089/wound.2014.0561

Trüeb RM, Gavazzoni Dias MFR, Dutra Rezende H (2023) Parasitic diseases and infestations of the hair and scalp. In: Dutra Rezende H, Trüeb RM, Gavazzoni Dias MFR (eds) Hair in infectious disease: recognition, treatment, and prevention, 1st edn. Springer, Cham, pp 261–290

Truong VL, Bak MJ, Lee C, Jun M, Jeong WS (2017) Hair regenerative mechanisms of red ginseng oil and its major components in the testosterone-induced delay of Anagen entry in C57BL/6 mice. Molecules 22(9). https://doi.org/10.3390/molecules22091505

US-FDA (2005) FDA authority over cosmetics: how cosmetics are not FDA-approved, but are FDA-regulated. https://www.fda.gov/cosmetics/cosmetics-laws-regulations/fda-authority-over-cosmetics-how-cosmetics-are-not-fda-approved-are-fda-regulated. Ascessed 5 Jan 2024

Vasanthan J, Gurusamy N, Rajasingh S, Sigamani V, Kirankumar S, Thomas EL, Rajasingh J (2020) Role of human mesenchymal stem cells in regenerative therapy. Cells 10(1):54. https://doi.org/10.3390/cells10010054

Vest BE, Krauland K (2022) Malassezia furfur. In: StatPearls [Internet]. StatPearls Publishing, Treasure Island, FL

Villablanca S, Fischer C, García-García SC, Mascaró-Galy JM, Ferrando J (2017) Primary scarring alopecia: clinical-pathological review of 72 cases and review of the literature. Skin Appendage Disord 3(3):132–143. https://doi.org/10.1159/000467395

Wagner T, Gschwandtner M, Strajeriu A, Elbe-Burger A, Grillari J, Grillari-Voglauer R, Greiner G, Golabi B, Tschachler E, Mildner M (2018) Establishment of keratinocyte cell lines from human hair follicles. Sci Rep 8(1):13434. https://doi.org/10.1038/s41598-018-31829-0

Welankiwar A (2013) Review: regulatory provisions regarding Cosmetics in India. https://www-pharmatutororg/articles/review-regulatory-provisions-regarding-cosmetics-in-india. Ascessed 5 Jan 2024

Welle MM (2023) Basic principles of hair follicle structure, morphogenesis, and regeneration. Vet Pathol 60(6):732–747. https://doi.org/10.1177/03009858231176561

Xia Y, Chen J, Ding J, Zhang J, Chen H (2020) IGF1- and BM-MSC-incorporating collagen-chitosan scaffolds promote wound healing and hair follicle regeneration. Am J Transl Res 12(10):6264–6276

Xie B, Chen M, Ding P, Lei L, Zhang X, Zhu D, Zou Y, Deng Z, Sun G, Li J, Nie H (2020) Induction of dermal fibroblasts into dermal papilla cell-like cells in hydrogel microcapsules for enhanced hair follicle regeneration. Appl Mater Today 21:100805. https://doi.org/10.1016/j.apmt.2020.100805

Xu Z, Wang Z, Yuan C, Liu X, Yang F, Wang T, Wang J, Manabe K, Qin O, Wang X (2016) Dandruff is associated with the conjoined interactions between host and microorganisms. Sci Rep 6(1):1–9. https://doi.org/10.1038/srep24877

Yen CM, Chan CC, Lin SJ (2010) High-throughput reconstitution of epithelial-mesenchymal interaction in folliculoid microtissues by biomaterial-facilitated self-assembly of dissociated heterotypic adult cells. Biomaterials 31(15):4341–4352. https://doi.org/10.1016/j.biomaterials.2010.02.014

Yin GOC, Siong-See JL, Wang EC (2021) Telogen effluvium–a review of the science and current obstacles. J Dermatol Sci 101(3):156–163. https://doi.org/10.1016/j.jdermsci.2021.01.007

Yoo HG, Chang IY, Pyo HK, Kang YJ, Lee SH, Kwon OS, Cho KH, Eun HC, Kim KH (2007) The additive effects of minoxidil and retinol on human hair growth in vitro. Biol Pharm Bull 30(1):21–26. https://doi.org/10.1248/bpb.30.21

Yoo B-Y, Shin Y-H, Yoon H-H, Seo Y-K, Park J-K (2010) Hair follicular cell/organ culture in tissue engineering and regenerative medicine. Biochem Eng J 48(3):323–331. https://doi.org/10.1016/j.bej.2009.09.008

You J, Woo J, Roh KB, Jeon K, Jang Y, Choi SA, Ryu D, Cho E, Park D, Lee J, Jang M, Jung E (2023) Evaluation of efficacy of Silybum marianum flower extract on the mitigating hair loss in vitro and in vivo. J Cosmet Dermatol 23:529–542. https://doi.org/10.1111/jocd.15978

Yum S, Jeong S, Kim D, Lee S, Kim W, Yoo JW, Kim J-A, Kwon OS, Kim D-D, Min DS (2017) Minoxidil induction of VEGF is mediated by inhibition of HIF-prolyl hydroxylase. Int J Mol Sci 19(1):53. https://doi.org/10.3390/ijms19010053

Zerbinati N, Sommatis S, Maccario C, Capillo MC, Di Francesco S, Rauso R, Protasoni M, D'Este E, Gasperina DD, Mocchi R (2021) In vitro hair growth promoting effect of a noncrosslinked hyaluronic acid in human dermal papilla cells. Biomed Res Int 2021:5598110. https://doi.org/10.1155/2021/5598110

Zhang X, Xiao S, Liu B, Miao Y, Hu Z (2019) Use of extracellular matrix hydrogel from human placenta to restore hair-inductive potential of dermal papilla cells. Regen Med 14(8):741–751. https://doi.org/10.2217/rme-2018-0112

Zhang J, Chen R, Wen L, Fan Z, Guo Y, Hu Z, Miao Y (2021a) Recent Progress in the understanding of the effect of sympathetic nerves on hair follicle growth. Front Cell Dev Biol 9:736738. https://doi.org/10.3389/fcell.2021.736738

Zhang Y, Huang J, Fu D, Liu Z, Wang H, Wang J, Qu Q, Li K, Fan Z, Hu Z, Miao Y (2021b) Transcriptome analysis reveals an inhibitory effect of dihydrotestosterone-treated 2D- and 3D-cultured dermal papilla cells on hair follicle growth. Front Cell Dev Biol 9:724310. https://doi.org/10.3389/fcell.2021.724310

Zhang Y, Ni C, Huang Y, Tang Y, Yang K, Shi X, Zhang Y, Li Z, Wang JA, Zhu Y, Li H (2021c) Hair growth-promoting effect of resveratrol in mice, human hair follicles and dermal papilla cells. Clin Cosmet Investig Dermatol 14:1805–1814. https://doi.org/10.2147/CCID.S335963

The manufacturer's authorised representative in the EU is Springer Nature Customer Service Centre GmbH, Europaplatz 3, 69115 Heidelberg, Germany. If you have any concerns regarding our products, please contact ProductSafety@springernature.com

Printed and bound by CPI Group (UK) Ltd, Croydon, CR0 4YY
01/10/2026
02225915-0005